A 1960s Global Odyssey

Around the World in 80 Months

David W. Skillan

After a lifetime of wandering the world, I cannot watch a plane taking off, a yacht setting sail or a train pulling out of a station without wanting to be on it.

– The Author

 FriesenPress

Suite 300 - 990 Fort St
Victoria, BC, V8V 3K2
Canada

www.friesenpress.com

Copyright © 2021 by David Skillan
First Edition — 2021

All rights reserved.

No part of this publication may be reproduced in any form, or by any means, electronic or mechanical, including photocopying, recording, or any information browsing, storage, or retrieval system, without permission in writing from FriesenPress.

All photos are from The David Skillan Collection.

ISBN
978-1-5255-9932-3 (Hardcover)
978-1-5255-9931-6 (Paperback)
978-1-5255-9933-0 (eBook)

1. TRAVEL, ESSAYS AND TRAVELOGUES

Distributed to the trade by The Ingram Book Company

*TO MY CHILDREN, GRANDCHILDREN
AND FUTURE GENERATIONS*

May you live in a world of peace, goodwill, tolerance, and understanding.

Contents

Acknowledgements vii
Chapter 1: In Sha Allah (If God Wills) 1
Chapter 2: Algerian Holiday 33
Chapter 3: Mosquitoes and Bananas 47
Chapter 4: Monkey Business 73
Chapter 5: Playing With Steel 113
Chapter 6: The Open Road 129
Chapter 7: Wandering through Africa 141
Chapter 8: The Bum's Way 165
Chapter 9: Relief at Last! 183
Chapter 10: Fuji, Sake and Donburi .205
Chapter 11: A Romantic Interlude 221
Chapter 12: A Pommie in the Outback 237
Chapter 13: Sheep, Mountains and Apples 265
Chapter 14: A South Sea Sojourn 281
Chapter 15: Thieves and Serenatas .293
Chapter 16: Pete Saves a Life 305
Chapter 17: Latins are Lovely 321
Chapter 18: My Life in their Hands 337
Chapter 19: Mariachis and Robbers .359
Chapter 20: Welcome to America! 369
Chapter 21: Down and Out in New York 385
Chapter 22: Where There's a Will 401
About the Author 417

ACKNOWLEDGEMENTS

My heartfelt thanks go to:
The many who, as well as befriending me, gave me a ride, a bed for the night, or an offer of employment. By doing so, they made my epic journey possible and brought such pleasure and happiness into my life.

And the countless others who were instrumental in making so many of my remarkable and sublime experiences take place.

It is with the deepest gratitude that I acknowledge:
My great friend **Pete**, for his inestimable companionship, without whom 'the trip' might never have occurred.

My mother, **Olive**, who despite poor health was always cheerful and full of sound advice.

My dad, **Arthur**, who would have done anything to have participated in my great adventure.

My sister, **Janice**, for patiently spending countless hours typing the original manuscript.

The late **David Wright**, veteran newspaper reporter and editor, for his friendly advice and encouragement without which my professional writing career may never have started.

My good friend, **Hugh Griffith**, for his never-ending enthusiasm for this project and constant prodding to get it done.

Last, but far from least, **Anne Ward**, who has given freely of her time and expertise and, through her painstaking efforts, helped get my manuscript to the finishing line. And her husband, **Brian Manyluk**, who has supported both her and me every step of the way. Their help and advice throughout the reading and editing process have been invaluable. I cannot thank them enough and will be forever in their debt.

CHAPTER 1:

In Sha Allah (If God Wills)

'Wherever you go, go with all your heart.'
— *Confucius*

Was it really true? I asked myself as I shaved in the mirror hanging on a nearby tree while beads of sweat trickled down my naked chest and back onto my shorts. *Me, an ordinary young man from England, now catching monkeys for a living in the heart of Africa.* I glanced down at my bare feet and legs and pondered the mosquito bites, scars and scratches. I looked up. Shadarak was mopping his brow as he boiled water over the campfire. Henry was dusting down the Land Rover. And Saidi, humming to himself, was squatting over the metal trunk which doubled as a larder and kitchen table, contemplating the day's menu.

I sighed with satisfaction as I surveyed the rest of my little domain. The monkey house. The orange and green tents. The yellow weaver birds flitting in and out of their round, upside-down nests. The dartboard rump of a solitary male waterbuck grazing peacefully with a small herd of zebra in a nearby clearing. And a female warthog, trailed by three tiny offspring, tails curled in the air, scurrying across the open grasslands in the distance. All everyday sights where I now lived.

And yes, it was true enough. What a far cry from my job as a chocolate salesman for the Nestlé Company in England's southwest counties!

The last few months had certainly been eventful for me and my three companions. Working in the mountains of Algeria had been interesting and fun, and the trip up the Nile had proved eye-opening, but it had been a tremendous wrench leaving the Land Rover behind in Cairo. Pete's illness had been touch and go, and the many problems and obstacles had been challenging.

Despite all the ups and downs, so far it had all been worth it, and I was already looking forward to whatever was coming next. Nothing ventured, nothing gained was my personal motto. I intended to try everything and do whatever presented itself in the months and years to come during what would turn out to be a marathon odyssey and the journey of a lifetime.

It had all started simply enough several months earlier on a clear, sunny English autumn day, when we slipped quietly out of London on the morning of September 5, 1962. As we drove cheerfully along the A2 bound for Dover and the English Channel, neither Pete nor I could ever have dreamt of the extraordinary incidents and events that we'd experience over the next several years.

'Valid for all foreign countries,' that's what our passports said. Algeria, Kenya, South Africa and Ethiopia; India, Thailand, Japan and Malaysia; Australasia, Colombia, Peru and the United States of America would be the scenes of some of our adventures. Not even in our wildest dreams could either of us have imagined some of our future jobs and relationships. And in our constant quest for adventure and new experiences, we would sometimes bite off more than we could chew.

My burning ambition to see the world had nothing to do with being bored with life in England. I wasn't fed up with my job, or the UK. I wanted to do something meaningful with my life and make it count as well as achieve my boyhood ambition. From an early age, I'd found seeing new places addictive and intoxicating. My inquisitive nature and adventurous spirit, coupled with my schoolboy travels, led me to being well and truly bitten by the travel bug.

Originally, it was to be just Pete and me, but Eddie joined us after realising we were deadly serious and deciding he didn't want to be left out. And Mike, too, at the last minute.

We'd all met during our National Service ('NS') in the 2nd Battalion Green Jackets, also known as the '60th Rifles' and later renamed the Royal Green Jackets, an elite light infantry regiment in which everything was done 'on the double'. Indeed, the many regimental duties and nonstop activities made everyone breathless and pissed off. But remarkably fit.

As young soldiers, we'd been thrown together and became good friends. Our hopes of serving abroad in some delightful exotic spot — such as Malaysia, Cyprus or Hong Kong — where the British Armed Forces served were rudely shattered. We found ourselves in County Down, Northern Ireland, performing numerous guard duties.

I was in Headquarters Company, where they put all the 'odds and sods'. As Battalion librarian for a time, I got to enjoy one of my favourite pastimes — reading. My boss was a dour, middle-aged Scottish warrant officer. He saw some potential in me and recommended me for promotion. Like Mike, I was promoted to lance corporal.

Regardless of the job you had, once per week we all had to do guard duty. 'Standby platoon' was the worst one of all. We stood by in full battle order armed to the teeth for 24 hours in case of an attack by the Irish Republican Army. In spite of the threats and spasmodic outrages of the IRA, we learned to live with them.

The unequalled camaraderie of service life was a major attraction. Also, Ballykinlar was a pleasant enough posting, even though the weather was often dull and wet. It was a tiny place consisting of a few houses, one grocery shop, a Sandes Home for Soldiers (akin to a seaman's hostel), a Protestant church, and our own Abercorn Barracks on a windswept peninsula. In between our military duties, we had a liberal amount of free time. Even so, some fellows hated the posting. But I thoroughly enjoyed it. There was more than enough to do with plenty of outdoor activities, great scenery, fresh air and good grub. What more could one ask?

Much of my spare time was devoted to rock climbing with a group of friends in the nearby and often misty Mourne Mountains. Like me, Pete seized every chance to get away and escape the monotony of barrack life. We enjoyed walking and climbing in the verdant hills where we sometimes spent overnight at the newly built Kinahalla youth hostel. A couple of times we made it to the top of Slieve Donard, the highest peak in the area. Eddie joined us occasionally, and Mike came along once or twice, but after developing very sore feet, he decided that the girls and pubs of the nearby town of Downpatrick provided a more agreeable way of passing his spare time.

As you will read, I spent most of my teenage years in Germany. There, at boarding school, I'd learned the basics of sailing and loved being around small boats. This prompted me to take an army course in map reading, ridge walking and climbing on Scotland's Island of Arran which I thought would be useful in the future. Upon completion of the course, I qualified as an Adventure Training Instructor. I started teaching other like-minded souls in the regiment the art of rock climbing in the Mournes as well as canoeing and sailing on Strangford Lough, a long coastal inlet teeming with birdlife. Outward Bound had not yet come into its own.

Being an NCO gave me the opportunity not only to enjoy the great outdoors, but also to pick and choose my companions. So it was that Pete, always gung-ho, sometimes Eddie, and more often than not, Rob accompanied me.

Rob Townsend was a lanky six-foot-four woodwork teacher in civilian life and store man in the army. He'd become a good friend and was a keen walker and hiker. On one of our many outings in the Mournes, Rob met Kathleen, a girl from Belfast, whom he subsequently married. Were it not for meeting Kathleen, he probably would have joined Pete, Eddie and me. We stayed in touch and exchanged correspondence on and off throughout the trip. And, indeed, to this day.

My other role in the armed forces, which I shared with Mike, was that of Company Orderly Sergeant, or COS, which gave me two more 'imaginary' stripes as well as a lot more responsibility. My main job was to rouse the company at 6 a.m. at the sound of Reveille by banging on the dormitory doors with my cane and trying to emulate the drill sergeant's voice with, 'Wakey, wakey. Rise and shine. Shake a leg. Thirty minutes to breakfast!' After supervising breakfast, I assembled the entire company — catering staff, store men, drivers — outside Company Headquarters office. To the command of 'Attteeeeeention!' I would bring them smartly to attention. I'd do roll call for the duty officer, who would then inspect them and give out the day's notices before dismissing everyone to carry out their duties. It was also my job to march the men to various guard duties at different times of the day; march them when necessary in front of the colonel; and, charge them for misdemeanours such as going AWOL or being drunk on duty. It wasn't the most popular of jobs, but somebody had to do it, and it allowed me to more or less come and go as I pleased.

My stay in Northern Ireland was made more memorable when I met a young woman from Terenure, a suburb of Dublin. It was the summer of 1960, and a good one weather-wise when I went with a couple of pals to a Saturday night dance in Newcastle, a popular seaside town at the foot of the Mournes just across Dun-Drum Bay. There, I met the girl of my dreams, a slim beauty about five feet, five inches tall with short, dark wavy hair. She had a small pert nose, large blue eyes, a wide generous mouth with even white teeth, and the loveliest smile you could imagine. A white knee-length dress showed off her gorgeous figure to the best advantage. The most beautiful woman I'd ever seen, she took my breath away. Her name was Evelyn. I would call her Eve.

After plucking up the courage, I asked her to dance and couldn't believe my luck when she accepted. When the last dance was announced, I practically ran across the floor to where all the girls sat, afraid someone would beat me to it. But to my great pleasure, I got the last waltz when we could hold each other close. It went without saying I could have danced all night.

After the dance, I walked her home to her aunt's place where I kissed her lightly on the cheek and asked if I could see her the next day. When she said yes, I thought I was walking on air. I seem to remember skipping down the street to the three-tonner which stood waiting until midnight to take everyone back to barracks. If you missed the ride back, it meant a nearly ten-mile walk along dimly lit roads. But the way I felt that night, I wouldn't have cared. I was in seventh heaven and felt like the luckiest man on earth.

Eve was 21, a year older than me and a civil servant in the Irish parliament. She reminded me of those gorgeous girls you would see in an open sports car with sunglasses and hair blowing in the wind and just wish … you knew her.

Right from the start, we got on famously. We wrote long, rambling love letters to one another and met a few times over cups of tea in little cafés. We wandered hand in hand around Belfast and the village of Ballykinlar. I also took her sailing, which thrilled her no end. As we bobbed around in the small dinghy, we declared our undying love for each other over and over again. On several occasions, I even serenaded her by singing 'La Mer' when we were alone. It was the only song in French I knew, something I'd picked up from listening to Maurice Chevalier on the radio.

Our favourite place was Tyrella Beach which bordered Abercorn Barracks, my military home. There, on what seemed like always warm, sunny, deliriously happy days, we'd lie among the sand dunes, our bodies aching with longing and desire, talking, laughing and kissing for hours at a time. It was a passionate but completely innocent relationship. We also took photos of each other with my little camera, perfect mementos of the times we had shared together.

I'd had a few casual girlfriends by then, but no relationship as intense as this. Over the best part of a year, she very quickly became the first true and unforgettable love of my life. I confided in her and told her something about my hopes and dreams for the immediate future, which at the time didn't include marriage. That was a big mistake. I should have kept those plans to myself.

After we'd been almost 18 months in Ballykinlar, the battalion was posted from Ireland to Berlin which I knew having lived there for several years. Eve and I stayed in touch by mail. Then, much to my dismay, she married someone else. I was crushed. When she broke the news to me in a 'Dear John' letter, I wrote back and wished both her and her new husband well. But there was no denying it — I was totally devastated. Like all first real romantic relationships, I would remember her for the rest of my life. It was young love at its finest, sweetest and most innocent. In the future, I would have some love affairs, but none would be as unforgettable as the one I had with Eve, my first love.

At the time, my parents still lived in Berlin having moved there some years before, allowing me the privilege of being the only single man of 600 to live out of barracks. As soon as the posting was announced, I was instructed to brief the entire battalion about life in Germany and specifically Berlin. I also contributed an article about it to the battalion newsletter, the first of many articles I would ultimately write in my lifetime. Around that time, both Mike and Eddie completed their NS and returned to England and civilian life.

Meanwhile, once in Berlin, Pete, Rob and I explored the famous city and my home for some years. By then, Berlin had become a hip and trendy place known for its many jazz clubs which we occasionally frequented. We used our meagre soldier's pay sparingly. During the day, we dressed in uniform, carrying out various duties in barracks and around the British Sector — showing the flag. Then, at night, we changed into civvies and hung out in the coffee shops, night clubs and bars. It was at one of these that Pete met a pretty young *fräulein* with whom he exchanged letters for a while.

Our battalion took it in turns with the other Allied Forces (American,

Soviet and French) to guard the notorious Spandau Prison where Rudolph Hess, last of the Nazi war criminals, was imprisoned. Another duty was guarding the British train (hauled by a Puffing Billy before diesel engines came in) that by Allied Occupation Agreement travelled at night with blinds drawn from the station of Charlottenburg, in Berlin through the Soviet Zone to first Helmstedt, a town in British Forces territory and then to the large city of Hanover. On arrival in the morning, everyone disembarked and went their separate ways while the train was prepared for the return journey to Berlin.

Once, I was assigned to the train as guard commander and, needing two riflemen to make up the full contingent, I recommended to my CO that Pete and Rob accompany me. I'd travelled on this train many times on my way to and from boarding school in Wilhelmshaven but now, grown up and a soldier, I saw the journey from a completely different perspective. I was responsible for the entire train and a couple of hundred passengers. Fortunately, the night passed without incident and there was no confrontation with Russian troops. After exploring the city of Hanover and enjoying a fine dinner of wiener schnitzel and beer, the three of us returned to Berlin as passengers in our own comfy berths that same night.

Three months after arriving in Berlin, our time as soldiers was up. NS was being phased out. I had spent two years minus six weeks serving Queen and country. Pete, Rob and I were demobbed, said 'Auf wiedersehen' to Berlin and returned to England and civilian life. Five months later, on August 13, 1961, the Wall went up.

My parents stayed in Berlin a few more years then retired and bought a B&B in Hunstanton, a beach town on the Norfolk coast.

Back in the UK, I rejoined the Nestlé Company, this time as a relief rep after a two-week sales course, covering the Counties of Surrey, Hampshire, Devon, Cornwall and Dorset. It was a job which I greatly enjoyed for it involved spending a lot of time travelling to new places and meeting new people.

Despite having passed my army medical A1, I failed the medical required for the Nestlé Company pension plan. So the company sent me for a second opinion to Dr Evan Bedford, a leading cardiologist in London's world-renowned Harley Street, whom I learned had been Sir Winston Churchill's specialist during and after the war. It was with great apprehension that I went to my appointment. After a thorough examination, Dr Bedford advised me that I had a heart murmur which might require surgery in the future. When I told him of my intentions to travel, he replied, 'As far as I'm concerned, you can go anywhere you like'. Those few words were music to my ears and made me impatient to see the world as soon as possible.

One day when making sales calls in the village of Burley in Hampshire's New Forest, I stopped to pick up a female hitchhiker. The middle-aged woman wearing a belted mackintosh and headscarf was Kathleen Newton, the third and last wife of Irishman Jim Phelan, professional vagabond, wordsmith, political activist and probably the most well-known tramp (hobo) in the British Isles. He was the author of several books, including *Tramping the Toby* which consisted of stories about people eking out a living on the highways and byways of Britain. Like him, Kathleen was an authentic tramp and blessed with the gift of the gab. She realised that one of the best ways to promote her husband's books was to the drivers who gave them a lift. From what she told me, she and Jim did a roaring trade as they hitched rides, selling about a dozen books each day. After listening to her spiel or as her husband Jim would say, his 'line of guff' as we drove, I bought a book about her husband's long and interesting life on the road.

Little did I know then that Jim and I had a few things in common, including itchy feet and a strong desire to follow our own destinies. I still have the book Kathleen sold me. It is entertaining and full of good, if somewhat dated, advice for those who for whatever reason find themselves spending long periods on the road, as I would.

Pete, Eddie, and sometimes Rob and I met frequently in London to reminisce and exchange news. On a few occasions, we drove to Snowdonia, in North Wales to hike and rock climb amid the beautiful, rugged Welsh hills. As the only one with a company car, my small Austin A40 came in useful for such pleasurable excursions. As had been drummed into us during our military service, 'Gotta think of your mates,' I made sure we all got to enjoy it.

It was in a London pub amid swirls of cigarette smoke and the downing of several beers in early 1962 that I put forward my proposition. Something that I'd long considered. 'Would you,' I asked Pete, 'be prepared to join me on a trip around the world?' For a second or two, he looked askance.

'You must be joking,' he laughed.

Despite his rather negative initial reaction, I carried on. The proposal seemed straightforward enough to me. 'Why not team up and try to work our way around the world? A combined effort would increase our chances of success,' I continued. Scepticism replaced laughter as he realised I was as serious as anyone could be. And the more I talked, the more interested he became. He'd spoken vaguely about immigrating to Canada one day, but so far had done nothing about it. Now, he was lost in thought.

It would take weeks for Pete to fully come around to my way of thinking. It was a big decision by any stretch of the imagination. In fact, it would be a colossal undertaking that would require great determination and perseverance. I was so excited that I'd completely forgotten that I was afraid of flying, had a very sensitive stomach, and got desperately seasick once out of sight of land — hardly the makings of a potential world traveller! But none of that mattered to me at the time. It would, however, become very significant and all too real in the months and years ahead.

For centuries, men had worked their passage on board ships to every imaginable port on earth, but not many had worked their way around the world the way we intended doing. Of the enterprising few who had travelled extensively, many had been well-off or sponsored in some way. What hope had we, unknown and unaided?

The next few weeks were rather trying for me, hunched over pints of beer and cups of coffee at frequent meetings with both Pete and Eddie. My reasoning remained the same, and my enthusiasm, always considerable when it came to something I liked and believed in, refused to be dampened. We could and should do it was my repeated mantra. Relentlessly, I pressed my case. Of course, there would be problems and difficulties, I agreed. But 'there would also be adventure and fun,' I said time and again. Then one day I threw down the gauntlet: 'What a challenge it would be!'

That did it for Pete. There was a glint in his eye as he considered the implications of my proposal, his options and precisely what I had in mind. Never one to refuse a challenge, he suddenly announced, 'OK. Let's give it a go.' Or something to that effect. I was elated. My efforts had paid off. Now we were two. The trip was on.

With Pete convinced and raring to go, it wasn't difficult to persuade Eddie to join us as he'd already heard my proposal a dozen times. A week or two later, as he inhaled the smoke of a Rothman's, he casually declared, 'You can count me in.' That brief comment meant we were now three. With that, it was full steam ahead.

Our first priority was to purchase an inexpensive, rugged all-weather vehicle with four-wheel drive. Scanning the ads in various London newspapers brought us luck in the form of a well-used 1947 Land Rover. We immediately contacted the owner, inspected the vehicle and paid a deposit for it on the spot. Our preparations had begun in earnest.

When I broke the news to my family that I was heading off to distant lands, my younger brother Tony, an RAF communications specialist, and older sister Janice, a secretary, were all for it. My dad, Arthur, had wanderlust in a big way like me and wished he could join us. He would follow our travels vicariously from the comfort of his favourite armchair while my mother, Olive, worried about my safety. She was, after all, losing a son for God knows how long.

I promised to write home regularly and would keep my word — writing monthly letters from some of the most obscure places. In all, I wrote roughly 100 letters which both parents read with great interest. My mother faithfully kept them, and my dad referred to them often as he tracked my route on a world map with coloured pins.

Over the next several months, the three of us met often to discuss what we now referred to as 'the trip', and at weekends we busied ourselves on the Land Rover. Already 15 years old, it required lots of elbow grease. It had to be overhauled, fitted out, and newly painted. But we didn't mind. It was a labour of love. There was much else to be done. Maps and camping equipment, including tents, camp beds, stove, cooking utensils, water bottles and rucksacks, had to be purchased; inoculations had to be arranged; and visas had to be applied for. We joined the Automobile Association and the Royal Automobile Association to obtain the necessary documents, including a carnet allowing us to take the Land Rover in and out of various countries customs free (like a passport for a motor vehicle). Finally, we applied for international driving licences. In short, we went about organising for our trip as if it were an expedition, which of course it was. I gave it a name: Adventures Unlimited.

Studying a world map, we decided to travel through Western Europe to

Spain, cross to Morocco, then drive down the West African coast to South Africa. Once there, we anticipated finding work to save enough money for the next leg, details of which would be planned when the time came. Our plans would always be vague and flexible to enable us to divert at will, remain in a place if we liked it, or move on quickly if circumstances warranted. We estimated that our marathon odyssey would take approximately three to four years. We decided on a date of departure that would allow us sufficient time to accumulate 50 pounds sterling each (the equivalent of 900 pounds in today's money) — which in those days was the maximum you could take out of the UK on any trip abroad. This money would be stretched to the limit and take us a fair distance. It would be a challenge to see how far such limited funds would take us before we had to look for work.

We were impatient to start and agreed to leave in early September to avoid travelling through Europe in winter.

In the beginning, there wasn't enough money in the kitty for picture taking, which I personally would come to regret. As the journey progressed, photography would become an important part of my travels — and my future career in the travel industry. But early on, we were all focussed on more important matters than picture-taking which, in retrospect, was a huge pity for we would miss some wonderful photo opportunities.

As it happened, 1962 was a memorable year, not just for me and my erstwhile companions: John Glenn orbited the earth in *Friendship 7*; Nelson Mandela was arrested for sabotage; the Cuban Missile Crisis took place; Marilyn Monroe, the world's most glamorous movie star, died of a drug overdose; the Beatles released 'Love Me Do'; Anthony Burgess's controversial novel, *A Clockwork Orange*, was published; the first James Bond film, *Dr No*, was released; Cliff Richard and The Shadows began their musical careers; Shirley Bassey recorded the album *Let's Face the Music*; the Rolling Stones rock band was formed, and the Beach Boys and Doris Day, my long-time film star sweetheart, were going strong with hit after hit.

One day, we drove the Land Rover on a test run to the Lake District, England's magnificent lake and hill destination, and a mecca for hikers and anyone who loves the great outdoors. Earlier, I'd suggested to Pete and Eddie that we ask Mike if he'd like to join us. As they both knew him well and it meant another person to share expenses, they happily agreed. So, on the way, we stopped in Birmingham and called on him. After explaining our plans in detail, he was visibly excited and didn't need much persuasion, agreeing almost immediately. In fact, by the time we'd finished talking, Mike was licking his lips in anticipation! It was obvious he regarded the invitation as heaven-sent and didn't want to be left out. Now we were four. However, as time would tell and events unfolded, two of the four would, for quite different reasons, fall by the wayside.

By the time we four met as young soldiers, I'd travelled the most and had a fair amount of real-life experience under my belt. As a youngster, I'd led something of a peripatetic life.

During the war, while my dad was away serving in the Royal Artillery, my mother, sister, and I, like many, were evacuated to the countryside for safety. We spent three years amid the stark beauty of the remote hills and dales of rural Yorkshire, not far from where the Brontë sisters grew up. For me, a young, active and curious boy, it was a fun-filled, carefree time enjoying the wide-open spaces and the natural world.

Although only a young boy, I clearly remember the rationing in the UK in wartime and a few years after when just about everything was in limited supply. Ration books had to be stamped for milk, butter, poultry and clothing. As a result, we more or less lived on bread and dripping, beans on toast, and jam sandwiches for weeks on end. We were always ravenous and couldn't wait for the every-so-often life-saving consignments of food parcels from South Africa, America, Canada and Australia. On many occasions during 'the trip', we'd relive this rationing experience — living on cheap, monotonous food for days on end.

While in Germany for five years, I regularly made the round trip of 600 miles to and from boarding school in Wilhelmshaven on the bleak North Sea coast from the family home in Berlin. How well I remember leaning out of the windows of the school train, pulled by a smoking steam engine and packed with boisterous, noisy kids as we passed towns and villages and ever-changing scenery. I loved every minute of it.

As well, I'd taken a few road trips with my parents to various European countries, and also hitch-hiked through France which gave me a taste for different places. Pete had gone once or twice to the Spanish island of Mallorca with a few pals, where he admitted to getting thoroughly drunk every night.

But neither Mike nor Eddie had yet left the British Isles. Not that it mattered. Now we were all in the same boat.

Between us, we decided that to make it all work, we all needed trip-related jobs. As Pete was a tool and die maker and worked with his hands, it was unanimously decided he should be the mechanic. He accepted this decision with enthusiasm, even though he knew little about cars. Eddie would be the navigator because of his military training in maps. As Mike was a good organiser, he would look after the provisions and do the bulk of the shopping while I being the most experienced driver — I'd been driving for two years while the others had only just passed their test — would be the main motorist and chief bottle washer. I would also keep a diary about our life on the road.

As four youthful, fresh-faced and fairly naïve young men, although quite different in many ways, we had much in common including being the same age, 23, born only months apart in the same year. We'd all had our share of military discipline, including two-minute cold showers, early morning runs, and square-bashing for hours on end until our feet hurt.

We were bursting with energy and optimism, ready and willing to take on the world. At that age, we were full of piss and vinegar with an air of invincibility and a devil-may-care attitude that most young soldiers tend to have. In reality, there would be times when we literally flew by the seat of our pants. None of us were back-slapping extroverts, but our military training would prove we were quite 'tough buggers' and could, if push came to shove, all take care of ourselves.

We knew the way ahead would be a formidable and challenging journey. Surviving and earning a living were going to be our greatest challenges but, thanks to pure luck and God's good grace, things would often just fall into place. There would be dangers lurking here and there, but as the saying goes, 'ignorance is bliss'; and, what we didn't know, we didn't worry about. All the same, this trip would be a life-changing experience for every one of us in so many ways.

Mike Paige was an insurance agent from Leeds. Tall, slim and fair-haired, he'd been kept busy as Headquarters Company clerk during our time with the Green Jackets. He was quick-witted, affable, and had a keen intellect and a mischievous sense of humour. Like many witty people, he always looked a little smug and pleased with himself. A Methodist minister's son, he'd rebelled when growing up and was the complete opposite of what you would expect of a churchman's son. He had a casual, nonchalant manner and 'couldn't care less' about a variety of things. Also, as the rest of us often laughingly reminded him, he fell quickly for almost every girl he met.

Edward Mathew (known to everyone as Eddie) was from Leytonstone in East London and was by occupation a technologist in the paint trade. He'd been in the Intelligence Section of the battalion whose main job was to interpret maps. Tall, well-built with a slightly ruddy complexion, he was naturally serious and unassuming. He blushed easily and his shyness was evident. It was only by speaking to him first or discussing a subject dear to his heart that he could be brought out of his shell. But he relaxed among close friends. Despite his impressive appearance, he was very nervous by nature and a chain smoker. A voracious reader, he would state his case whenever contributing to a discussion, then categorically defend his statement with the words, 'Because I read it!' As if that proved his point. Quite the opposite of Mike, he'd never had a girlfriend in his life which as we would see in the future made him extremely vulnerable to the charms of the opposite sex.

Pete Barrington, another Londoner like Eddie, hailed from Willesden Green, a working-class district near Edgeware Road. As a qualified tradesman he would easily find work. Lucky for him, his skills would prove to be an open sesame when it came to job hunting in the future. Good-looking with dark hair and a gleaming smile revealing almost perfect, even white teeth which he took great pains to maintain, he'd been a signaller in the battalion and knew Morse code. A slight five-foot-five, he was in fact as tough as nails. A typical Londoner, he was also garrulous and friendly.

He was also very independent, tenacious and fiercely competitive, as I was to find out. Thankfully, he turned out to be the perfect travelling companion, always considerate and cooperative. He was as honest and as trustworthy as they come. Unlike Eddie and me, who were prolific readers, Pete admitted to only having read half a dozen books in his life for which we teased him unmercifully. And which he took good-naturedly, like everything else. Well-read or not, he was to prove himself successful in a variety of ways. It was Pete's habit to walk quickly with a jaunty bounce, often breaking into a trot. He disliked dressing up, always preferring to be in shirtsleeves. Ultimately, as events unfolded, he would graduate from being an ordinary London lad to become the quintessential world traveller.

As for myself, I was of average height, pale-faced, thin and wiry. I had my dad's naturally curly hair — which I didn't like and did my best to plaster down — and his prominent nose, which made me very self-conscious. Fortunately, my self-confidence was usually quite high, and as I went through life, I often forgot about my not-so-appealing physical attributes. This was just as well; I had much more important things to do and think about.

I was born three months before the outbreak of World War II in the family bungalow on the outskirts of the town of Romford in Essex which was, when I was a boy, well known for its open-air Saturday market (with cattle, sheep, pigs, fruit and vegetables and just about everything else). It also had the Star Brewery, a popular dog-racing track, and a Third Division football team.

I'd been a delicate, sickly child, subject to frequent bilious attacks which would come out of nowhere, and sometimes prevented me from attending school. The doctors never diagnosed the precise problem, but my mother put it down to 'nerves'. Also, as a boy and much to my mother's consternation, I was the fussiest of eaters. That would all change during military service and travelling the world when at times I'd be so hungry I'd be happy to eat anything!

Like most young boys, I was something of a little rascal, always up to some mischief. More than once, the neighbours predicted I would break my neck. As my mother would often tell me, I was a happy kid, so much so that she often called me 'Cheerful Charlie'. I loved being out of doors fishing or catching frogs and tadpoles and spent almost every minute of my free time, especially in summer, exploring and listening to birds, including the unmistakable sound of the cuckoo echoing across the surrounding fields and woods. Naturally restless, I had 'ants in the pants', according to my mother.

As something of an idealist and dreamer, I was eager for experiences of any kind and determined to live life to the fullest. I was qualified only in enthusiasm to achieve what I wanted most.

Like many people, I was a complete contradiction — sometimes shy and introspective, other times full of confidence. I learned to conceal my shyness from fear of missing out. Despite my rather reserved disposition, I was seldom at a loss for words, especially on matters I knew something about and subjects I enjoyed. I could also be very persuasive, which was to come in handy at times.

I knew I was going places when as a nine-year-old, I set off on a midsummer morning with my best friend, Alan Surrey, bigger than me and one year older, to cycle the roughly 26 miles from our homes on the outskirts of Romford to Southend-on-Sea, our nearest seaside town. It started as a dare to one another to see how far we could go in a day. Around eight o'clock in the morning, we told our mothers we were going for a bike ride. Before long, we were on the A127 major highway to the East Coast. There, on the conveniently built bike lane with traffic flashing past us, we pedalled like mad for several hours before arriving at our destination. The town of Southend, with its pebbly beach and numerous fish 'n chips shops, also had the famous Kursaal, a huge fun fair which, when first built at the turn of the century, was the largest amusement park in the world. It also had the longest pier. Once there, we spent all our pocket money — mine was a mere sixpence at the time — on rides and ice cream.

The long return journey home turned out to be pure hell. My second-hand, battered and bruised Raleigh, which I cherished and was all I could afford, couldn't compete with Alan's newer model. Three or four times we had to stop to fix a flat. Finally, after repairing at least three punctures and realising the inner tube had had it, we took it in turns riding each other's bikes the rest of the way. It must have been around nine o'clock when we rounded the bend of Jubilee Avenue and Marina Gardens because it was getting dark. By the time we arrived home, we were totally exhausted and I was 'green in the face', my sister later told me. Our mothers were waiting with anguished expressions for us to turn up, about to call the police, they said, as they enveloped us in their arms.

The episode became a well-told story in our family for years to come. After all was said and done, it was quite an accomplishment for two young boys to cycle a distance of well over 50 miles in one day. Neither I nor Alan was any the worse for our adventure, and I was pretty chuffed with myself. However, I also promised my mother not to put her through anything like that again! Little did I know then that for various reasons I would not keep my word.

In the summer of 1951, one month after my 12th birthday, and at the height of 'the Cold War', I moved with my family to join my father in Berlin, then known as 'The Divided City'. A former London bus driver, he got his big break when stationed in Germany immediately after the war. My dad had transferred from the British Army to the newly established Control Commission of Germany in the former capital which was shared between the British, French, American and Soviet military governments. We arrived in West Berlin after an overnight trip by troopship from Harwich to Hoek van Holland, then travelled the rest of the way by train.

It was while travelling for this first time by boat that I experienced my earliest bout of seasickness, something that was to rear its ugly head time and time again in the future. Indeed, it would be my Achilles heel. But I would be in illustrious company. Such luminaries as Lord Nelson, Agatha Christie and Charles Darwin, all at one time or another, shared the same awful affliction. Something that many people to this day still describe as 'absolute torture' or 'a fate worse than death'.

As events would unfold, my family's change of fortune and going from a semi-rural area of England to a nicer, grander lifestyle in Germany was due to one ironic piece of luck. We were on the winning side of the war! And as a result, we would lead a life of relative privilege.

No sooner had we settled into our spacious new home at 32 Miquelstrasse in the well-to-do, leafy district of Dahlem, than my older sister and I were packed off to Prince Rupert School (better known by pupils and staff as PRS). A unique co-ed, comprehensive British boarding school for children of British Armed Forces personnel, it was situated on the outskirts of the town of Wilhelmshaven, on Germany's bleak North Sea coast about 150 miles west of Hamburg, the much larger, better-known port. The town was an important wartime base for U-boats and the pocket battleship, *Tirpitz*, as well as the German navy. And we kids got to take it over, and live and play in their former barracks and sports grounds.

It was at boarding school where life was spartan and discipline was strict that I joined the school sailing club. The unpredictable weather and wind-swept location made it the perfect place to practise this exciting sport. Under the direction of my portly and bearded maths master, Mr Robertson, head of the Combined Cadet Force and former naval officer, I learned essential nautical skills. As well as teaching me to sail, he gave me extra lessons in maths, my weakest subject, for which I would be eternally grateful. When free time allowed, I sailed as often as I could, enjoying the fresh, salty air in my lungs. Sailing soon became a passion of mine. I had no problem finding my sea legs in and around coastal waters in small boats but sailing in larger ships on the ocean would be another story!

It was also at PRS that through my classmates I first became acquainted with Ian Fleming's James Bond books which I devoured as soon as they came out. Around the same time, in 1954, I heard my first rock-and-roll music when I listened to Bill Haley and the Comets singing 'Rock Around the Clock'. A year or two later I would hear Elvis Presley sing 'Blue Suede Shoes' for the first time, which gave me and all the other kids something different and exciting to write home about.

During school holidays, I explored much of Berlin on foot or on my trusty second-hand blue bike or sailed at Wannsee Lake with the RAF sailing club. Almost every Sunday, the entire family piled into my dad's black Volkswagen Beetle for a drive around the city. The VW was the government vehicle dad used in his job as transport supervisor delivering classified documents to the various Allied offices. Together, we explored all the districts under the control of the British, French and American allies and also went to Soviet-occupied East Berlin, which had been almost totally destroyed and, unlike West Berlin, was very slowly being rebuilt. We visited Neukölln, home to students and artists; Schöneberg, famous for its KaDeWe store and amazing deli; Mitte, where the imposing Brandenburg Gate stands; Charlottenburg, where the British Headquarters was based; Spandau, known for its notorious prison; Grunewald with its lakes and forest and where I learned to ski; and Tempelhof, scene of the life-saving 1948 Berlin Airlift — always stopping, when time permitted, at Café Berkel at the head of Kurfürstendamm, Berlin's wide and fashionable avenue, for the most delicious ice cream.

Twice during the long summer holidays, we drove to Switzerland, Italy and France. Heady stuff for a teenager. All of which left an indelible impression on me.

After leaving school at 17 — I was a hopeless scholar — and after much thought and in discussion with my mum and dad, I decided to try my hand at law enforcement. So I applied to join London's Metropolitan Police Force as a cadet. While I waited to hear if my application was approved, my dad got me a temporary job in Berlin as librarian at the British Headquarters housed at the Olympic Stadium. This modern complex of buildings was built by Hitler for the 1936 Olympics and included an impressive swimming pool where I learned to swim.

Once my cadet application was accepted, I took my first ever and very bumpy flight from Berlin's Gatow Airport to RAF Northolt, West London, in a DC-3 courtesy of the RAF (which my dad managed to wangle). After passing all the requirements (physical, written examination and interview), I was accepted. Alas, halfway through the twelve-week course, I came to the conclusion that a policeman's lot was not for me. And definitely not as glamorous as I first thought. So I quit. The CO of Hendon Police College, a chief superintendent, tried to talk me out of it, but my mind was made up. Naturally, when I broke the news to my parents in a long-distance telephone call, they were very disappointed.

Shortly after that, I got a job at the Nestlé Company, the internationally known food and confectionery company. As assistant to the art director in the design and photo department, I learned the basics of taking pictures, the ins and outs of plate cameras, and development of both black and white and colour photos in the darkroom. The work also involved a few trips to pretty places like Corfe Castle, Lulworth Cove, Polperro and Penzance. I would accompany Mr Geoff Cardew, my immediate boss, who wore double-breasted suits and colourful bow ties and came to work on a motorcycle fitted with sidecar. Under his watchful eye, I learned the art of composing and shooting photos. Using both plate and Rolleiflex cameras, we shot castles and picturesque rural and seaside scenes to be used on chocolate box covers. Such stimulating outings in the field were only for a few days at a time but, with nice hotel accommodation and all expenses included, I enjoyed them immensely. *Anything to get outdoors and away from the studio and lab*, were my thoughts at the time.

I was now 18 and a half and around then, not long before I would be called up for my compulsory two years of army service, I became friendly with Les, an artsy type employed by the company as a store window dresser. Les was a couple of years older than me and quite worldly. Together, we'd sometimes take the Tube to the West End and wander around SoHo and peer into the numerous strip joints and nightclubs. On one occasion, we saw the newly released and much-talked-about movie, *And God Created Woman*, starring Brigitte Bardot, the original 'sex kitten', directed by her lover and soon-to-be husband, Roger Vadim.

In the summer of 1958, Les and I decided to hitchhike to the French Riviera, the Côte d'Azur, roughly 540 miles as the crow flies from Dover. Coming from austere Britain in the late fifties, there was something very exotic about the south of France, the haunt of numerous French movie stars – the palm trees, expensive yachts, and glamorous people who paraded arm in arm along the promenade. And the always clear blue skies and lovely warm sunny climate. We carried rucksacks with our sleeping bags and a small collapsible tent. As we thumbed our way through France, we ate in cheap restaurants and camped in fields and campsites along the way.

After passing through Nice, Antipodes and Saint Tropez – the hangout of France's beautiful people – we arrived at Toulon. There, having already read up about our next destination, we took the 20-minute ferry ride to the Île du Levant, a tiny island and nudist colony which was then and still is shared by the French Navy. It was an interesting experience for a couple of virile, inquisitive young men who admired the opposite sex and the female form, especially when exposed in all its naked glory. At first very self-conscious, we got used to nude sunbathing on the beach and wearing what was loosely called *le minimum*, little more than a G-string. We liked the island so much that we camped and stayed there for three days and nights.

Prior to NS, I'd been sharing the rent in a nice, three-bedroom Victorian duplex in Acton with two other fellows, Hugh and Keith, both of whom worked for Nestlé as chemists. When I left the army, I moved in with my widowed aunt Mabel, my dad's oldest sister. Mabel's home was close to Richmond Park with its red and fallow deer herds. Wimbledon was close by and at times there were crowds and lots of traffic coming to see tennis greats like Lew Hoad, Rod Laver, and Pancho Gonzales. It was in the driveway at my aunt's Kingston address at 129 Robin Hood Way that we sometimes worked on the Land Rover and from where our great adventure would begin.

This living arrangement suited both my aunt and myself perfectly for, although I was away most of the time, it meant occasional company for her and it was easy for me to get into London either by local transport or in my little company car. There Rob, Eddie, Pete and I usually met at a pub or coffee bar. Once in a while, we took in a show and, every so often, a dance at either the Lyceum Ballroom or the Hammersmith Palais where we would mostly stand on the sidelines admiring the girls. That said, Eddie seldom joined us at the dance halls; he was much too shy to step onto a dance floor, or for that matter have anything to do with the opposite sex.

As the date of our departure drew closer and as much as I enjoyed the variety of work then as a sales rep with Nestlé, the working world tour now consumed most of my interest. 'The trip' would fulfil a boyhood ambition. As a teenager, I would often daydream about exotic and exciting places far away. Such places as Australia, India, South America, and especially Africa were the constant subjects of my thoughts, and I knew I wouldn't be satisfied until I saw them for myself.

I'd seen a fair number of wildlife documentaries, including David Attenborough's *Zoo Quest* series, produced by the BBC. Alan Whicker's dry humour and astute observations on his weekly ITV travels to various exotic destinations further whetted my appetite and made me want to set off on my own travels as soon as possible.

Apart from Alistair Sim, the actor who played Scrooge in *A Christmas Carol* whose autograph I got when he visited Berlin to promote the film, I'd never met or seen anyone famous until I saw Russian cosmonaut Yuri Gagarin, the first man to fly in space. He was making a goodwill tour to England when he waved to admiring crowds from his motorcade during his triumphant tour of London and Manchester in July, 1961, three months after making his historic flight around the earth. As I stood among the many well-wishers I couldn't help but be thrilled on seeing a modern day explorer in the flesh which further added to my spirit of adventure, particularly as he was quite youthful and dashing, and someone I could vaguely relate to.

Like Eddie, I had an appetite for books of all kinds; particularly true adventure stories, such as *The Ascent of Everest, Great True Tales of Human Endurance, Scott of the Antarctic, Accounts of Livingstone's Journeys*, and *The Voyages of Captain Cook*, to mention a few. I was also fond of biographies and learning what drove famous people and made them tick.

I particularly liked quotes, particularly this one by Lord Byron, which I took to heart and which would often be appropriate, 'Young men should travel if but to amuse themselves.' The famous poet could also be credited with, 'Always laugh when you can. It is cheap medicine.' Both were well worth remembering and applying to everyday life.

Of the four of us, Pete and I were kindred spirits with the most in common. Our main interests were a love of adventure and an earnest appreciation of the great outdoors. Throughout our military service, we were encouraged to use our initiative. Now it was about to be put to the test. Over the next several years, we would be living mostly by our wits so we would have to be adaptable and flexible at all times.

Three of us were smokers at the time, and Eddie was a particularly heavy one. Mike and I liked an occasional pipe, but Pete didn't smoke at all, which created some friction in the group. Sometimes we'd argue over small stuff such as whose turn it was to make dinner and other trivial things.

As former soldiers, we were all familiar with the most profane parade and barrack-room language and could all swear as only soldiers can, but generally we kept it pretty tame to a few daily expressions, the most common of which were: 'Get a bloody grip.' 'Silly sod.' 'Wanker.' 'Pull your finger out!' But nothing malicious or rotten. If there was an argument about anything, it usually ended with one or more of us expressing his disapproval with, 'Bollocks', 'Get stuffed', or 'Fuck off'. The nastiest expletives were reserved for officious officials or rude and obnoxious people we might encounter.

We didn't always see eye to eye on everything, but all things considered, including living very close together for weeks at a time and enduring the usual bullshit all good male friends experience, we'd get on pretty well. As members of the armed forces, we'd been through a lot together and knew how to cooperate to get things done.

We were a well-equipped little party, or so we thought. But in our haste and impetuosity, we'd not considered all the implications that such an undertaking involved, which would only reveal themselves as our journey progressed. In retrospect, this was just as well for had we pondered too deeply about the obstacles that we might face, we may never have set off at all! Indeed, by any stretch of the imagination, the journey would be a lot more than we bargained for.

Perhaps the greatest problems that we'd encounter would be due to our distinct lack of contacts. The only people we knew abroad amounted to one uncle in South Africa, another in Australia, and an aunt in the United States.

All mine. We also carried no letters of introduction to influential people to secure us assistance and work when the time came. No one to rely on if we became ill, or for help if and when needed. In short, we were to discover the world by ourselves, the success of which would depend on our own initiative and good fortune.

At last the great day dawned and, apart from a sleepless night and a bout of early morning nausea brought on by apprehension and excitement, I was as ready as the others for whatever lay ahead. The die was cast. There was no going back. We were on our way, heading into the unknown and everything that would come with it.

Ahead lay numerous foreign countries, countless strange cities, hundreds of new faces, and a variety of unusual jobs, not forgetting the natural elements of extreme heat and intense cold. Incredibly, protected by our innocence and ignorance, and what could only be described as divine intervention, luck would be on our side most of the time.

At Dover, having booked the late afternoon ferry, we surrendered our tickets and drove on board. Only when we had pulled away from the quayside and the white cliffs began to recede did it sink in that our journey had really begun!

While not excessively rough, the crossing to Calais was choppy enough, and I was soon forced to abandon our game of cards and lock myself in the WC. Fortunately, I didn't know then that crossing the English Channel was nothing compared to some of my future sea voyages.

It was drizzling and dark when we drove onto French soil. After travelling a few miles 'on the wrong side of the road', we stopped to camp in a wooded clearing. There, we heated up and drank some hot soup, climbed into our sleeping bags and fell fitfully to sleep. It was still raining when Mike was the first to rise the next morning. Dressed in T-shirt and pyjama trousers and clutching an umbrella, he was an amusing sight as he bent over the Primus stove preparing coffee.

We were delighted on reaching Paris, particularly as some friends had predicted that we wouldn't get that far! After overnighting at a campsite, and downing a couple of glasses of cheap cognac at a sidewalk café to celebrate being on our way at last, we were off again, driving in the fast-moving Parisian traffic around Place de la Concorde three times while trying to find the road leading south.

Initially, there would be many times we strayed off course and had to retrace our steps. Pete openly confessed to a hopeless sense of direction. Nevertheless, after weeks and months of travelling, both he and I developed a knack of orienting ourselves quickly with our surroundings and bearings no matter how strange or different they were. In fact, amazing as it may sound, I don't recall in all my travels ever getting lost or feeling out of place. I was 'at home' wherever I roamed.

It took several days of leisurely driving to reach the Italian Riviera via Switzerland. By then, we had already begun to lose track of time, savouring the pleasures and delights of our new-found freedom and doing what we liked when we chose. Mostly, we stopped to admire a particularly attractive view, buy some food, and camp. As time passed, we became more organised and once the two tents were erected and chores completed, we had time to take walks and explore the surrounding terrain. If there was a hill nearby, as we all had more energy than we knew what to do with, it was always climbed. Although born and bred in the city, Pete was fond of animals of all kinds. He was very interested in snakes and was forever levering up rocks or peering into bushes in search of the infernal things. But it was the constant activity in the back streets and alleys of towns and villages that fascinated and intrigued us the most. Although we were to hear of others being mugged and robbed in such environs, our curiosity to see things with our own eyes was always to get the better of us. Hopefully unscathed.

When on the road, we sat two in the front and two in the back, eyes glued to the road, afraid of missing anything. We seldom stopped to eat but made and munched cheese sandwiches and an apple or orange as we drove. Only in the evenings did we prepare a proper meal, taking it in turns to perform the role of duty cook. Most meals were the same and consisted of a fry-up — usually eggs, canned beans and chips. Fortunately for us, Mike took special delight in cooking and, when his turn came, he invariably came up with something new. Actually, if it hadn't been for his culinary prowess and aptitude, we all might have starved rather than face such dull repetition.

Having grown up during wartime, we had all learned, by necessity, to become financially creative so we were thrifty to the point of frugality in our spending. Being young and healthy, we were always hungry, so we ate lots of fresh fruit, bread, salami, and cheese, and occasionally treated ourselves to a bottle of cheap wine. Prior to buying anything, we debated if it was worth it and when we considered it was, lashed out the equivalent of a couple of shillings.

Soon, faint wisps of stubble appeared on our respective chins as we began to grow beards, and a friendly competition quickly developed as to who could produce the most luxurious growth. But, despite going unshaven, we developed almost a fetish about keeping ourselves respectable, preferring to take pride in our appearance and determined never to be labelled as 'scruffy'. This attitude was to pay off.

Thrilled with the life we had chosen, we delighted in our ever-changing surroundings. Our appetite for the new and unusual seemed endless. Every day was different from the last and everybody and everything was interesting to us as we constantly enthused about our exciting new way of life. Wherever we drove, people waved and every time we stopped, we were quickly surrounded by curious, eager spectators enquiring about our destination, 'Where are you going?' Our reply, 'Around the world!' invariably caused a stir as our audience murmured their wonder and approval to each other. But it was the Land Rover that attracted the most attention. Newly painted bright blue, laden with equipment, and sporting a tiny Union Jack, it provoked curiosity and excitement at each stop. For our part, we were rather proud of it and maintained it as well as we could.

Although we were to encounter just about every nationality throughout our wanderings, we never found communication to be a really serious problem. Between us, we had some basic German and schoolboy French. English was becoming increasingly international and we could always rely on facial expressions and gestures to make ourselves understood. None of us were linguists but, occasionally, through sheer necessity we would have to learn a few elementary foreign phrases and words. We all made a point of learning the local equivalents of 'please', 'thank you', 'hello', and 'goodbye'. Although the outcome would sometimes provide some tongue-twisting and amusing moments, it was a sure-fire method of showing our appreciation and friendliness.

In San Remo, we basked in the warm sunshine and enjoyed a few dips in the Mediterranean, but the sun soon got the better of Mike, Eddie and myself. Being very fair-skinned, we got badly sunburnt and suffered at night. Since reaching the shores of the Mediterranean, it had warmed up and the 85°F heat seemed quite hot to us. That was nothing compared to the temperatures we would face before long.

Monte Carlo, St Tropez, Nice and San Raphael were next on our itinerary. Before long, we had crossed the Pyrenees and were in Spain. Never one to hide his emotions, Pete got quite excited when his brand-new passport was stamped for the first time.

Pete was still at the wheel when we arrived in Barcelona during the rush hour. It is a large busy city and suddenly there was dense traffic on all sides. He'd soon had enough of 'those crazy Spaniards' and suggested someone else take over, which I did. He could hardly be blamed for feeling nervous for the traffic was fast, furious and frightening and he had only recently passed his driving test.

Not long after encountering the frenetic Barcelona rush hour, Pete suffered the first of what was to be several mishaps. We had pulled up and made camp and he was busy preparing a meal. All at once, there was an almighty bang, and Pete was sent reeling. We rushed to pick him up. 'You alright?' 'What the fuck was that?' We fell over each other with questions and attempts to help. It turned out that the butane stove had exploded as he changed canisters near the fire. He'd accidentally punctured it, creating a leak, followed by an explosion. His right hand and arm were painfully burnt. Eddie hurriedly got out the first aid box and, with my help, swabbed the burns with iodine and bandaged them up. 'Bloody lucky it wasn't worse,' piped up Mike as we all agreed Pete was fortunate not to have burnt his face. Weeks in bandages were to elapse before his burns completely healed.

That incident reminded us to be very careful about everything we did, for nobody wanted to return to England because of an accident or ill health. On that note, we got out and took the first of many doses of anti-malarial tablets in preparation for our entry into Africa.

In Gibraltar, we camped on the beach at Catalina Bay and looked across the straits at the dim, hazy outline of the North African coast. Leaving one person to guard the vehicle, we clambered up the base of the Rock to see the famous Barbary macaques.

Feeling rather reckless, we then decided to eat out and soon discovered the cheapest place in town, Smokey Joe's, a favourite place and hang out for both locals and travellers like us. We were to come across many similar 'cheap and nasty', 'greasy spoon' cafés around the world, all located in the city centres, and offering a home away from home atmosphere for the travelling fraternity.

Eating out often caused problems for me. Much as I liked food of all kinds, and over the course of several years would taste just about everything under the sun, I, more than any of the others, would pay dearly and suffer greatly from an upset tummy all too often in the future. Normally, I was as healthy as a horse but ordering the cheapest dish on the menu or some of the places where I ate, not surprisingly, would make me as sick as a dog. But all that was ahead of me and of no concern then.

For us, these cafés were always good places for a reasonably priced wholesome meal and to swap ideas, compare notes, and obtain information about the route ahead. Inside, over large mouthfuls of thick, juicy, succulent steak, we chatted to two bearded and flamboyant German travellers, one of whom wore a Stetson with a bullet hole in it. They had been on the road for more than two years; they lost no time in telling us. 'Two years,' we exclaimed suitably impressed for it seemed an eternity to us, even though we anticipated being away longer. We listened wide-eyed and mouths open hanging on every word as they recounted tales of their adventures in West Africa and the Canary Isles. We would have many such encounters both with fellow travellers and people of all nationalities, ages and walks of life as we travelled the world, most of which would be enjoyable and memorable.

We didn't bother to erect the tents that night — and fell asleep in our sleeping bags on the beach. Alas, we experienced our first rain for more than a week and got soaking wet. It didn't dampen our spirits though, and early in the morning found us en route to Algeciras to catch the ferry to Morocco.

So, after nearly a month, our journey through Europe had come to an end. None of us had expected it to be particularly eventful or thrilling, but somehow we felt Africa was going to be a very different story. As we made our way to the docks and the ferry to North Africa, little did I know that this would be the beginning of my passionate, lifelong love affair with the African continent.

A 1960s Global Odyssey

After a pleasant, two-hour millpond crossing of the Straits of Gibraltar, we landed in Morocco and camped alongside the main road to Tangier. We awoke at dawn to the sound of a flute. An Arab in djellaba and turban was squatting on the ground, playing music that sounded distinctly out of tune to us. As we watched, others appeared out of what seemed to be caves in the hills while one leading a pack donkey strolled past acknowledging us with a graceful Arab salutation, hand to brow and heart while uttering the words, 'Assalamu alykum' (Peace be with you). A universal greeting in that part of the world.

From that moment on, we knew we were in a different world. Africa, an amazingly varied, dynamic and beautiful continent with all its different sights and sounds and even its own distinct smells, lay ahead of us in all its magnificence waiting to be explored.

An unforgettable drive, weaving between camels, donkeys, goats, and people in djellabas on foot took us slowly to Tangier where we promptly pulled up outside the central Kasbah (souk or covered market). Within the ancient, thick fortress walls was a maze of labyrinthine alleyways, with silversmiths, leatherworkers, cobblers and cloth makers hard at work. People dressed in every imaginable colour and outfit milled around, shouting and bartering, oblivious to the pungent odour of fish, meat, incense and herbs.

A blind beggar hopped along on crutches; bearded, ferocious-looking hill tribesmen with gleaming daggers tucked in their belts marched imperiously by; and water vendors rubbed shoulders with women mysteriously hidden behind veils. We gazed in awe at the milling crowd gliding past the tropical fruits including lemons, oranges, melons, pomegranates, quinces, nuts and dates; and past brass artefacts, exotic perfumes, toilet paper, hand-woven baskets, and gorgeous Arabian carpets, all displayed for sale. It was a fascinating riot of colour, people and noise.

Outside Casablanca, best known for the movie of the same name (which was actually shot in a Los Angeles film studio), while looking for a quiet spot to consume our sandwich lunch, we drove onto the glorious white beach and immediately got bogged down. Our wheels were stuck up to the axles in soft, white sand and, despite engaging four-wheel drive, the vehicle wouldn't budge. As we stood discussing our dilemma, a group of Arabs gathered around. Using sign language and gesticulations, we enlisted their aid. After two or three momentous heaves, the vehicle was soon freed, but our thanks were not enough. They made it clear they wanted *baksheesh* (a tip). There were too many to pay, so we gave them a friendly wave of thanks and hurriedly, and somewhat guiltily, drove off.

Two or three times, we had punctures which Pete and Eddie usually fixed. On one occasion we broke down and Pete, never afraid of getting his hands dirty, sorted it out. It was an electrical problem which after a lot of head-scratching and trying this 'n' that was eventually fixed.

Heading south, we stopped frequently to swim. The waters of the Atlantic were biting cold after the warm Mediterranean. We dried our tingling bodies playing leapfrog in the strong sun. The magnificent white, sandy beaches were deserted except for us. But they wouldn't be for long. Two years later, massive concrete hotels would be erected to accommodate the tourist boom.

The smell of fish in the port of Safi was overpowering, and we stopped only long enough to buy some bread. Fortunately, our senses were to become accustomed to many strange and often unpleasant smells, noises and sensations over the years. As we headed hastily out of town, we heard the distant, melancholy wails of the muezzins (Muslim priests) calling the faithful to prayer.

The 1960 Agadir earthquake had resulted in untold ruin and 20,000 deaths. However, when we arrived, much of the debris had been cleared and work had started on new homes and offices using prefabricated materials. In due course, that entire section of coastline became a tourist mecca.

A young French couple named Jean and Isabelle approached us at a filling station wanting to chat and to ask us where we were headed. After greeting us in limited English, they invited us to their home for dinner. An enjoyable evening followed, sipping cognac and trying to converse. We were to lose count of such invitations, and the generosity and friendliness displayed to us by complete strangers never failed to impress us.

Our route now lay south through the Sahara and, after filling every available jerrycan with petrol and water, we set off into the largest and one of the most forbidding deserts in the world. It was typical Lawrence of Arabia country and any minute we expected to see hordes of Arabs on horses and camels appearing out of nowhere and firing shots into the air. After previously driving on reasonably decent roads, it was a strange sensation driving across flat, sandy, barren waste where the track was rough and only occasionally signposted with empty, 44-gallon petrol drums. The sun beat down mercilessly with a dry, searing heat, and when we stopped, all was strangely silent. There was neither a breath of wind nor the slightest sign of life. Yet, we agreed, it was beautiful and impressive in an awe-inspiring way. As we all reflected on the quiet beauty of the place, Eddie summed it up nicely, 'Shit, I'd hate to be stranded here!'

At the Moroccan border with Algeria, a tiny army post, beyond which stood a sign saying 'No Man's Land', the sentry on duty beside the barrier refused to allow us to pass, insisting that to proceed further required written authority from the area commandant. Out came our passports and other official-looking documents in an attempt to impress him but he remained stone-faced.

'He's only carrying out orders,' Pete commented as our spirits sank.

After a huddled discussion, we drove back to Guelmim, a scorching hot, Arab town set in dusty hills. There at the military headquarters for the district, we requested an interview with the commanding officer, a middle-aged, uniformed brigadier. He listed attentively then informed us it was impossible to cross that border 'for security reasons' — due to a Moroccan-Algerian dispute at the time. He made it very clear that his decision was final, and no amount of persuasion would make him change his mind. 'Fuck it,' muttered Mike, voicing everyone's sentiments as he exhaled smoke from one of his cheap Arab cigarettes.

That evening, despite the bitter cold of the desert and the chattering of our teeth, we sat on our camp beds around the fire discussing the problem well into the night. The only other route south, through the Spanish Sahara, was also closed, as we had learned earlier from fellow travellers. We considered sneaking over the border under cover of darkness but decided against it, realising it to be foolhardy and fearing the repercussions if we were caught. We finally came up with another plan.

We'd return to Casablanca, Morocco's largest port, to search for a freighter heading south. If unsuccessful, we would alter our entire route, drive across North Africa to Egypt, from where we would try to reach Kenya (then still a

British Colony) where we felt sure of finding work. 'In Sha Allah,' muttered Eddie, thus concluding the discussion and confirming that we were on our own in every way. One thing was for certain: we all agreed, come hell or high water, we weren't going home.

CHAPTER 2:

Algerian Holiday

> '*The purpose of life is to live it,*
> *to taste experience to the utmost,*
> *to reach out eagerly and without fear*
> *for newer and richer experience.*'
> — Eleanor Roosevelt

Back in Casablanca, we visited various shipping offices and combed the docks to no avail. Every ship we checked out was either remaining in port for some time or heading in every direction except the one we wanted to go. So, we dropped that idea and drove to Marrakesh, a beautifully laid-out Arab city of palm trees, mosques, palaces, and exotic gardens to acquire visas for our new route across North Africa. There, at a camping site in the heart of the city, we took full advantage of the facilities, which included tap water and a magnificent view of the snow-capped Atlas Mountains.

A day later, we were on the road to Meknes, another vibrant, fascinating walled city full of interesting Moorish architecture and Muslim history. Stopping to brew a hot drink, we were approached by an old man and his young granddaughter. As they stood watching us, Mike offered them a cup of cocoa which the old man accepted, his lean, dark face creasing into a wide, friendly grin. Murmuring a few words to the little girl, she suddenly ran off and quickly disappeared. She soon returned, however, clutching two newly laid eggs, which she insisted on presenting to us in return for the drink.

Just outside the city of Oujda, we crossed the Algerian border uneventfully, and drove to the large town of Tlemcen (pronounced Klemcen), known for its many classic Moorish buildings and 11th-century Grand Mosque. As we stood around the vehicle, a young Algerian said "'ello'. As he knew a little English, we asked if he knew a convenient camping spot. Much to our surprise, he invited us to stay at his home, the second of innumerable such invitations.

Fardaheb was his name, and he was a teacher who occupied a small stone house next to the local primary school. After introducing us to his attractive wife, we spent a most entertaining evening over dinner, which consisted of vegetable soup, goat stew and fresh grapes followed by strong, sweet black coffee. Fardaheb had apparently once spent a most enjoyable holiday in the United States and felt obliged to pay back, in this case to us, some of the hospitality and kindness he had received. He also wanted to practise his English, like countless others we were to meet.

During the conversation, he explained the current situation in Algeria. Ahmed Ben Bella, former freedom fighter who led the bloody movement for independence, was the newly elected Prime Minister. We knew the name because, in typical Arab fashion, it was plastered everywhere — on the sides of houses, bridges and even buses and trains. Independence had finally been attained from the French some three months previously but, because of the large number of 'Algeria for France' fanatics and sympathisers and the regular border disputes with neighbouring countries, armed troops maintained vigilant patrols to prevent things getting out of hand. Fardaheb himself had been imprisoned by the French for his political activities for a period of three years, but amazingly felt no bitterness or resentment towards his ex-captors. 'My people are now 'appy,' he declared, in his French accent. 'Now is the time for rejoicing, not remembering the past.'

As I sat guarding the vehicle in the centre of Tlemcen the next morning, the others marched off to shop for groceries. No sooner were they out of sight, when a slightly built European in a check shirt and dark pants approached me. 'Spotted the flag from across the street,' said the fellow, pointing to our tiny, already tattered Union Jack, 'so decided to come over and buy you a drink.' Fortunately, there was a pavement café nearby from where I could keep an eye on the Land Rover, and when the others returned, they were invited to join us.

Paul Roland, an Englishman in his mid-thirties, worked for Service Civil

International (SCI), a Swiss volunteer service. He was the leader of a group of workers from various European countries engaged in welfare work and the construction of a village for Arab refugees. For the next half hour, Paul, a writer and something of a raconteur, described the life of the volunteers who lived in an old Foreign Legion fort in the mountains. He concluded by saying, 'Come and visit us for the day, and if you like, you can stay longer.'

With nothing to lose and anxious to seize the moment and any unusual opportunity that presented itself, we accepted Paul's invitation and with his directions set off there and then for El Khemis, a village and base for the volunteers, some 50 miles due south. The road through the mountains was winding and treacherous, full of sharp bends and steep gradients. Two hours later, after a hair-raising drive, we spotted our destination at the end of a long, green valley, surrounded on all sides by steep, craggy mountains. It was to be our home for the next four weeks.

As we drove slowly along the gravel road up to the gates of the fort, we passed a herd of goats and a cluster of tents close to the huge stone entrance. Once inside the fort, we pulled up at what appeared to be the largest and most important building. A young, fair-haired woman came out immediately to greet us. 'Hello,' she said, 'I'm Glynis. Welcome to El Khemis. Come and have a cup of tea.' Inside the building which contained a kitchen and communal dining cum common room, we met several other mostly young people of different nationalities, all of whom worked for and represented the SCI.

At the time, there were already some volunteer organisations such as the American Peace Corps and the British Voluntary Service Overseas (VSO) where one could sign up to do good work abroad and, at the same time, see how other people lived. It would be many years before 'voluntourism' as such would become part of the international travel scene.

'How long are you staying?' a bearded Welshman asked. 'Where are you from?' a red-haired French girl enquired. We told them of Paul's invitation and our world trip. 'Vell, in dat case,' said Theo, a well-built blond Dutchman, 'you must stay a vile.'

As we talked, we learned something about the volunteers' work which most of them described as a working holiday. Around thirty volunteers represented professions as diverse as their nationalities. There were nurses, teachers, architects, bricklayers and carpenters from Holland, Germany,

Switzerland, France and the British Isles. All had one thing in common. They had temporarily left their respective jobs and countries of origin to work for nothing to assist the destitute people in the region. Most of the young men were employed in building a permanent village for the refugees, while the young women were responsible for their health and welfare. All were very friendly and welcoming, making us feel immediately at home.

The fort itself stood in beautiful, rugged surroundings, perched on the top of a large hill. Old and small, it had obviously seen far better days. Originally built and occupied by the proud French Foreign Legion, it had been abandoned a few months earlier; the only evidence of their once glorious stay being a small, faded tricolour left sadly hanging at the top of the flimsy wooden lookout tower. Within the high, thick stone walls stood a few dilapidated stone buildings, with roofs made of corrugated tin. All had been commandeered by the volunteers, the girls occupying some and the men, the others. At the foot of the hill was a shanty town and a mile away stood a village of dirty, brown tents housing Algerian refugees, who had been displaced by the newly ended Algerian war for independence.

Our first night was spent in the garrison prison, a large musty dungeon then used for storing cement. The next day, however, we transferred our belongings to more comfortable and congenial quarters, into what had once been the Legionnaires' bar. Furniture and all other amenities were sparse and at a premium, but we didn't mind a little discomfort now and again, having chosen the 'rough life' and grown used to being cooped up in the Land Rover for fairly long periods at a time.

Despite the prospect of not getting paid, the work appealed to us, and after deciding to stay, at least for a while, we rolled up our sleeves and went to work, adapting very quickly to our new routine.

Each morning after breakfast, the camp leader or his deputy would issue notices and instructions, details of the day's duties which were allocated by the camp committee the evening before. We four were separated into pairs; and while Mike and Eddie joined the rock party, Pete and I went off to load sand.

Our work party included Big Peter, Little Sid and Mohammed. Big Peter was an English postman who had forfeited his holiday to work, while Sid, a student at the London School of Economics, had taken a year off.

Mohammed, the Algerian driver, was one of the few paid employees. Tall, lean and olive-skinned, he boasted a pencil-thin moustache and a dozen gleaming gold teeth.

We worked all day shovelling sand from a pit into an old Dodge truck, pausing only for tea breaks and lunch. As to be expected, we were quite relieved when the day came to an end. The next day, it was our turn to load rocks while the others loaded sand, and we were even more thankful when that working day ended. After a while, though, despite the days being long and the work strenuous, we enjoyed our volunteer jobs. Indeed, the harder and longer we worked, the fitter and happier we became.

Up at six, we finished breakfast by seven and assembled outside the common room, awaiting transport to our respective worksite. Early in the morning, as at night, it was bitterly cold, and the short drive to work in the back of the Dodge had us huddling close to keep warm. Fortunately, as we worked and exercised our limbs, the cold soon passed, and by 10 a.m., when the sun was warm, we had taken off our shirts and were reduced to only jeans and boots.

Back at the fort by six, we sat down to dinner. It was nothing special but, as we were always starving from the fresh air and exercise, we found it plentiful and nutritious. Prepared by the girls, who took it in turns to cook, it invariably consisted of stewed or roast goat meat and assorted vegetables, followed by fresh oranges and dates, and swilled down with cheap red wine and black coffee. An hour or two of lively discussion and letter writing usually followed, before turning in. Once in bed, we all slept soundly, and even the occasional howls of the odd jackal that prowled around the fort walls failed to disturb us.

One day Heinrich, a German plumber, completed the water pipeline he had been working on for some weeks; and when the water gushed out the villagers were ecstatic, singing and dancing and clapping each other on the back. Previously, they had to walk a mile to the nearest stream to fetch water. Now, they had to walk little more than a hundred yards.

We had been at El Khemis about a week when Mike confessed to feeling lousy, the first to succumb to stomach pains and diarrhoea, commonly referred to as the trots. Pat, an American Quaker nurse, examined him and diagnosed amoebic dysentery, a serious and very unpleasant illness, and gave him some pills. For three days, he lay in bed, his tummy wracked in pain, wanting to relieve himself constantly. Soon, he wasn't alone in his suffering for Willy, a Swiss, and Hans, a German, were affected the same way and before long, others too.

While they certainly had everyone's sympathy, it was an hilarious sight, seeing them race across the square in a mad dash to reach one of the two outside toilets, hoping against hope that they'd make it in time! But it wasn't funny. Dysentery is a serious illness if not treated. As we would find out, it's one of the common problems when living or travelling in third world countries where sanitary conditions and the handling of food leave much to be desired. And it was one which I was to personally experience in no uncertain way in another and very different exotic part of the world.

Around the same time as the dysentery outbreak, I was restricted to 'light duties' at the fort for a different reason. While lifting rocks from a nearby hillside, I sprained and bruised my ribcage. As it remained painfully sore for a few days, I was advised by Anne, another nurse, to have myself X-rayed. So Paul drove me to the main hospital in Tlemcen in what we referred to as the 'Bumblebee', a tiny red and black Fiat used as a taxi service for the volunteers. Inside the hospital X-ray room, I was instructed by a fat, bald, chain-smoking orderly in a white smock to strip to the waist and lie down. Two minutes later, my examination was over, but instead of being allowed to leave, I was curtly told to be seated while the next patient was X-rayed. Next in the queue was an old man, who was seized roughly by the arm and injected with a three-inch needle. 'Follow me,' the orderly then instructed, escorting me along a corridor to the operating theatre, in which an operation was actually in progress. As I stood anxiously in the doorway, fearing the worst, the surgeon in mask and with bloodied gloves casually looked up and remarked, 'You 'ave no fractures. Take azpirin for ze next few days.'

For some reason, Paul was in a hurry heading back and drove as though his very life depended on it, refusing to allow anything on the road to get the better of him. After careering and skidding round hairpin bends and narrowly missing two stray donkeys, I was very relieved to reach the fort in one piece, feeling I'd had more than enough for one day!

An evening or two later, Pete, a keen fisherman, decided to go night fishing. Donning three sweaters and anorak, he collected some bread and corned beef from the kitchen and, after a cheerful wave, set off alone into the freezing night. His destination was a large lake midway up the valley. When morning dawned and he hadn't returned, a search party was organised. With every available volunteer, we searched every inch of the lakeside, shouting and whistling. But to no avail.

After a few hours, we began to fear the worst. Perhaps he had fallen in and drowned, somebody suggested. 'But what about his reel and line?' someone asked which had us shuddering. Surely if he'd slipped down the steep banks into the water there would be some sign or trace, we told ourselves. But there was nothing. Our anxiety mounted as time dragged past, and he still failed to appear. We returned to the fort, feeling thoroughly depressed.

It wasn't long, however, before Paul returned from an overnight trip to Tlemcen and loudly announced to everyone present that Pete had been found and was safe and well. Relief engulfed us as Paul related the story, and soon we were laughing at what had actually happened.

Early that morning, Paul had been called upon by the local police to identify a foreigner in gaol. He duly reported to the police station, where he found Pete behind bars. Apparently, on his lone night walk to the lakeside, Pete had been stopped and questioned by police. Being a foreigner and not in possession of his passport, they were immediately suspicious, and he was taken into custody to the Tlemcen gaol. There, after trying twice unsuccessfully to escape, he spent the night. When Paul arrived and identified him, he was immediately released.

However, not content to return to the fort empty-handed, Pete insisted on their way back that he be dropped at the lakeside to accomplish his original purpose. Several hours later, he returned triumphantly to the fort, grinning from ear to ear, holding two large barbel fish proudly in the air!

Yet another incident involving Pete occurred late one Sunday afternoon, when he and I decided to do some rock climbing in the nearby mountains. After negotiating a fairly high, steep face, we split up to wander around at the top, arranging to meet back at the fort. An hour after my return, he arrived, a little the worse for wear with a cut and swelling on his forehead. He'd fallen on a boulder and knocked himself out. Fortunately, he had no other injuries, but his independent nature would reveal itself frequently in the future, for which he was to pay quite dearly. It was beginning to look like he was accident-prone. At that time, it was unanimously felt that everything was happening to Pete, which it was, but my turn was to come much later on.

Before long, the days were racing by and three weeks had passed. Since recovering from his illness, Mike had spent most of his time in the kitchen with the women, where everybody admitted, he excelled — at cooking and flirting. We other three continued with manual labour and, despite the hard menial work, actually enjoyed it. Above all, we adored the local environment and people, and we knew when the time came, we'd be very sorry to leave.

Occasionally after dinner, we took it in turns to give English lessons to little Mohammed, a small, orphaned Arab boy who had more or less been adopted by the volunteers. Like all Arab children, he was very handsome with large dark eyes and an engaging smile. He reciprocated by teaching us a few words of Arabic. Like millions of other poor and underprivileged children throughout the world, toys were totally unknown to him, and he would amuse himself endlessly playing with rubber tyres and listening to our transistor radio.

One evening, Eddie called us outside to confirm what he thought to be a shooting star. We traced its flight for a few miles before deciding it must have been a satellite. How it stirred our imagination, for it was difficult to comprehend such scientific progress when compared to our primitive surroundings. This was obvious that same night when Bernard, a Canadian architectural student, discovered a snake in his bed and killed it instantly with one well-aimed swipe of his T square.

Saturday was a normal working day and Sunday supposedly a day of rest. Even then, however, 'volunteers' were requested to do extra work. One Sunday afternoon, we drove to another old, deserted fort where we spent hours demolishing the buildings with pickaxes and sledgehammers to utilise the timber and stones. Surrounding this fort was a high wire fence, behind which the French had planted mines. We spent an enjoyable hour blowing them up by carefully taking aim from a distance and gleefully throwing stones. It then dawned on us that there were unmarked minefields throughout the area. Several months after our departure, we learned by letter that Pierre, a French bricklayer volunteer, had suffered terrible injuries, the loss of both legs, when he accidentally stepped onto a hidden mine.

By day, the weather was usually warm and sunny, but once a sudden gale swept through the valley and blew off two of the fort's tin roofs. Theo nearly followed them as he bravely weathered the fierce wind in an attempt to batten down the remainder. Fortunately, he was successful.

The previous evening, Theo, Eddie and I had visited the one and only coffee shop in the village. It consisted of a dilapidated wooden shack; a roughly hewn timber counter; three woven mats on the uneven ground on which to sit; and one ancient, unpredictable Primus stove. On November 1st, the anniversary of the start of the Algerian War of Independence, most of the volunteers, including us, trooped into the same coffee shop for a quick Arabic coffee prior to participating in the festivities. These entailed a feast of falafel, tabbouleh, pita and hummus followed by all night singing and dancing with the elated villagers. After a few hours, most of us had had enough and went to bed.

A job which I always watched with great interest was the daily distribution of milk. At 11 o'clock in the morning, Renée, a young French woman, together with her Arab assistant assembled buckets of powdered milk at the fort gates and supervised the milk ration to what seemed a never-ending queue. A similar practice occurred when a fresh shipment of donated used clothing arrived.

Mail call was a relatively exciting event, when everyone clustered around Paul, hoping to hear their name called. I wrote home regularly to family and friends, describing our unusual and eventful new life with boyish enthusiasm.

We sometimes used the Land Rover, but as two old American trucks, and a Volkswagen minivan and the Bumblebee were at the volunteers' disposal, its services were not often required. Once or twice, we gave some Arab women and children a ride, much to their obvious delight, their own method of transportation being donkeys and sturdy feet.

Another time, we took Jean-Jacques, a French teacher, to his newly adopted village, a ten-mile drive to the east. Two years before, the French had razed it to the ground, and now it was merely a collection of temporary tin shacks. Jean-Jacques had volunteered to spend two years teaching in this village, a fact that we couldn't help marvelling at. We left him surrounded by some 40 young children outside his school, which was little more than a tumbledown shed.

In every country that I was to visit, thanks to my inquisitive nature, I made a point of learning some interesting facts — from the name of the head of government to the size of the population, the climate, and various customs and traditions, etc. Among other things about Algeria, I learned at the time that 82% of the population was illiterate and the infant mortality rate was 25%, typhoid and malnutrition being the main causes.

All too soon, we had been at El Khemis an entire month and reluctantly decided it was time to move on. Our stay had been enjoyable, and we had no regrets. On the day of our departure, the Land Rover was a little heavier than when we'd arrived with a considerable supply of Aspirin, a huge can of instant coffee, and a couple of pounds of milk powder generously given to us by the volunteers.

We were also presented with a letter of introduction from Paul addressed 'To whom it may concern.' It was headed 'The Fort, El Khemis, Commune of Seb-dou', and was signed by Paul Roland, Camp Leader.

In appreciation of our services, Paul gave each of us a five-pound note. 'Bonne chance. Au revoir. We'll miss you,' he said as we all shook hands. Then, rather glumly, the four of us climbed into the Land Rover and exited the gates of the fort for the last time. As we drove slowly along the now-familiar road, we kept glancing back to see little Mohammed and the girls waving wildly from the walls. They kept waving until they faded from view.

All of us hated farewells. Throughout our travels, I intensely disliked goodbyes, and I would never get used to it. However, I consoled myself on such occasions by telling myself that perhaps one day I'd be back. Fortunately, we were to discover that with each move, the grass was invariably greener than before. And there was always the thought of what we might have missed had we not carried on.

Later that day, we passed through Sidi Bel Abbes, once headquarters of the French Foreign Legion. The great fort and training centre had recently closed, and the Legionnaires transferred to French Somaliland and the Island of Corsica.

In the city of Oran, we stayed for a couple of days in a seventh-storey apartment overlooking the blue Mediterranean as guests of two Swiss volunteer nurses who paid routine visits to the new village and fort at El Khemis where we had worked and stayed.

We all liked Algiers, despite sleeping one night in a disused garage and another under a bridge. While in Algeria's capital, we spent two frustrating days to obtain the necessary visas for Libya and Egypt. We also met up with some other travellers with whom we exchanged information. Eager for any advice or tips, we quizzed them about the route ahead. Almost all our travel tips, road conditions, border requirements, and reasonably priced restaurants and accommodation originated from fellow travellers and, when in the position, we happily did the same for others. Occasionally, a fellow traveller dismissed our efforts with something like, 'You'll never make it.' This was slightly maddening, but no doubt spurred us on. We were reasonably confident of success, but others did not always share the same opinion.

On the road again, we fell back into our accustomed routine, stopping only to pick oranges and lemons from the trees alongside the road. And once, I picked a prickly pear. Ouch! A big mistake for I spent hours with tweezers, trying to extract the minute prickles.

Every time we stopped, no matter for how long, the vehicle was immediately surrounded by curious people who appeared out of nowhere. A few men tried to befriend us by linking hands, an Arab custom that we didn't particularly appreciate, especially when some of them led us along the street to show us off to their friends! Theirs was a different culture, we told ourselves, but we went along with it, strange as it seemed.

At the Tunisian border, immigration officials welcomed us to their country by offering us a cup of tea, declaring at the same time that very few foreign visitors passed that way. Future mass tourism would change all that. Before long, hundreds of tourists would take advantage of the glorious white beaches with swaying palms and ancient Roman ruins.

Two days of nonstop motoring later, we approached Tripoli, capital of Libya, known for its crusader fortress, the Citadel of Raymond de Saint-Gilles.

After Tripoli, we came across the impressive ruins of Leptis Magna, which would later be declared a World Heritage site. It is one of the best-preserved Roman sites in the Mediterranean, and was a major trading post, rivalling Carthage and Alexandria nearly 2,000 years ago. After a sandwich and a quick look around, for none of us were history buffs, we continued on our way.

Taking the only road east, we drove for hour after hour, through endless miles of nothing but sand. Along this rutted and mercifully straight and narrow road, a few remnants of the Western Desert Campaign of the Second World War remained: parts of trucks, tanks and broken jerrycans, rusted and half-hidden in the enveloping sand. It was here that Rommel's troops and tanks fought Montgomery's famous 8th Army 'Desert Rats'. Every so often, we passed the remains of isolated clay brick houses, damaged or destroyed by shells and bombs. On some of them, we noticed occasional handwritten signs, preserved despite the years. Typical of such were, 'You will not laugh if Gerry strafes, safety first, keep dispersed', 'H.B. Taylor 1942', and 'Kilroy was here'.

Stopping to stretch our legs, something caught my eye sticking out of the sand. I picked it up. It was an unexploded tank shell! It occurred to me if I dropped it on a rock it would blow me to kingdom come. I gingerly put it back where I found it.

As we continued along the coastal desert road under a cloudless sky of azure blue, we passed other famous places such as El Alamein that saw fierce fighting during the war. Once in a while, we stopped at a small roadside store to fill our jerrycans with water and, as it was unbearably hot, to take a swim in the Mediterranean.

Just outside the massive fortress town of Tobruk, we visited the British Commonwealth cemetery where, in the beautifully maintained gardens, we stooped down among the hundreds of white crosses to read some of the epitaphs. Most of the fallen, 'Whose names shall liveth for evermore' were about our age, and we left with great lumps in our throats feeling tremendously sad. 'Bloody waste of young lives,' was everyone's thought at the time.

Driving east, we'd only gone a mile or two, and ours was the only vehicle on the desolate road when police motorcyclists with arms waving and sirens blaring bore down on us, signalling us to pull over. Seconds later, the elderly King Idris himself acknowledged us with a regal wave from his large, black American limousine as his royal entourage swept past. Just as we were unsure of our future, so was he. He was soon to be deposed by some of his junior army officers, led by Moammar Qaddafi, an arrogant young Libyan officer and despot who was to prove far less benevolent and much more tyrannical than the king.

Usually, as darkness fell, we pulled off the road and made camp deep in the desert under the stars. Like all deserts, the western desert was extremely hot by day and bitterly cold at night. Nevertheless, we slept well, for our sleeping bags were warm and our camp beds comfortable. But before falling asleep, we discussed the day's events as we stared into the star-filled, clear sky. As we chatted, we watched shooting stars as they fell. Up at the crack of dawn, we made it a habit of first checking our footwear for snakes and scorpions.

As we busied ourselves around camp late one afternoon, we noticed through the desert haze a solitary tent a mile or so away. An hour elapsed before we spotted a lone figure silhouetted against the sky, heading our way. Another good 15 minutes passed before the figure took on the appearance of an elderly, bearded Bedouin. When he neared the camp, he took up a position leaning nonchalantly against the Land Rover.

To begin with, we ignored him and resented the intrusion. Mike told him once curtly to 'piss off' and Eddie followed a minute or two later with 'bugger off.' But he just stood there silently watching us. After observing our antics for what must have been ten minutes, he casually produced from under his djellaba five chapattis and nine hard-boiled eggs, which we assumed were for sale. Not wishing to buy, we motioned to him to be on his way. But either he didn't understand or had no intention of moving for he just stood there quietly smiling and watching. Finally, as it seemed we'd never get rid of him, Eddie offered him some money as payment for the food. With this gesture, he politely handed over the eggs and chapattis, but surprisingly refused the money. We all then grouped around him and insisted he take it, but to no avail. He was adamant. Still smiling, he then bowed; touched his brow and heart in salute; and slowly turning around wandered back in the direction he had come.

Only then did it dawn on us that sharing with strangers is the law of the desert; something we should respect and certainly not forget.

CHAPTER 3:

Mosquitoes and Bananas

'Adventure is the champagne of life.'
— G. K. Chesterton

As far as I was concerned, there was an indefinable thrill about entering a foreign country or a distant land that one has always yearned to see, and a peculiar tingling sensation of excitement would often run up and down my spine. Egypt was no exception.

We entered the sweltering heat of the ancient land of pyramids and pharaohs after crossing the border only hours before our Libyan visas expired. Extremely strict and despite our complete innocence and protests voiced by me of 'Nothing to declare,' the Egyptian customs officials made us unload everything inside and on top of the vehicle. This aggravating process took over an hour and annoyed us greatly, but there was nothing we could do except comply. After satisfying themselves that we weren't smuggling arms, drugs or contraband, the officials gave us the 'OK' and waved us on our way.

Driving slowly through the jam-packed, frenetic streets of Alexandria, the second-largest city after Cairo, three of us were in the front of the vehicle while Mike was in the back, preparing cheese sandwiches. As we slowed to turn a corner, a small, scruffy street urchin snatched the cheese from his hands and raced off. Mike immediately leapt from the vehicle and gave chase, catching up with him 50 yards down the street. After cursing the boy, he then boxed the kid's ears and retrieved the cheese! Our first knowledge of what happened occurred when a friendly shopkeeper frantically signalled us to stop the vehicle, and we saw Mike puffing up behind us. 'That'll teach him.

The little bastard!' said Mike nonchalantly, his one and only comment about the incident.

After passing the hot and dusty single-street town of Mersa Matruh, we sighted Cairo from a steep escarpment and quickly braked to admire the view. After so much barren, brown desert landscape, the Nile delta was very lush and inviting. Stopping for cool drinks at Giza near the great pyramids — one of the great wonders of the world — we were quickly besieged by jostling Arabs intent on selling us camel and pony rides.

Cairo was pure bedlam, a chaotic, pulsating mass of people, animals and traffic, infinitely worse and more hair-raising than other cities we had so far encountered. The streets were jammed with donkeys, camels, flocks of skinny sheep and goats and old cars that hooted all the time. The dilapidated buses amused us to no end with so many people hanging on outside.

It was early evening by the time we located an empty piece of land slightly out of the city. We had just started to make camp when a policeman approached with a poker-faced expression and revolver drawn, curtly motioning us to move on. There was no point in arguing, so we packed up and drove close to the pyramids. Two hours later, dead tired, we were no sooner dozing off when two friendly Arabs shook us awake and indicated that it was unsafe to stay there the night. To emphasize their point, they drew their forefingers across their throats! So we moved again, a few hundred yards this time in the vicinity of the night watchman's beat, where, after befriending him over a cup of cocoa, he assured us in sign language, all would be well.

By this time, we had camped in some rather weird places, but that site turned out to be one of the most impressive to date. The next day, when the early morning mist lifted, the view over the great Nile valley was simply breathtaking.

Back in Cairo to acquire more visas, we were invited into several different shops. Although we had no intention of buying anything, the shopkeepers insisted that we 'look around'. Each time we accepted, we were led into a small back room, offered peppermint tea or *araq* (an anise-flavoured distilled spirit), then were immediately asked, 'You wanna change dolla or sterling?' or 'You wanna buy souvenir?' In fact, almost every time we spoke to anybody, the subject of money came up. The little boys, though, were the worst. Scruffy and dirty, they milled around us with hands outstretched demanding, 'Baksheesh, baksheesh.' As we were dressed in shorts, our legs were the object of everybody's stares, and brought giggles from some of the lovely Egyptian girls. Like many of their Arab counterparts, they were mostly Muslim, wore hijabs, had large brown eyes and were beautiful and sultry.

At the Sudanese consulate where we applied for visas, we encountered a setback. A Nubian consular official explained that it was necessary to pay a hefty financial 'bond' for the vehicle, as a guarantee that the Sudanese government would not be liable for payment in the event of a breakdown. This news was disheartening, to say the least. We had been so sure of ourselves but were now facing a problem which we didn't want to hear. There was no way we could afford the necessary bond. As we discussed our plight, we counted our money. But we knew it was no good. We had only enough to cover the cost of petrol and food for a few weeks. We realised we would have to leave the Land Rover behind. 'Sod it!' said Eddie, taking a long drag on his cigarette and speaking for all of us as we faced yet another big dilemma.

Some ripe barrack-room language pierced the air, venting our disappointment and frustration among ourselves. After discussing and weighing the pros and cons of our predicament, it seemed there was no alternative but to leave the vehicle somewhere safe until we could afford to come back and get it. Nobody liked the idea for we'd become used to our home on wheels that had faithfully carried us this far. But as we all agreed, there was no point in flogging a dead horse.

Soon after, at the Nile steamer ticket office, we made enquiries about the route south and tickets for the Aswan-Wadi Halfa steamer. There, we learned if we missed the next boat, we'd have to wait a further three days for another. Time was money; we couldn't afford to hang around. We needed to keep moving.

Back at the Great Pyramid of Giza, we set up shop and placed some of our older, well-worn things on the ground and opened our own bazaar. Soon, we were surrounded by Egyptians only too happy and eager to barter for 'good English clothes'. We then hauled out our rucksacks and packed only what we considered to be necessary essentials; the contents and weight of which, at least for two of us, were to remain more or less constant for the next few years.

We then drove to the well-to-do suburb of Heliopolis, where a friendly

British consular official, having learned of our problem, had kindly agreed to garage the vehicle until we came back. Little did we know that would never happen and we'd never see it again. The idea of doing without it appalled and saddened us, for it had served us well over the past few months. Because we were limited to what we could carry, we left many things behind, including, foolishly, our mosquito nets, which would have come in very handy down the road.

After deciding to hitch-hike to Aswan to conserve our limited money, and taking a crowded local bus to the outskirts of the city, we split into pairs. Eddie and Mike had already decided to travel together. This arrangement suited Pete and me, for we knew we'd always have each other's back. Leaving the other two to be picked up first, Pete and I walked along the main road heading south for about half a mile. There we sat by the side of the road, the first of countless times, talking about our immediate plight and later, our hopes and dreams for the future.

At first, our 40-pound packs full of essential gear seemed impossibly heavy and uncomfortable, and we had to rest from time to time. But gradually we got used to them. That would be the first of hundreds of times we'd hoist those packs containing most of our worldly possessions onto our backs.

We didn't have to wait long before a large truck appeared. The driver stopped, inviting us with hand signs to hop in beside him. We were to depend a lot on lorry drivers, who seemed much the same around the world. Invariably helpful and friendly, they represent a certain breed. In this case, we made frequent stops when the driver insisted on buying us glasses of sugar cane. When we tried to reciprocate, he tut-tutted, wagged his finger and refused.

We jumped out at the town of Beni Mazar, as the truck was going no further, and bought some dates and bread rolls, surrounded by what seemed the entire local population. We were rescued from the crowd of curious onlookers by a young man who introduced himself as Abdul, a local schoolteacher. 'Perhaps I could be of some assistance?' he asked, anxious to use his English. 'Would you please join me for some mint tea?' We did and, in the course of conversation at a small outdoor café, explained to him our imminent destination. 'I will help you,' Abdul said. 'Please follow me.' Minutes later, we found ourselves at the town hall where Abdul requested permission to see the mayor.

Beneath a huge portrait of a smiling Gamel Nasser, soldier-president at the time, Abdul repeated our story to the local mayor, who then picked up one of his two telephones and spoke briefly to someone in Arabic. After putting it down, he murmured a few more words in Arabic to Abdul, then extended his right hand and wished us 'Assalamu alykum'. Outside, Abdul explained that the mayor had arranged a lift for us at the nearby military check post.

As we trudged beneath the weight of our packs, Abdul suggested that we spend the night at his home. He emphasized that if we stayed, he would willingly provide a 'lover', presumably an attractive young woman. This was our first offer of this kind but would not be the last! Pete and I exchanged knowing glances as our imaginations got the better of us, and we relished the enticing suggestion. But much as we liked the idea, we refused, intent on reaching our destination on time.

At the barrier, the duty sentry was waiting for us. He had received instructions to stop the first passing vehicle and order the driver to take us as far as he was going. The driver would drop us at the nearest check post at the end of his journey and relay the instructions to the next driver. As there were numerous military check points throughout the country, it seemed like a great plan. What luck, Pete and I agreed, grinning at each other!

Shortly after, we were sitting on top of a truck laden with sugar cane. That night, we reached Menina, having seen no sign of the other two on the road. Standing in the centre of this strange town, we were unsure where to stay. We couldn't afford the price of a hotel, so we decided to make our way to the central police station to deposit our packs for safekeeping and to enquire where best to spend the night.

A constable met us at the main door and introduced us to the duty officer. A young, handsome lieutenant, resplendent in khaki and Sam Browne belt, he immediately clicked his heels and saluted us, then bid us relax with the words, 'You. Sit down.' Then dispatched an orderly to fetch some tea. So began a conversation of sorts. His English was rather limited, but like so many other people we were to meet, he was very keen to practise it. Our conversation was soon interrupted when duty called — to deal with that evening's lawbreakers.

As we sat sipping tea and eating banana sandwiches, several *fellahan* (peasants) were brought in, including one in handcuffs and two with bloodied faces. One by one, the lieutenant shouted at them, and they were led away to be locked up. What dreadful fate lay in store for those unfortunate miscreants we never found out, but we assumed it would be a cooling-off period in gaol. As darkness was approaching and we had nowhere to stay, the lieutenant invited us to remain at the police station. We slept that night in our sleeping bags on tables in an adjoining room.

From then on, we would often utilise the facilities of police stations in Africa. If we needed information, somewhere to leave our rucksacks for safekeeping, or shelter for the night, we automatically resorted to the police as they seemed only too pleased to see us, especially in remote regions. And glad to help out.

Although fairly frequent, our rides, mostly in trucks, were quite slow. Once, to speed things up, we caught a bus. After elbowing our way past several peasants, some chickens and a couple of kid goats, I seated myself next to a respectable-looking fellow wearing a red fez. As the bus proceeded, neither of us spoke. Sometime later, we stopped in a village square. Leaning casually out of the window, my seat companion called out in Arabic to a small boy who was selling oranges nearby. He purchased two and smilingly handed one to me. I felt very humble as I thanked him, but he simply put his forefinger to his lips as if to say, 'That's enough. Don't mention it.' With that, we continued on our way without speaking as before. With this small act, we learned to regard most Egyptian people, like so many other nationalities we would ultimately meet, as gracious and welcoming.

It was then that Pete started to complain of stomach pains, which grew worse as we travelled on a truck laden with tiles. We slept on the tiles as the truck rumbled through the night, then in the early morning transferred to another loaded with crates of dates. Finally, we boarded a truck which was destined for Aswan and, upon learning this information, we were happily convinced that we'd reach our destination well ahead of time.

After driving along the Nile for mile after mile and passing the famous ruins of Luxor, city of ancient monuments and gateway to the Valley of the Kings and Queens, which I would visit and explore many times in years to come, we entered the eastern desert.

The road soon deteriorated, and what was once macadam became only gravel and sand. Approaching a steep hill, the driver changed into the lowest gear and we started to climb. Alas, the engine wasn't powerful enough, and we ground to a halt then slowly rolled back. Three times our driver, named Ahmed, attempted to negotiate the hill but failed. On the third try, we rolled back too fast and hit a partially hidden rock, then heard a grinding noise as the rear axle snapped. 'Shit! That's done it,' I vaguely remember saying to Pete.

The three of us waited in the cab about two hours for another vehicle to appear in the hope of getting help, but the road remained empty and silent. We were in a tight spot and knew it, but it wouldn't be the first time. We sweated profusely in the ferocious heat of the desert sun as we contemplated our predicament.

Ahmed then made a decision. He gave us a loaf of bread and some water, bade us stay where we were, and in faltering English said something like, 'I shall be back as soon as I can.' With a half salute and half wave, he set off on foot to God knows where to seek help. How far he walked, we shall never know, for there were no villages in sight and no sign of life. The desert was completely barren, with nothing but sand and rocks as far as the eye could see; a more inhospitable place you could not imagine. How long he'd be gone was anyone's guess, but we desperately hoped it wouldn't be too long.

It was around 11 o'clock in the morning and already blisteringly hot. Pete complained about stomach pains again. Whenever he spoke, it was through clenched teeth as if to ease the pain. I could tell he was getting worse but, unable to treat his complaint, I could only sympathise, and insist that he see a doctor when we reached Aswan. The hours stretched on and the sun grew higher and hotter. Sitting inside the cab of the truck made the uncomfortable circumstances worse. So we sought and found some slight relief in the shade of some huge rocks nearby. There, Pete lay down and tried to doze off. More hours passed, and we finished our bread and most of the water, and still there was no sign of any other traffic. By now, we were becoming increasingly concerned.

Eventually, nightfall descended and as it turned bitterly cold, we fell asleep under the tarpaulin canopy in the back of the truck. We woke up in the middle of the night to the sound of a truck pulling up. Unfortunately, it was too full to take us. But we made ourselves understood. Despite feeling rather sorry for ourselves, we felt we could hardly abandon the vehicle without consulting Ahmed. More bread and water were handed to us, and we were at least relieved to know that the road was, in fact, used once in a while!

Early the next morning, to our great relief, Ahmed returned out of nowhere and told us he'd arranged for a mechanic and repair truck that would be along soon. As we waited patiently, a petrol tanker stopped, and Ahmed and the other driver exchanged conversation. We were then told to change vehicles. Previously listless, we suddenly came to life as we clambered into the cab. But then we realised there was only room for two. What about our driver? 'OK,' said Ahmed with a thumbs up, reading our thoughts. With that, we drove away. We had been stranded in the desert for a little over 24 hours.

Finally, in Aswan, we chatted to two young American archaeologists, named Al and Bill, whom we met in a small café. One look at Pete, whose eyes were now slightly jaundiced, and Al immediately suspected hepatitis. He advised him to consult a doctor at once. With that, I accompanied Pete to the nearest local hospital staffed by a German doctor and Swiss nuns. He took his turn in the queue of townspeople. After examination, the results came back positive. Pete had, as Al had thought, hepatitis. The doctor prescribed a specific diet and lots of rest, but Pete, proud and independent as always, and intent on reaching Kenya as soon as possible, dismissed the advice.

At the Grand Hotel — a deceptive name for the place was nearly falling apart — where our American friends were staying, they smuggled us into their room, where we enriched ourselves with long-overdue showers. As a final show of generosity and kindness, Al and Bill treated us to a typical Egyptian meal of rice, beans and lentils which made a refreshing change from what had become our regular diet of bananas and bread and jam. They then took us by taxi to the waiting steamer. There, we were reunited with Mike and Eddie, who had arrived uneventfully the day before.

After buying third-class tickets, otherwise known as deck class, we made our way on board the steamer. For the next 36 hours, our method of transport was desperately cramped, alive with people, children and chickens. Somehow, we managed to find enough deck space for our sleeping bags and laid them out.

Our river journey included a brief stop at Abu Simbel to allow everyone to look around the famous temple which would soon be carved up and moved to higher ground to avoid being flooded. This monumental task was organised by UNESCO. Lake Nasser would be created when the multimillion-dollar Aswan High Dam, then under construction, was completed. When our boat emitted three blasts of its siren, everyone trooped back on board.

Everyone, that is, except Pete. Only when we had cast off and started to head upstream, did we notice his absence. As we wondered where he was, we heard a shout from the shore. There he was, pelting along the bank, waving and shouting to attract attention. Fortunately, the mate spotted him from the wheelhouse, and, with a clang of bells, we hove to. The gangplank was dropped, and he came on board completely out of breath. Despite feeling lousy, he had been so intrigued by the inside of the temple that he failed to hear the steamer blast. Once again, he'd had another narrow shave; he could have been stuck there for days! We all laughed at the incident and pulled his leg about it, but we also shuddered at the possible consequences.

At Wadi Halfa, an insignificant desert town, we transferred to a crowded train. Inside, we stretched out on the luggage racks in an effort to be more comfortable and to evade the chickens, sheep and goats which occupied the aisles. We travelled that way for the next day and night — covering the entire length of the Nubian Desert, a formidable mass of barren land by any stretch of the imagination.

Arriving in Khartoum, capital of the Sudan, it was stinking hot. The temperature registered 102°F and Pete was affected drastically. As I walked, he dawdled behind, unable to keep up. It was the last straw. He had to have urgent medical treatment. A passer-by directed us to Harper Hospital, a private British nursing home, where the Welsh director, a Doctor Jones, examined him, then ordered him immediately to bed. By this time, Pete could hardly walk. The pain he described was like a hot knife thrust into his belly, and he was certainly in no position to have his own way. Meekly and thankfully, he did as he was told.

The prognosis was not good. According to Doctor Jones, Pete had an advanced case of hepatitis and, had he not received immediate medical attention, would have ruined his liver for the rest of his life. 'Don't worry,' the doctor told us, 'What he needs now is good wholesome food and proper rest.' He also said that Pete was not the first to arrive in such a sorry state. 'I see a few pass through here,' he said, 'travelling on a shoestring and not eating properly.'

Relieved at last that Pete was in good hands, we three others moved into a dormitory at Khartoum University for a small fee, at the invitation of a Sudanese student we had met in town. Here, we delighted in the luxury of clean sheets, real beds and the use of a swimming pool. We visited Pete every afternoon during our five-day stay, and each time we saw him, he looked and felt decidedly better. His one complaint, however, was that he couldn't afford the fees of five pounds per day, and he mentioned this to Dr Jones every time he made his rounds. 'Look Doctor,' he would say, 'Regarding the bill, I just can't afford it now. I'll send you the money later.' And each time the good doctor replied, 'Don't worry. Just rest.'

Meanwhile, there was a steamer leaving from Kosti for Juba in southern Sudan within a couple of days. Apart from flying, which we couldn't afford, this was the only regular river link south through the infamous Sudd, a vast area of dense, inhospitable marshland and relentless, muggy heat. The steamer service, unreliable at the best of times, only operated every ten days. With our funds dwindling rapidly, we didn't have to think twice. It was essential that the three of us catch it.

We discussed the situation with Pete, who characteristically urged us to keep going — easing our conscience by telling us he'd be discharged in time for the next one. We left him behind reluctantly, after doing our best to encourage him and cheer him up, but we felt we had no alternative. We could have hunted for work, but none of us relished the idea of working in oven-like temperatures for what would probably be a tiny pittance. Besides, we'd need work visas, which would have been impossible to get.

Another problem then reared its ugly head. In Egypt, we had been assured that United Arab Republic (UAR) currency was negotiable anywhere, but we soon discovered this was far from true. When changing traveller's cheques, we would change only what we estimated we needed to cover our expenses in each country. Anything left over, we changed at the borders or banks. But in this case, we had over calculated. We tried changing the UAR at several banks, a few hotels, shops, and the British embassy — all to no avail. Even the Egyptian consulate didn't want the lousy stuff!

The actual sum was equivalent to about 20 British pounds, not much to most Westerners but then a princely figure to us. Fortunately, we met a German traveller in a little restaurant and, as he was heading for Egypt, he agreed to a swap. Had we not been successful in changing that 'useless' Egyptian money, we might well have had to visit the nearest British embassy and been repatriated. Having come this far, the last thing we wanted was to be sent home at taxpayer expense. The prospect of repatriation appalled us, but it was to hang over our heads more than once.

After again purchasing third-class train tickets, stocking up with tinned food, and queueing up at the station for the train to Kosti, the long wait and the intense heat proved too much, and we collapsed several times into nearby café chairs where we weakly clapped our hands — Arab style — for service and cool drinks.

Like all Egyptian trains, our train, was pulled by an old British steam engine. It finally crawled out of the station five hours late. Having forced our way on board through hordes of shouting, jostling Arabs, we sat next to three students who offered us homemade jam and sweetmeats throughout the journey. The train with its open windows and packed to capacity took 12 hours to reach the town of Kosti — a distance of only 100 miles! As we coughed dust out of our lungs, we thought we'd never reach our destination, for we stopped at every village and hamlet along the route. Finally, we boarded the *SS Tagool*, an ancient paddlewheel steamer which was to be our home for the next ten days and nights.

Once aboard the *Tagool*, we commandeered some space and looked around. As third-class passengers, our sleeping quarters amounted to throwing our sleeping bags on deck, a space we shared with a couple of hundred Egyptians and Sudanese. 'Home, sweet home,' remarked Eddie jokingly, and, although none of us were exactly impressed with the conditions, we could hardly complain, the fare being only the equivalent of about five pounds each. There were half a dozen other international travellers on board, including three or four Europeans and two young South Africans who were heading home from London to Johannesburg by motorbike. Drawn together by a common bond, we quickly became friends.

Initially, we were all quite excited to be travelling upstream on the famous Nile River. After the first day, however, we quickly became bored, for there was little to do, even less to see, and it was oppressively hot. Consequently, when someone produced Alan Moorehead's relevant and absorbing book, *The White Nile*, it was regarded as a godsend and was passed eagerly from hand to hand. As was John Gunther's *Inside Africa*, one of the few detailed reference works about the African continent then available.

As the steamer entered the virtually unexplored and forbidding Sudd, the river was immensely wide in some places, but most of the time it was quite narrow. For mile after mile, the banks were lined with tall, dense papyrus reeds, too high to see over for any view. The current was so strong that the helmsman negotiated it by purposely hitting the banks and bouncing off in an attempt to maintain course. In addition, the water was full of small, floating islands, making it sometimes difficult to make any headway at all. Every so often the giant wooden paddle wheels got clogged with dense reeds, bringing us to a halt and making it necessary for all hands to clear the reeds manually with the aid of machetes.

Once or twice a day, the monotony was broken by the sight of a dead, bloated antelope floating past. The few Nile perch that we saw were enormous, sometimes three or four feet in length and weighing 300 to 400 pounds, which made excellent eating, like the other well known, but smaller, tilapia, found in profusion in those parts. We also noticed huge crocodiles on the banks basking in the sun and, occasionally, hippos playing in the water. But most of the time, we saw nothing. And the only sound we heard was that of the boat's powerful engine, which droned on day and night.

It was about that particular section of the river that the British explorer, Sir Samuel Baker (also known as 'Baker of the Nile') had written, 'All is wild and brutal, hard and unfeeling'. Not much, we noted, had changed over hundreds of years. It was as nature intended, with virtually no sign of life. Slowly, the days dragged by, and in our boredom, we paced up and down the deck, restless like caged lions.

Just as most days were uneventful, they were also intolerably hot, and we were grateful to be in the shade. Such as it was. Because of the intense heat, we constantly drank bottled water and took frequent river water showers in our shorts and T-shirts.

There was no relief from the muggy conditions at night, and the king-size mosquitoes had a field day swarming and feasting with impunity on everyone. It got to the point when just one buzzing around my head would keep me awake all night. No matter where I went on the boat, they would dive bomb me, targeting my arms, ankles, neck, and ears. Despite being well covered with repellent, they still attacked me. It reached the stage when, in desperation, I muffled myself up in two sweaters and thick socks. However, I soon began to sweat and was forced to strip. Slipping into my sleeping bag brought me no relief, for they followed me there too. So I got up and wandered around deck. Finally, at about 3 a.m., I would fall asleep totally exhausted.

When dawn broke, the activity on board made it impossible for anybody to sleep in peace. With only two or three hours sleep a night, our tempers frayed and like Eddie and Mike, I became irritable and fed up. Hardly surprising under such extreme circumstances.

Just about every second day on our river journey, our boat pulled alongside a remote village to pick up or discharge cargo and passengers. At each stop, we all went eagerly ashore, a very welcome opportunity to stretch our legs and restock with food. The one and only local shop like all the other 'buildings' in such an isolated area was usually little more than a mud and straw hut. Our purchases were always the same — bread, jam, cheese and bananas — our staple diet! But we soon tired of the same food, especially bananas, which we only bought because they were so plentiful and cheap.

Often waiting to welcome us stood the entire local population, most of whom were very black and quite naked except for coloured beads around the waist. Meeting the steamer was always an exciting occasion for them, for it provided their only real contact with the outside world. We too regarded each stopover as a blessing. Frequently, we were invited into mud huts by some of the natives, who had been educated by missionaries. They talked quite freely about the brewing revolution and their resentment toward their Arab 'masters,' who lived predominantly in the north but who controlled the entire country. Years later, South Sudan was to fight a protracted war against the north, finally winning their independence and a seat at the United Nations.

Christmas 1962 dawned unobtrusively. Our fellow European travellers joined us to celebrate. Before leaving home, my mother had insisted I take a large tinned Christmas cake and pudding which, despite cursing the weight, I'd dutifully carried that far. Now was the time to haul them out.

Amazingly, we each contributed something for the special occasion and between us, produced some biscuits, tinned ham, and even a bottle of sherry

— and, of course, bread, cheese and bananas. A banquet fit for a king, we all gleefully thought. Tony, an artist from England, rose and proposed a toast. Plastic beakers and tin mugs were raised as 'A happy African Christmas' was declared. Our celebratory meal was followed by singing carols into a tape recorder. It was my first of several Christmases abroad in exotic places.

As we were all chatting away and celebrating that special day, a shocking incident marred the happy event. A sudden high-pitched scream rent the air. To a man, we all jumped up. A toddler had fallen overboard, and his mother was screaming hysterically. We looked over the side and stared into the murky depths of the river, hoping against hope to see the small figure, but there was no sight of him. All too quickly, he had disappeared. The current was swift and treacherous — he never had a chance. He had quickly drowned or been snapped up by a croc. Nevertheless, the captain chose to search up and down the river for nearly two hours, but there was no trace of the unfortunate victim. We hardly slept at all that night, lost in thought as to the little boy's dreadful fate, the tragedy emphasized by the terrible, incessant wails and sobs of his distraught mother.

On Boxing Day, we decided to buy cool drinks. Purchasing a Fanta or Coca-Cola became part of our daily routine on the boat. Apart from quenching our perpetual thirsts, we had an ulterior motive. We could escape our highly uncomfortable conditions and find relief from the unbearable heat in the first-class dining salon for a couple of hours.

By then, we were ravenous and desperately hungry for a decent meal. When the big coal-black chief steward sat down at a table next to us to eat lunch, Mike and I stared wistfully at his plate piled high with meat, potatoes and vegetables. After taking no more than one or two bites, he could stand our stares no longer. 'Help yourself,' he said as he beckoned us to finish it up. Needing no second bidding, we descended on it like wolves, and voraciously polished it off to the last morsel!

After ten long and boring days and nights, we came to the end of our river journey and docked at Juba, much to my own great relief. Next time, I vowed, it would be first or second class for me, with a proper bunk, fan, mosquito net, and good food. But time was to erase my miserable memories quicker than I thought.

Soon after docking and plodding out of the sprawling port town, which held little of interest except for a few large and lovely old stone buildings, we all sat by the roadside waiting for a lift. We sat there for some hours in the sweltering heat before a truck gave the three of us and Tony a lift all the way to the Ugandan border. There, across the frontier barrier, we chatted to an African student who had been barred from entering the Sudan because of his financial position. He told us that he was considered destitute for possessing only the equivalent of about ten pounds sterling. Imagine our feelings when we counted our money and realised we had little more than him.

Most governments have a rule stipulating all visitors must either be in possession of an 'onward' ticket or 'sufficient funds' to provide for themselves during their stay. Without this, you could be refused entry. We had neither, but were confident and optimistic that we wouldn't become a responsibility to anyone. Fortunately, as Tony was with us and relatively well off, we borrowed some money from him to enter Uganda and returned it to him after our passports were stamped. Once again, we split up to make getting rides easier. Eddie and Mike stayed together while Tony and I teamed up and walked ahead.

After crossing the equator, which was well signed with a huge map of Africa, and experiencing three breakdowns in two cars, we all finally arrived safely in Nairobi, capital of Kenya. There we met up and compared notes. It was New Year's Eve 1962 but, with less than 20 pounds between the three of us, we were in no position to celebrate.

Our first night, we slept at the main police station where the duty officer, a youthful English inspector, was kind enough to allow us to occupy one of the cells. Ironically, it was right next door to the classy Norfolk Hotel, the finest hotel in town — where I would stay years later as a paying guest. After a chance meeting with the Chief of Boy Scouts, a big Welsh fellow named Clive, we stayed at the Boy Scout Headquarters, also in the centre of town, for a couple of days.

Our immediate objective then was to find long-stay, inexpensive accommodation and, even more importantly, work. Tony was headed to South Africa and anxious to be moving on, so he left us within a few days and headed south alone. As we waved him off outside the Thorn Tree café, quite by coincidence as I scanned people's faces — as I now always did in case I saw someone I knew — I spotted a vaguely familiar face in the crowd, Captain Robin Leach.

During our NS days, Robin had been Signals Officer of the battalion (Pete's boss). After hearing about our adventures and expressing his surprise at meeting us, he suggested that we call on others we had known, who were then attached to and serving in the Kenya Regiment. Shortly after, we were exchanging news and reminiscing with a handful of once familiar army friends. During our conversation, we were invited to utilise two of the large army tents which sat erected on the lawn adjacent to the barracks and parade ground. We were so pleased that, as former members of a sister regiment and to show our appreciation, we joined the famous regiment there and then.

The Kenya Regiment at that time was a voluntary, multiracial regiment which entailed only weekend soldiering. It also had a splendid reputation gained during the Mau Mau emergency (1952–1960) and the Second World War. Upon signing up, we were issued with khaki uniforms, which included Australian type bush hats, shorts and knee-length socks, and on the few occasions we went on parade we felt quite smart. With our accommodation problem solved, our main intention was now to find work — much easier said than done.

I took it upon myself to make up an ad which advertised our combined services in the *East African Standard* and *The Nation*, Kenya's principal English newspapers. Our ad said we were 'willing to do anything and go anywhere'. We got one response from a Mrs Sampson, the co-owner with her husband of Samcax Book Services, a bookshop dealing mostly in encyclopaedias and various reference works. She immediately offered us jobs as sales reps. As we were just about completely broke by this time, we couldn't afford to be fussy, so we accepted her offer and began work on a commission basis.

Meanwhile, Pete arrived, found us and related what had happened to him after we had left him in Sudan's capital. He'd remained in the Khartoum nursing home for a period of ten days, each day pressing his financial status case before Dr Jones. And every day Dr Jones told him to relax and not to worry. On the day he was discharged, the good doctor shook his hand.

'Good luck and forget the bill,' he said as he pressed an English five-pound note into Pete's palm!

The next day, following in our footsteps and the route we had outlined to him, Pete took the train and steamer and made the trip without incident, like us teaming up with other fellow travellers and picking up the various messages that we had promised to leave at strategic intervals. Now, never one to let grass grow under his feet, he immediately found a job in a little factory in the industrial area of Nairobi. Amusingly, he ate the same meal of eggs, sausage, chips, and beans almost every day at the same small Indian-owned restaurant. He had taken Doctor Jones's recommendation to heart about eating well, even though it was not the healthiest of foods! His sobering illness had in one fell swoop put an end to his pub-crawling days and, while he didn't give up booze completely, he greatly reduced it.

The modern New Stanley Hotel was centrally located on the corner of Kenyatta Avenue and Kimathi Street. The Thorn Tree pavement café was part of the hotel. Named after a thorn tree sapling planted in the middle of the café, which over many years would attain the height of eight storeys, the café had become a favourite rendezvous for us, as it was for many travellers. There we indulged, money permitting, in Longonot sandwiches (a triple-decker sandwich named after a famous volcano in the Great Rift Valley).

One day, seeing two empty seats at the bar next to two vivacious young women in their twenties, I sat down beside one with Mike next to me. As I pondered what to order, the young woman and I smiled at each other. 'Would you like a chip?' she asked in a charming Italian accent. I accepted with pleasure. The next thing we'd introduced ourselves and were talking like old friends. And Mike was chatting up her sister. During our brief talk, I learned that 22-year-old Anna-Maria Peccioli was a secretary at the Italian Embassy and Christiana, her younger sister, worked for Alitalia Airlines. Their parents had sent them to Nairobi from their Mogadishu home in Italian Somaliland, as it was then called, to improve their English. They shared a small flat not far from Nairobi city centre and operated a little white Fiat 600 as their runabout.

Thus, began brief but fairly intense and passionate relationships for both Mike and me. Mike dated the quiet and more demure Christiana while I went out, when in town, with the more outgoing, and somewhat voluptuous Anna-Maria. A free spirit if ever there was one.

One evening shortly after we'd met, the four of us went out for dinner and afterwards the sisters took us back to their flat, which wasn't far from the Kenya Regiment Barracks. By this time, we were all starved for intimate female company, and what happened next both Mike and I eagerly embraced. No sooner had we entered their premises than Anna-Maria led me into her bedroom, closed the door, gave me a long lingering French kiss then slowly and deftly undressed and seduced me while Cristiana did the same to Mike in the bedroom next door. 'Stone the crows!' exclaimed Mike, licking his lips and grinning like a Cheshire cat as we later left the darkened flat in the dead of night, and made our way on foot back to the army tents where we were staying. No doubt about it, we happily agreed, it had been quite a night!

Several weeks later, for my 24th birthday, Anna-Maria gave me a Swiss army knife which came in useful and which I would carry with me on boats, planes and trains all around the world. Like my little Optima camera, it became my constant companion wherever I went, until one day years later, it was confiscated at Cape Town Airport after security at all airports was tightened.

Like other travellers of the time, we hung out a lot at the Thorn Tree, swapping information with other hitchhikers and like-minded souls. The café boasted a large noticeboard permitting anyone to post notes about where they were staying, where they were headed, and their onward address. And because it was adjacent to the New Stanley Hotel, more than once a specially equipped open safari vehicle would pull up by the kerb and out would step some Hollywood movie star accompanied by their white hunter, followed by onlookers and admirers with mouths agog. Once, I spotted Ava Gardner with Jimmy Stewart. Another time, Deborah Kerr and Stewart Granger. It was still the time of hunting safaris, and Percival Safaris and Ker & Downey safari vehicles were seen everywhere. I had no idea then that well into the future, I would not only be occasionally staying at the New Stanley Hotel but also running my own tour and safari company.

It was also at the Thorn Tree café where we became acquainted with a young white Kenyan named Gavin MacHutchin. Gavin wore specs, had a mop of curly black hair, owned a Land Rover, spoke fluent Swahili, and lived with his parents in a Tudor-style mansion in the exclusive suburb of Karen. His father was Vice-President of Pepsi-Cola, which was well established and very popular throughout the continent. And, as I would discover, could be found in the most remote places!

Born and bred in Kenya, Gavin was an experienced bushman, a year or two younger than us. No sooner had we all met than he invited us on a number of excursions in his Land Rover. One such outing was to Nairobi National Park, just outside the city, the first designated park in Kenya. Covering an area of 45 square miles and surrounded by a high wire fence with a wide gap to allow the animals to migrate, it had all manner of game, including Cape buffalo, wildebeest, hartebeest, hippo, baboon, impala, giraffe, spotted hyena, waterbuck and even rhino, leopard and lion. There, we whiled away hours of enjoyable game viewing.

On another occasion, Gavin invited us on a weekend safari, the first of several. And for one specific reason, that particular weekend remains vividly in my memory.

It was a Saturday, and there were six of us in the party: Mike, Pete, Gavin and Gavin's parents plus myself. After driving out of the city well beyond the Ngong Hills, we camped beside a stream lined with acacia trees not far from Lake Magadi, a huge salt lake and an intensely hot place. We finished lunch and left Gavin's parents taking a nap in the shade of some tall fig trees while we drove off in search of game.

We had already seen a fair number of animals, including dozens of Grant's and Thomson's gazelle, giraffe, hartebeest, topi and buffalo. We stopped to watch the antics of a herd of wildebeest, also called gnu and known as 'the clown of the plains' because of their weird appearance and particularly their behaviour during mating season. We had just passed the half-eaten carcass of a zebra trussed to a tree as lion bait, the other half no doubt eaten by hyenas — hunting was then still permitted and would not be banned until 1977 — when we began driving alongside a stream, usually a good place to sight and flush out wildlife. Gavin was driving; Pete was sitting next to him; while Mike and I stood in the back of the open Land Rover.

As we drove alongside the stream where the vegetation was fairly dense with thorn trees and wait-a-bit bushes, Mike and I were constantly ducking down to avoid getting scratched. Suddenly, Pete shouted, 'Baboon!' Simultaneously Gavin shouted 'Heads!' In looking for the baboons, I failed to bob my head. All of a sudden, I felt something dig into my face. 'Stop!' I shouted, throwing my hands instinctively up to my face, immediately feeling the warm stickiness of blood. 'Oh God,' I groaned to myself. 'What the hell have I done?' Ugly razor-sharp, two-inch-long thorns had torn into the left side of my face, narrowly missing my left eye. And I was bleeding profusely.

Mike had witnessed the accident and immediately pressed a handkerchief tightly to my injured face in an effort to stem the flow of blood while Gavin quickly turned the vehicle around and sped back to camp. There, I lay on a camp bed while Gavin's mother soaked a towel in water and wrapped it around my face. When the bleeding stopped, she washed my face gently, and then applied iodine lotion to prevent infection and to soothe the nagging pain. That put an end to our weekend adventure. Within an hour, the camp was packed up, and we headed back to the Nairobi suburbs.

For a while, the left side of my face was a mass of ugly scabs, which brought stares from many people and made me very self-conscious while Anna-Maria was both horrified and sympathetic and tried not to notice. Those wounds, superficial as they were, would take weeks to heal, and I was to bear the scars of the accident for a few years. Eventually, after travelling and living mostly in sunny climates, they gradually faded completely. Early on, though, I came to see the funny side of this incident, often joking that I'd been involved in a fight with a baboon! I also viewed it as a warning. I was lucky not to lose an eye and would have to be more careful in the future.

Living in the tents, while not unpleasant, meant dampness and mosquitoes. On our river journey, I had realised just how much they liked me more than the others. I couldn't sleep for the mosquitoes. We were all grateful to be allowed to move into the main block of the regimental barracks. With a bathroom right next door, brushing teeth, taking a shower, and relieving ourselves were all now so much more convenient.

We continued to meet new people, including three American nurses named Barbara, Pat and Carole, all of whom worked at the Aga Khan Hospital and were around our age, in their early to mid-twenties. After some discussion, the seven of us rented a house together. Soon after we moved in, I got a job on safari that involved being away a lot. But I heard from the others that the house was the scene of many parties.

It was there that Eddie met another Carol, a white South African office worker. Right from the start, she had her eye on Eddie. She was his first real girlfriend, and he succumbed quickly to her charms.

From our previous experience in the Green Jackets we knew that soldiers have to be active or seen to be active. Someone suggested to the CO that we tackle Mount Kenya, Africa's second-highest mountain after Mount Kilimanjaro. One weekend we joined members of the regiment to climb this long-extinct volcano of slightly over 17,000 feet located on the equator. After driving from Nairobi in a convoy of the regiment's Land Rovers to the township of Naro Moru, we turned off into the Aberdare Forest where we established base camp and erected several large sleeping tents, a kitchen tent and a smaller mess tent. We watched sunset over the iconic mountain and sipped our 'sundowners' — alcoholic beverages like whisky and soda, gin and tonic or local beer — before turning in early.

The next day, rising at dawn, we set off on foot with laden backpacks through dense bamboo. Along the way, we spotted a couple of grey duiker; several eland, largest of the antelope family; and some mountain zebra. After struggling for several miles over open alpine moorland along a rough, sodden track made by elephant and buffalo, we reached the snowline which for the majority of us was the absolute limit! I had extremely sore feet from my new army issue boots, and most of the others also had valid reasons to call it a day.

No sooner had we set up camp within view of the mountain's three peaks than the mist rolled in. After spending a freezing, sleepless night at 15,000 feet in the small, flimsy tents, we were greatly relieved when the officer in charge called off the climb. This pleased almost everyone. Typically, though, Pete was one of the keen three or four who left to climb Point Lenana, but they were unable to reach the summit due to lack of time and proper equipment.

We also conscientiously turned out on the firing range to test our rusty skills, and shortly before the regiment disbanded, the following item appeared in the notes of the regimental newsletter, written by the adjutant:

> *I remember being struck by the sudden appearance on the firing point of four extraordinary apparitions, dressed up in an odd*

assortment of semi military attire and winkle pickers [Mike wore those]. On closer examination, these proved to be our four round-the-world volunteers hot from our parent regiment, but not having paid a visit to Bwana Kalele [the name of the firing range]. The entry of these chaps into the platoon so soon after leaving one of our parent regiments is, I think a unique event and it is a pity that Dave, Pete, Eddie and Mike could not have had more time with us.

Meanwhile, after about two months of selling books, I for one felt I knew every paving stone in Nairobi as Mike, Eddie and I made sales calls at shops, offices and lots of small factories in the industrial area, mostly owned by polite, well-educated Indians, in our quest to sell books. The job itself was totally uninteresting. To make matters worse, Mrs Sampson (married to Colin, a quiet, mild Englishman) was demanding and a bit of a sourpuss, a real taskmaster who lectured us how to conduct a sale as often as she could. Eddie and Mike weren't enjoying it, and my heart just wasn't in it, so we collectively resigned. The job had served its basic purpose, and we had eked out an existence, but that was all. By then, however, we had made a number of helpful friends in Nairobi. Within a day or two of quitting the books sales job, Eddie found work with Singer, the sewing machine company, and Mike got an offer from an insurance company.

To be employed on safari was one of my greatest ambitions. I made enquiries for such a job and one day contacted Carr Hartley, a well-known white hunter and trapper. At the time of our meeting, he was concentrating on capturing rhinos for zoos and the film *Hatari* (Swahili for 'danger') for which he was a consultant. He captured them with the aid of lassoes from fast-moving vehicles. Like many of his profession, his reputation preceded him. He had a very loud voice, and his arms and legs were covered in hunting scars. He owned a ranch named Rumuruti in Laikipia District where he kept a variety of animals as pets, for movies as well as for zoos. During my brief interview, he proclaimed, 'I either want to die at 90 murdered by a jealous husband or be gored to death by a rhino when I'm 95!' As he was something of a braggart, I wasn't sure if I wanted to work for him. As it transpired, he had no job vacancies at the time.

Thanks to our new-found friends, we visited the most important places of interest in and around Nairobi. We visited the acclaimed Coryndon Museum, later renamed The National Museum of Kenya, and admired the countless stuffed birds and animals as well as the exquisite drawings and paintings of Kenyan flowers and tribesmen in traditional garb by Joy Adamson. She had already found fame as an artist. Soon, her book, *Born Free*, would be published and become a best seller. A couple of times we also spotted her driving around Nairobi in her white convertible.

Almost every weekend we went somewhere different, either in Gavin's Land Rover or for short distances, in Anna-Maria's little Fiat, which had its limitations on the mostly dirt roads.

At Lake Naivasha, a paradise for birdwatchers, we'd take a boat ride on the lake to see hippos and a variety of waders: herons, egrets, darters, cormorants and pelicans. We'd then celebrate with a Tusker beer at the town's Bell Café, or order afternoon tea with cake and sandwiches on the lawn of the Lake Naivasha Hotel as marabou storks watched silently on one leg, hoping for scraps.

Gavin took us to the Kenya highlands, known by the local whites as 'up-country', to the market towns of Nyeri and Nanyuki. It was in Nyeri where Lord Baden-Powell, founder of the Boy Scouts, and his wife, Olave, were buried. It was also home to the famous Treetops Hotel where, 11 years previously, in February 1952, Princess Elizabeth had been holidaying with Prince Philip when she became Queen. There, we admired the lovely tropical gardens at the adjacent Outspan Hotel and enjoyed a delicious Sunday curry lunch, complete with naan and mango chutney. In Nanyuki, we enjoyed a few beers at the Sportsman's Arms Hotel with its fine views of Mount Kenya in the distance.

Twice we drove the 320 miles along the main unpaved highway via Mombasa to Malindi, a sleepy one-street resort town. There, we camped on the beach, swam and sunbathed on Kenya's sun-drenched Indian Ocean coast where they were just beginning to build all-inclusive resorts in anticipation of a massive tourist boom. It was not uncommon to see plenty of game, including huge herds of elephant and buffalo; large numbers of zebra, wildebeest, hartebeest, topi and giraffe; and at least two or three rhino crossing the main road connecting Nairobi with Mombasa, the country's biggest port. In those days, like many African countries — such as Northern/Southern Rhodesia and Bechuanaland (all soon to become independent and renamed) — Kenya was still teeming with wildlife. Regrettably, this would change dramatically due to poaching and loss of habitat.

On another trip, we passed Menengai Crater, a vast, mostly dormant volcano, en route to Lake Nakuru where we gazed in awe at the thousands of lesser and greater pink flamingos and white pelicans. Yet another time, we ventured into Hell's Gate, a narrow canyon with furnace-like temperatures, so devilishly hot we didn't stay long. But we were lucky enough to catch a fleeting glimpse of the huge and relatively rare lammergeyer vulture catching the thermals high in the sky. As well, there were buffalo, zebra and baboon along with several hyraxes scurrying around the rocks. Prior to leaving, we spotted our first klipspringer, a small dainty antelope that makes its home among rocky outcrops.

On one memorable weekend, we finally saw our first *simba*, Swahili for lion. Close up. It was my turn at the wheel as we raced through the wild and rugged open country, slowing down only to get the lay of the land and to peer through binoculars. The climax came as we slowly negotiated a rutted track. Suddenly, about 30 yards ahead, we spied a full-grown lioness. I jammed on the brakes and we all sat in silence, watching one of Africa's most dangerous and majestic animals with bated breath. Graceful and lithe, she was 250 pounds of quivering muscle. As we watched in awe and some concern, for this could be a very dicey situation, she stood still for a few seconds and sniffed the air. Then, with barely a glance from her yellow eyes in our direction, she silently padded across the track and disappeared into the tall elephant grass. With that, we all let out a sigh of both relief and admiration. It was one of many such thrilling wildlife encounters in Africa that I was to experience in the days ahead.

It wasn't long after that the Kenya Regiment was disbanded in May 1963. I would be away on safari, but Mike, Eddie and Pete all marched in the final parade through the streets of Nairobi. They were very pleased to participate and wear the uniform one last time. Judging from the photographs, they looked very proud and smart. We all kept our regimental Australian-style, broad-brimmed hats, which would come in very useful in the many tropical countries that lay ahead.

By then, my friends and I were eating good wholesome food. Regardless of past events, we still all enjoyed bananas. And with so many exciting things to see and do, it was not surprising that most of our previous, unpleasant experiences were soon forgotten.

CHAPTER 4:

Monkey Business

*'On safari, each breath you draw gives pleasure,
(and) you wake with a new sense of wonder.'
— Elspeth Huxley*

Amazingly, my dream for a safari job was to come true. Pete had heard that Air Spray, a crop spraying company, required a field assistant but, as he was gainfully employed at the time, he kindly suggested that I apply.

Telephoning the office at Wilson Airport (for small planes), I was told the vacancy had already been filled, but there was need for a man 'on the monkey side'. If interested, I should call on Mr Noon, a director of the company. My ignorance of what the job might entail did little to stifle my interest. The following morning bright and early, I was at the front door of Mr Noon's beautiful home in the well-to-do suburb of Karen where Gavin lived, several miles from Nairobi city centre, with its English-style mansions and vast, sprawling well-maintained gardens.

A slim, dapper man with greying hair and a matching moustache, Alec Noon was a well-known East African personality who had served with distinction as a squadron leader and spitfire pilot in the Royal Air Force during the war. Middle-aged now, he was active in local flying circles. As the co-owner and founder of Air Spray, he had recently established Primates, a new company, described on the letterhead as suppliers of animals for medical research.

Despite my obvious lack of experience, he sensed my enthusiasm and concluded the short interview on the doorstep with, 'Well, old chap, we'll give you a try.' I was thrilled and didn't have to think twice about accepting. Nearly everyone dreams of going on safari in Africa but only a wealthy few then could afford it. The prospect of running my own camp for an extended period excited me enormously. It would also mean remaining longer in East Africa, a part of the world I had already come to love.

A few days later, in Mr Noon's office, the job was explained in detail by his assistant, an elderly Scottish woman. The plan was as follows: Together with Joseph, an African driver, I would drive an old Bedford truck to a campsite located close to the Great Ruaha River, some 600 miles south of Nairobi in what was then Tanganyika. (In April 1964, the Island of Zanzibar would amalgamate with Tanganyika; the new country would be named Tanzania.) On arrival, we'd make contact with Curly McPherson, a young white Kenyan who was already trapping. He would teach me the ropes for two or three weeks before returning to Kenya. Joseph would drive the Land Rover, which Curly had been using, back to Nairobi. Once I took over, Curly would hitch a ride back in one of the company aeroplanes when they picked up the monkeys.

Our journey down the Great North Road took two and a half days and was uneventful except for encountering occasional game, such as impala, zebra, ostrich, wildebeest and giraffe dashing or leaping across the road ahead of us. Not long after leaving Athi River, a tiny township known for its abattoirs, where scavenging vultures patiently sit on the flat-topped acacia trees waiting for their next meal, the bitumen ended. The thick, red dust quickly filled our eyes, nostrils and ears — an occupational hazard driving on Africa's undeveloped roads. I already knew that travel on African roads is arduous, and with the washboard effect, always bone-jarring. One can get used to the African bush and the extreme heat but never to the dust, which penetrates everywhere and everything.

> *Winding for more than 1,000 miles through rain forests, game plains and mountain stretches, the Great North Road may well be the world's worst international highway. Along its flat stretches, the road is little more than a trail of treacherous sand or slide mud (depending on the weather). Right of way is often usurped by rhino, elephant and lions.*

This was a quotation from *Time* magazine, which I cut out to send home. I realised why certain stretches of this famous highway were incorporated into the annual East African Safari, the popular annual motor race often referred to as 'the toughest race on earth'. It was invariably won by the very durable French Peugeot 404, then widely regarded as 'the finest car' in Africa.

After passing through the small main towns of Arusha and Dodoma — the latter would become the capital of Tanzania — we left the murram highway about 60 miles north of Iringa and turned off into the bush. A couple of inverted rusty spears in the ground were the only sign that indicated a rough track going anywhere. As we followed it, our progress was reduced to a crawl for about eight to ten miles. An hour had elapsed when we stopped to ask a local man the whereabouts of the camp.

Just then Curly drove up. 'Been expecting you,' he said, staring suspiciously at my recently scarred face as we introduced ourselves. Then, without more ado, he told us to follow the Land Rover. Suddenly, the bush opened up to reveal a long flat dried mud clearing, a natural airstrip. Driving across to the other side, we pulled up at the camp — such as it was. Curly had been too busy organising the Watu (local people) to concentrate on creature comforts, he told me, but I was welcome and told to make myself at home.

The camp consisted of a long, roughly built, open-sided, grass-roofed hut containing rows of elevated cages, some of which contained monkeys. Surrounding the immediate area, cooking utensils, camping equipment and other items were strewn around. Extreme disappointment was my immediate reaction; I wondered what I had let myself in for. Mr Noon had said living conditions would be satisfactory for me and my staff. They would be later but certainly weren't at the time!

It was early evening by the time I had looked around and been introduced to Shadarak and Henry, two African camp helpers, employed respectively as general factotum, and driver/interpreter. Then, after helping myself to some bread and tinned ham, I laid my sleeping bag on a mattress, chatted to Curly, and with the advent of darkness attempted to sleep.

Within minutes, mosquitoes started buzzing about me, and I knew I was in for a bad night. As it was mid-April, late in the rainy season, they were still around. With no mosquito net, I was fair game. Throughout the night, they bit relentlessly. I hardly slept at all and was relieved to greet the dawn.

Fortunately, after a few days, no more rain fell during my stay, revealing cloudless blue skies. The trees, bush and grass changed from dark green to a dirty, dry brown.

Curly was a fair-haired, muscular fellow of medium height, and, although only 21, he had spent a great deal of time on safari and was familiar with bush life. He would show me what had to be done. But before learning about the job, I took a few matters into my own hands. Mr Noon had instructed me to organise the camp properly, and that was exactly what I planned to do.

After choosing a more convenient site at the end of the 'runway', Shadarak and I cleared the area of tall grass and acacia bushes, using razor-sharp *pangas* (machete-like tools used for cutting, slicing and digging). We then erected the tents, which were still lying in bundles, rolled up. Ashes went down to deter termites and ants.

My tent was erected first. *Almost too good to be true*, I thought. Like a small house, it was spacious, with mosquito-proof windows, built-in groundsheet and even a mosquito-proof veranda. My delight increased as we unpacked the furniture from wooden chests and trunks: two camp beds with mattresses, pillows, linen and blankets, two collapsible canvas armchairs, a table, and even a kerosene-operated refrigerator. There was also an identical tent, for guests, a larger one for the staff, and two smaller ones for use as shower and latrine. Within a few days, the camp took shape.

Once it was fully operational, I joined Curly and Henry on their rounds. Off we would go, bumping and rattling along a rarely used and barely discernible track to make contact with the trappers. Stopping at an isolated village, Curly would speak in Swahili to the headman, requesting and sometimes urging him to dispatch his men to set traps while I dutifully listened and watched.

The procedure was always the same. The trappers assembled around the vehicle and hauled out the monkeys from dirty, torn sacks. As the monkeys screeched in fright, the trappers gripped them firmly by the back of the neck and tail to avoid being bitten or scratched. Curly then examined them for injuries, disease or other telltale marks. Only perfect specimens were accepted. And we did not take pregnant females or mothers with young. Henry transferred them into freshly sterilised bags and loaded them gently into the vehicle. The trappers were then paid — five East African shillings per monkey — while the rejected ones were released on the spot, scampering into the bush. Sometimes, business was conducted on the banks of a river, or beside the track and more often than not, at a tiny, remote village. At each meeting, prior to dispersing, the next date was arranged.

The monkeys were vervets, a common species of primate found all over Africa — quite small; 15 to 24 inches in height; rarely exceeding 12 pounds; greyish green in colour with black faces, hands and feet, and tails longer than themselves; and with a life span of roughly 12 years. They are omnivores who live in family groups of 20 to 30 called troops, feeding on fruit, berries, insects, and farmers' crops, when they could get their hands on them. Like baboons, they sleep high in trees at night, safe from predators. The males are often referred to as 'blue-balled' monkeys because of their distinctive turquoise blue scrotums. Completely wild and naturally afraid of humans, they could be vicious. And would bite, claw and scratch. I didn't touch one for some weeks, preferring to learn the technique of handling and inspecting them by observation. After that, despite my confident dexterity, I still occasionally got bitten and scratched.

To capture them, traps of branches and twigs were laid and sprung, baited with sweet potato or maize. The trappers set and checked their traps every day. As each monkey was destined for medical research — as told to me by Mr Noon, to make polio vaccine — it was essential that a high standard of humane treatment, from capture to ultimate destination, be maintained. The trappers learned this the hard way, by not getting paid for any less than healthy animals.

I was, of course aware, as I went about my job, that not everyone liked the idea of conducting experiments on animals. But when I stopped to ponder that it was for the benefit of mankind and was therefore a worthy cause, I seldom thought of it again.

Early during my stay, an aircraft appeared, circled overhead, landed on the strip, and taxied up to the camp. Out stepped the pilot, Mr Noon. I entertained him over a cup of tea while Curly supervised the loading of the monkeys. Impressed by the camp layout, he suggested I charge guests per visit. With that, he handed me 40 shillings.

'How's it going?' he asked, questioning me at length about the job and asking me for my comments. 'Couldn't be happier,' I told him, 'except for one thing.' I needed a better vehicle. The old Bedford was impractical, with no four-wheel drive, and it occasionally broke down. 'Fair enough,' he replied. He suggested I accompany him back to Nairobi that day.

The aircraft was a 22-year-old Airspeed Oxford, a twin-engine monoplane virtually made of paper and wood, similar but much smaller than the DC-3. Hundreds were built primarily to train air crew and transport troops during the Second World War. I later learned that it was the same type of aircraft in which aviator Amy Johnson went missing in 1941. Although it was mechanically sound, the engineers of Air Spray swore among themselves never to fly in that particular plane. Once we were airborne, I understood why.

As we revved up and slowly moved off, it seemed to me that such a flimsy machine loaded with 75 monkeys wouldn't get off the ground. But with the throttle open to maximum, we gradually built up enough speed to soar over the surrounding bush with what looked like inches to spare.

Throughout the two-hour flight, the noise of the twin engines was deafening, and they shook everything — including me as I sat on a wooden box behind the pilot's seat, hemmed in by cages. Mr Noon, meanwhile, sat nonchalantly at the controls, puffing on a cigarette and periodically chatting into the mic. I wanted to share his indifference, but I was uncomfortable and afraid of flying at the best of times. To make matters worse, two monkeys escaped from their cages and pranced around the fuselage. It was with considerable relief that I stretched my rather unsteady legs when we finally touched down.

After two days in Nairobi, seeing Anna-Maria and exchanging news with Gavin and my three pals, I headed back to camp, this time in a new long-wheelbase Land Rover with the company name, Primates EA, painted on both front doors.

Curly remained for another week, patiently explaining everything, then left on the next plane. With that, I was on my own. Curly's departure meant that the job and everything related to it was exclusively mine, including a skinny, half-blind white and brown mongrel stray, nicknamed Lassie. My inheritance also included a native cook, named Saidi, whom Curly had engaged while I was away. This meant my staff of regular helpers now numbered three.

Shadarak was the eldest. Slim, middle-aged, with tribal scars etched deep into his face and grey hair balding on top, he was quiet and reserved. His principal task was care of the monkeys: feed and water them, and sterilise and clean the bags and cages. Plus all the odd jobs around camp. He was also a *fundi* (expert) at improvisation. Apart from learning about the monkey business and a great deal about humility and patience from my dealings with the local people, I picked up a lot from Shadarak's expert tuition. He had, after all, lived in the bush all of his life.

Saidi was of medium height and build, genial by nature, laughing at the slightest whim. He wore a faded crimson and dog-eared fez so much that I became convinced that he slept in it. He cleaned my tent and, as camp cook, baked the finest bread. And, considering he had only basic utensils (pots and pans) and an open fire, his culinary prowess was superb. He could produce roast beef, goat or chicken with all the trimmings in next to no time.

Henry had been educated at a mission school and spoke quaint but passable English. He was short and well-built and often played a handmade pennywhistle, usually as darkness fell.

All three were resourceful and a tremendous help to me. They wore old, tattered shorts and shirts until I issued them new ones. All boasted gleaming white teeth, which they conscientiously cleaned twice daily, using ashes from the fire and small sticks. Although Curly had spoken fluent Swahili, I was handicapped in this respect. So, to communicate, I mostly spoke to Henry, who translated on the spot. Often, I attempted (with sometimes amusing results) a mixture of broken English interlaced with gesticulations and the odd, newly acquired Swahili word. I referred to them by name or *rifiki* (friend), while they called me, 'Bwana (Mister) David'.

Every day was long and full of activity. Rising at sunrise, Saidi would fill the canvas water bowl outside my tent. By the time I had washed and shaved, breakfast was ready on the table. When I'd finished, it was time to check the monkeys and their cages. My greeting of 'Jambo' acknowledged ('Hello' — used universally in East Africa at any time of the day), and my tour of inspection completed, I issued instructions to Shadarak and Saidi for the day. Then, with a shout of 'Kwenda' (Let's go) to Henry, we'd move off in a whirl of dust.

All day we made the rounds, driving extremely slowly. Moving at 20 miles per hour, we rocked and rolled, swayed, dipped and plunged like a ship in a gale, through a sea of varied vegetation — including wait-a-bit thorns — and up and down steep dongas. In all, I would cover more than 12,000 miles in nearly six months. Initially, at the end of each day, I was a worn-out wreck of sweat, dust and fatigue. Although it was tiring seated behind the wheel for hours at a time, I soon came to enjoy it.

At first, everywhere looked more or less the same; but as my experience increased, so did my observation skills. In fact, every sense in my body became more keenly aware. The routes we followed were seldom more than tracks, which often petered out and left us to bump, rattle, stop and start continuously. But after a while, I found I could relax and still concentrate on the surroundings and where I was going. On every trip, we encountered game — usually giraffe, cheetah, topi, eland, and Grant's and Thomson's gazelles. As I watched in wonder, zebras streaked away, drumming the ground; wildebeest cantered off with ungainly grace; and impala leapt over one another in their haste to get away. Some were so close we could lean out of the vehicle and almost touch them. Henry occasionally relieved me at the wheel, but instead of negotiating around the bush, he tended to drive straight through it.

Frequently, we found ourselves cautiously skirting a huge herd of elephant or buffalo, the herds then consisting of hundreds of animals. Sadly, in coming years, those massive herds that I knew as a young man would vanish, never to be seen again.

We often spotted small groups of kudu, beautiful, shy creatures with huge spiral horns, and occasionally, just for fun, we raced ostriches when the landscape was dead flat. Frequently, we came across the bloody scene of a kill. Every kind of vulture flapped its wings and tore at what was left of the poor beast — head, hooves, bones, and even excrement — while spotted hyenas, their bellies full, skulked off to the next feast.

In the sweltering heat (usually around 95°F), my assistants and I wore only sandals and shorts, and I carried a small hunting knife at my side. The trappers dressed the same way, except that their footwear was cut from old rubber tyres and, instead of a hunting knife, they carried their pangas, either in their belts or hands. All wore shorts of different sizes and colours, and a few wore second-hand, shapeless jackets with nothing underneath, while others wore torn shirts or singlets. Altogether, some 50 to 60 local men volunteered to cooperate in trapping monkeys. They fell under my jurisdiction and looked to me for guidance. There was no district officer responsible for this area so, at the tender age of 24, I found myself not only part-time employer, but also guardian, teacher, and sometimes medical orderly.

As the company was a commercial enterprise and, presumably, a lucrative business, it was also my job to increase the number of monkeys captured. This I did, with Henry's assistance, by persuading the trappers to lay more traps. But it was difficult. Most trapped only when they felt inclined, preferring to work on their tiny *shambas* (cultivated plots of land with usually corn, sweet potato, a papaya tree or and a few scrawny chickens), help with a neighbour's new hut, or just loll around.

For my efforts, I was paid a reasonable basic salary. Fresh foodstuffs arrived with the weekly plane and, all in all, working at the camp provided an excellent opportunity to save. The fact is I would have willingly worked for almost nothing — I adored my new way of life and felt like a king.

Back at camp at the end of the day, the dust and sweat were showered away and, while Saidi prepared dinner, I grabbed my stout walking stick, whistled up Lassie and went for a walk. I often stood still momentarily, trying to identify the different sounds and watching the sun go down.

Later, while seated at the table, which also served as a desk, Saidi brought dinner on a wooden tray. It usually consisted of tinned soup, then roast beef or mutton with fried or boiled potatoes and a vegetable, followed by fresh fruit, mostly oranges and bananas, and coffee. It was plain but delicious. Breakfast was equally satisfying: bacon and eggs, toast and marmalade and tea or coffee. The food not brought in by plane was purchased in Iringa or, more often, locally from Somali *duka-wallahs* — bush shopkeepers — selling Coca-Cola, razor blades and Nestlé's tinned milk in the remotest places. Daily living in the bush was cheap enough. Six oranges cost a penny and a whole sheep seven shillings.

It became my routine to join my assistants around the fire after dinner while they prepared their meal. As Shadarak and Saidi took it in turns to stir the *posho* (pounded maize and water), Henry and I discussed the next day's work. However, very soon, I would have to retreat to my tent to escape the onslaught of mosquitoes while the others, lucky fellows, remained oblivious and seemingly immune. Most evenings and some mornings, I'd bathe my legs and ankles in a bucket of disinfectant which was soothing and stopped me from scratching.

After completing my daily report and accounts, and sometimes reading in the light of the hurricane lamp, I turned in. Often I lay awake for a while listening to what gradually became familiar sounds: the distant trumpeting of a lone elephant; the 'whoop whoop' of hyenas; and the incessant chirping of crickets. Sometimes, a nearby interloper would step on a dead twig, causing me to sit bolt upright, every nerve and muscle in my body tense and quivering. Even though I became used to these sounds, I never failed to remind Shadarak to build up the fire before he went to bed.

By then, I'd heard grizzly stories of lone lions and scavenging hyenas with their bone-crushing jaws biting off the heads and feet of unsuspecting human victims lying on cots in their tents — not only at night but also in broad daylight! I made sure the tent flap was tightly zipped at night. At least, it provided me with a false sense of security.

Twice, leopard prowled their way around camp in the dead of night. Partial to monkeys and dogs, they were an obvious menace. Silent, stealthy creatures, they were extremely dangerous and could tear man or beast to ribbons in a matter of seconds. 'Over there,' I shouted the first time as two large green eyes stared back at me when I shone my torch into the bush. Another night, it was Shadarak who shouted the alarm. Both times, our well-aimed fire sticks had them scampering away.

Many nights, I heard a lion roaring in the distance and its mate roaring back. Once I noticed lion tracks through the camp, which had me concerned, but I never saw the beast and it never returned. Over time, I learned to identify numerous animals by their sounds: the cough or sawing sound of a leopard, the barking of baboons, the squealing of warthogs, the yelping of hyenas, and the snorting of buffalo. I took appropriate steps to avoid the most dangerous animals, giving them a wide berth. There was a pack of 16 wild dogs in the vicinity that came and went. Considered one of Africa's most efficient hunters, they pursue their prey until it tires, then tear it apart. But they never bothered us at all.

Although the monkey house remained lit with kerosene lamps throughout the night, occasional predators (usually hyenas) infiltrated the camp area, scavenging for food. The monkeys themselves were their own best defence, screeching frantically and rocking their cages at the first sign of danger and quickly waking the camp.

The days sped by. I didn't need a watch. Here time meant nothing, and I wasn't going anywhere. Apart from the daily collections, cages and bags had to be repaired, water had to be transported from the river, and the airstrip had to be freshly painted with whitewash. Often, after delivering instructions to Saidi and Shadarak, I grabbed a panga and helped build a new section of the monkey house or clear fresh space to extend the camp. We also built a garage which consisted only of four sturdy poles topped with a grass roof.

Frequently, en route to a rendezvous with trappers, we encountered native people heading the same way on foot. We always stopped to give them a ride. From time to time, I was presented with gifts, usually fresh fruit and, once, six newly laid eggs. I always refused, but they insisted. Some waited regularly beside certain tracks; one was a young, attractive girl named Charity. Not content with climbing into the back of the Land Rover, she insisted on snuggling between Henry and me.

Once we stopped for a woman with what appeared to be a bundle of old clothes on her back. She seemed exceptionally distraught. The bundle turned out to be a dead baby; she was taking it to the local mission station to be buried. Encounters such as these commanded my admiration and respect. Tragedy and danger are ever-present in the African bush. Despite the constant threat of drought, fire, famine and flood, as well as serious illnesses such as malaria or sudden mysterious deaths of undetermined causes, the people with their simple, unaffected ways seemed to be far happier than so-called civilised peoples. *Perhaps it was because they lived closer to nature or maybe it was just plain ignorance*, I thought at the time. They were content with what they had. I came to greatly admire them for they were always cheerful and stoic no matter what. Indeed, some unknown, profound part of me felt totally and blissfully at home in this natural environment.

More than once I was reminded of Robert Ruark, author and Errol Flynn look-alike, who wrote in his 1962 highly controversial and absorbing epic, *Uhuru* (Swahili for 'independence'):

Africa grabs a piece of your heart, and never quite lets go.

Earlier, I'd managed to get my hands on his 1953 classic, *Horn of The Hunter*, as he followed in the footsteps of Ernest Hemingway, whom he greatly admired. Sadly, he was to die relatively young, aged 49, of cirrhosis of the liver, while Pete and I were still wandering awestruck throughout Africa.

Although I was now living in one of the remotest parts of Africa, I wasn't lonely, and the isolation didn't affect me except for occasional homesickness when writing or receiving letters. I was, for the most part, too happy and busy to think or worry about the 'outside world'. After only a matter of weeks, the so-called 'civilised' world seemed almost non-existent. And living out of doors under almost constant sunshine had made me leathery tough and as brown as a berry.

The African dawns and sunsets were unforgettable. As the dawn chorus heralded the start of a new day, the crimson red sun rapidly rose. At the end of the day, no sooner did the massive fiery red ball of sun sink below the horizon than darkness fell, and it immediately became night. There was no twilight, but when the sky was clear, which was most nights, it was full of glistening stars that seemed close enough to touch. Magical! Naturally, in such an environment, I became something of a stargazer. Indeed, I would be lucky to see amazingly clear skies at night as well as remarkable sunsets in many different countries. Perhaps, not surprisingly, it became my favourite time of the day.

Anna-Maria and I had remained in touch by letter and, to celebrate her 23nd birthday, I took a couple of days off to drive to Ngorongoro Crater with her. We met in the tiny township of Namanga outside the small Namanga Hotel and Bar. 'Ciao! How do I look for the safari?' sang out Anna-Maria in her Italian accent as she climbed down off the bus and jauntily approached me, clutching a tiny holdall.

'Well, you certainly travel light.' I laughed, eyeing her luggage. I had to admit that she looked very becoming in her brief white shorts, dark blue halter and orange bandana wrapped around her long hair. But hardly prepared for the occasion. Luckily, I'd brought some extra shirts and sweaters just in case, and remembered to shove a mattress in the back of the Land Rover at the last minute.

After enjoying a buffet lunch in the town of Arusha, a long, arduous drive over a very rough dirt road and steep escarpment brought us to Ngorongoro Crater, not far from the vast plains of the Serengeti. As one of the greatest natural wonders of the world, it is a spectacular caldera, 12 miles across, renowned for its great abundance and variety of wildlife.

It was early evening, getting dark and already freezing cold by the time we reached the park gates, halfway up the 8,500-foot crater, only to find them locked for the night. So we drove to a nearby sisal estate where we asked the Scottish manager if we could make use of his garage during the freezing night. After consulting his wife, he kindly invited us to utilise the guest house. 'You're welcome to join us for dinner,' Mr MacGregor added in his Scottish accent, 'but ye'll have to take potluck.' Potluck turned out to be a very welcome, steaming hot stew which we ate by candlelight and the flickering, glowing warmth of an open fireplace. That night, dead tired and greatly relieved to be in a safe and cosy haven, we fell asleep to the sound of hyenas yelping nearby.

The next morning, after a hearty breakfast graciously provided by our Scottish hosts, we slowly descended 2,000 feet in four-wheel-drive through thick swirling mist. The steep, winding one-way track had claimed more than one vehicle over the years. Not long before our visit, I'd heard about a terrible accident involving some Canadian tourists in a Land Rover which had gone out of control. The vehicle went over the edge, killing one person and leaving three or four others badly hurt and in agony for hours until help arrived.

There was little traffic in those days, so not much chance of getting help if needed. So I was particularly careful driving both down and later, up the perilous one-way track. At the bottom from the safety of the Land Rover, we observed a mass of different species of African wildlife, all of which made their home on the savannah and plains of the vast crater floor. Our bumper game-viewing day produced no fewer than 22 lions, several elephants, numerous spotted hyenas, a golden jackal, two bat-eared foxes and numerous zebra, buffalo, wildebeest, and too many Thomson's and Grant's gazelles to count.

As well as the great numbers of animals, we saw lots of birdlife including dozens of beautiful African crowned cranes which had migrated from some far-off place like Siberia and were feeding on the lush grasses of the crater.

After our thrilling visit to Ngorongoro, we headed to Lake Manyara National Park, a densely forested area on the banks of the saline lake where, after watching a troop of baboons playing, we cautiously approached a herd of elephants out in the open. I switched off the ignition as we quietly sat and watched them at close quarters for several minutes. An old bull then eyed us warily and for some reason resented our presence. As we watched, he flapped his huge ears, threw up his trunk, let out an almighty bellow — and then headed our way! A distance of no more than 40 yards separated us when I tried to switch on the ignition. No response. I tried again. Our elephant friend was coming closer, too close for comfort. Three times I tried to start the engine, each time with no response. Simultaneously, Anna-Maria was shaking me while frantically telling me to 'do something!' Suddenly on the fourth turn of the ignition, the engine roared into life and with my foot hard on the accelerator, we shot forward out of the path of the not-so-friendly jumbo, pulling up a few hundred yards later when we realised we were out of danger.

'Phew, that was a close one,' I exclaimed with enormous relief as Anna-Maria and I fell into each other's arms, shaking and laughing with emotion. Soon after, we turned off the track and into a secluded clearing. As it was a beautiful hot, sunny day and there were no other vehicles in sight, we hauled out the mattress, placed it on top of the Land Rover and stripped off to sunbathe. And as the sun shone on our youthful bodies, we embraced and kissed, and made passionate love, watched by curious impala, giraffe and zebra.

Our all-too-brief but memorable safari ended in Arusha, where we said 'Arrivederci,' and Anna-Maria hopped on a long-distance bus bound for Nairobi, while I headed back to camp and the business of catching monkeys.

Back at camp, I was in my element, especially since the avian life was so colourful and prolific. Both red and yellow hornbills hopped around boldly in search of scraps. There were African hoopoes, bee-eaters, flocks of Fischer's love birds, tiny colourful parakeets, and — what would become one of my all-time favourites — the lilac-breasted roller, an exquisite specimen with a bright lilac breast and pastel shades of every colour. The many different francolins and helmeted guinea fowl made good eating if you could catch them. At night, I sometimes heard the hoot of an owl and learned that for some people spotting one brought luck and a successful safari.

It was at the river and streams where I came across the greatest profusion of birds — carmine bee-eaters, blacksmith plovers, Egyptian geese, African jacanas (better known as lily pad walkers), darters, cormorants, pelicans, egrets, spoonbills, herons and ducks of all kinds.

I also often saw vultures — hooded, Egyptian, Nubian and white-headed — circling high above a kill, then swooping down and landing close to the carcass. They flapped their huge wings and screeched and hissed as they fought for a piece of the pie — any part, no matter how large or small. As I watched, scavenging hyenas and silver-backed jackals would appear out of nowhere to snatch their share.

Often, with my binoculars, I spied birds of prey such as osprey, hawks and eagles. I particularly liked seeing the martial eagle, largest of the eagle family; the resplendent red-footed bateleur eagle, and the very common tawny eagle. But my favourite raptor was the African fish eagle with its distinctive white head and tail, and chestnut belly. It would sit patiently on a branch of a tree overlooking a river, lake or stream, watching for fish. Its haunting cry echoed over the acacia and fig tree woodlands and across the water in the early mornings or late afternoons. It was the unmistakable sound of Africa.

Water fetching was a major event, and a double occurrence in the space of a week. Loading the Land Rover with all available jerrycans and 44-gallon drums, we drove the mile or so to the river, picking up villagers en route. Once there, I foolishly ignored the dangers of bilharzia, a parasite contracted from African rivers and lakes and carried by a species of minute water snail which transfers to humans through the pores of the skin or, more easily, via cuts and bruises. The parasite usually lodges in the liver, causing haematuria or worse if not treated. After a quick look around, I'd go for a swim, keeping a wary eye open for hippos and crocodiles. The river was fairly shallow but quite fast-moving, and I recall one day seeing a tall, proud Maasai warrior pay another tribesman a shilling to help him across.

Periodically, a lone villager or trapper wandered into camp complaining of pains or nursing a slight injury. With Henry's assistance, I would attend to the patient by dishing out Aspirin or dressing the wound, and twice I had to clean some nasty panga injuries. Fortunately, no severe cases came my way; had they done, I would have driven them to the nearest mission station staffed by a nurse, ten miles north.

Each day was different — like the time when I was sitting in the latrine tent and a tiny hole suddenly appeared near my left foot, followed by a slim snout. Snake! Standing ever so slowly and at the same time hauling up my shorts, I shouted for Shadarak. The urgency in my voice brought him running to my aid, panga at the ready. As the offender slithered out of the hole, Shadarak decapitated it with one blow. That was the first and last time I was caught with my pants down.

Shadarak revealed another technique while I was sitting in the small canvas bath and felt a peculiar squirming sensation beneath me. Standing up, I emptied out the water and there was another snake staring at me with its cold reptilian eyes!

'Shadarak!' I yelled. The snake coiled and prepared to strike. Before it could, Shadarak ran over clutching a hefty stick, and without a second thought, dealt the snake a paralysing blow to the head. He then cremated it with some kerosene and a match. Like most Africans, he feared snakes and killed them on sight.

Those scary incidents made me think. Being so isolated, I couldn't afford to be ill, involved in an accident, or attacked by any of Africa's many dangerous animals. There were nicer ways of dying than being trampled by an elephant, bitten by a black mamba, or eaten by a lion. But these terrifying occurrences did happen to local people, as I sometimes read in the Nairobi newspapers. Had I known too much about the potential dangers lurking in African lakes, rivers, mountains, and bush, I might never have signed up for the job.

Actually, it was and is a mystery to me why I never got malaria, yellow fever, sleeping sickness, dengue fever, or one of the many other tropical diseases endemic to Africa. Any one of them could have made me very ill, or even taken my life. *Someone must be looking over me*, I thought time and time again.

The large animals concerned me less than the creepy crawlies

— scorpions, *dudus* (flying insects) and the various one-inch ants, many of which had a nasty sting. I couldn't wait to strip to shake them off. Once, I was stung three times by the same bee; but that was retribution after Henry and I had disturbed its hive in search of honey.

After a while, I became quite familiar with the area, and sometimes out on a drive, we'd be flagged down by a white hunter with clients in their specially equipped safari vehicle going in the opposite direction. There, in mid-track, we'd stop to chat, and I'd be asked for directions, such as a good place to camp, or if I'd seen any sign of elephant, rhino or kudu.

Around the same time as I was running my camp, Jane Goodall was also living in the bush, at Gombe on the shores of Lake Tanganyika about 400 miles to the northwest, researching chimpanzees. We never met, but I was well aware of her interesting and valuable work. Years later, after she became famous, I was standing behind her in the queue at the British Airways counter at London's Heathrow waiting to check in for a flight to Vancouver. I thought later I should have introduced myself for we had a few things in common, including a love of Africa and the natural world. But I didn't want to intrude.

As I laid whitewash at the far end of the runway late one afternoon, a strong dust devil suddenly appeared. Racing to the Land Rover to warn the camp, I was too late. In a flash, the grass roof of the monkey house and garage had sailed away, and, in mere seconds, the camp was a shambles. Luckily, no one was injured, but some of the monkeys suffered from shock when their cages fell to the ground. The remainder of that day was spent clearing up.

Every second Saturday, my assistants took it in turns to accompany me the 60 miles via a steep escarpment to Iringa, the nearest town, set on high cliffs with a view overlooking the Ruaha River valley. Those trips were always regarded as a great treat. There, we purchased sacks of sweet potato and a few groceries and telephoned the Nairobi office to let them know how many monkeys we had. A decision would then be made which aircraft was available and suitable for the operation. After a quick Western meal at the Greek-owned Whitehorse Inn — the only hotel in town — we headed back to camp. This became a regular, once-fortnightly routine.

But for me, Sunday was the most important day: the weekly aircraft was due. Prior to its arrival, I made a second tour of the camp to ensure that everything was shipshape. At about lunchtime, a distant drone would be heard, and Lassie would leap around and bark. The aircraft would appear and circle the camp. Not long after, the pilot was seated in one of our movie director-type, canvas chairs relaxing over tea and biscuits brought by Saidi.

Whenever an aircraft landed, it was immediately surrounded by local people. It was a source of wonder to them, this great flying machine that made such a thunderous noise. They would stroke and pat it while emitting 'Ee-yehs'. More often than not, they admired the Land Rover the same way. After loading was completed, the plane took off with 75 to 100 monkeys bound for Nairobi, where they would be closely examined, fed and watered, and then dispatched to the laboratories of London, Tokyo and New York City.

It was late on a Wednesday afternoon when Christopher Noon, Alec Noon's 18-year-old son, touched down in the four-seater company Cessna. Taught to fly by his father almost as soon as he could walk, he already had his pilot's licence. He explained that he was passing by and dropped in to say hello.

No sooner did we exchange greetings than he said, 'Would you care for a joy ride?'

'Would I?' I exclaimed. 'I'd be thrilled.' That was an understatement. For the next 20 minutes, I held my breath as we banked, zoomed and swooped over and around the camp area in the small aircraft, barely missing the tree tops and scattering game in all directions. It was my first experience in such a small light plane. At first, it scared the wits out of me, but after a few minutes, I found it thoroughly exhilarating, and it certainly helped ease some of my fear of flying. Years later, I was to fly so much it seldom bothered me, no doubt due to the several small plane flights I took in those early travels.

Interestingly I was to run into Chris Noon several years into the future when I was a passenger and he was the Captain of a British Airways VC-10 en route from Nairobi to London via Entebbe to pick up the last of the Asians unceremoniously expelled from Uganda by military dictator Idi Amin.

It was Chris who told me about Curly's illness. Not surprisingly, he'd contracted bilharzia at the Great Ruaha Camp. It had taken some time for this particularly virulent but relatively common tropical disease to manifest itself. The treatment made him desperately ill and he nearly died. That story made me think twice about wading or swimming in the river as Curly had so often done!

Over a period of some weeks, several American and Japanese doctors came to spend a weekend at camp. Involved in medical research in New York and Tokyo, respectively, they were particularly keen to observe for themselves how the animals were treated at 'the other end.' They eagerly joined me for long walks in the area; took it in turns to accompany me on my rounds; and expressed a deep interest in everything they saw. Doctor Yamaguchi said he'd never enjoyed himself so much in his life. They all said how fortunate I was to be leading such a life. Not that I needed to be reminded, for I knew only too well I was living many people's dream.

After I'd been supervising the Great Ruaha Camp for about three months, Mr Noon flew in with a new, exciting proposition. He intended expanding operations, he explained, and wanted me to establish another camp in Kenya on the slopes of Mount Kilimanjaro. Permission to trap in the area had been granted and, he already had someone to take over from me, he elaborated. 'Report to the district officer when you arrive,' he said as he handed me my new trapping permit.

I didn't sleep very well that night, too excited at the idea of setting up camp close to one of the great natural wonders of Africa. I could hardly wait to get going. After teaching all that I knew to my successor, a big German fellow named Karl who knew the basics of animal husbandry, the Land Rover was loaded. I reluctantly said goodbye to Henry, Saidi and Lassie — who was now much fatter and healthier — and set off with Shadarak for the small township of Loitokitok, 5,500 feet up in the foothills of the world-famous mountain. Once again, I'd be in big game country, following in the footsteps of hunters, explorers and adventurers I'd heard or read about.

There was no mistaking where we were headed. Soaring majestically above the clouds and subtropical plains of East Africa, Mount Kilimanjaro personified Africa — seductive and alluring and visible for miles. I'd seen it before, of course, but only from a distance. On a clear day, it dominates the surrounding landscape for well over a hundred miles, and even when not so clear, the foothills can easily be discerned — except when low cloud or rain squalls blot it out completely. A long-extinct volcano, it lies approximately 200 miles south of the equator and its summit is perpetually covered in snow. At 19,340 feet, it is the highest mountain in Africa and regarded by some tribes as sacred. It consists of two peaks. Kibo is the higher, most photographed, and graceful snow-capped crater shape while Mawenzi, a rugged, pointed peak, rises out of the upper slopes to slightly over 16,000 feet.

Wild animals roam throughout the area, and elephant bones have been discovered as high as 15,000 feet. 'The Gods are smiling at us,' murmured Shadarak, beaming. Indeed, as Africa's most celebrated and beautiful mountain drew closer, it seemed to embrace us with open arms.

It was late afternoon when we reached our destination in the shadow of the mountain. A red hue glistened on the mantle of snow and ice high in front of us. We drove slowly up Loitokitok's single dirt street which had been reduced to a quagmire of mud after a brief but torrential downpour. We passed a Shell pump station, a couple of Indian-owned wood and metal stores, a small shabby hotel, and a police station before pulling up outside the DO's office. I could hardly contain my excitement at the breathtaking scenery as a rainbow suddenly appeared above the purple and white jacaranda trees lining the street.

As I went to enter the office, out stepped the district officer, a transplanted English civil servant named John Tigwell. A fortyish, dark-haired, handsome man, he looked smart in neatly pressed safari khakis, highly polished boots and a bush hat with a *'Dieu et mon droit'* badge. We exchanged introductions, and I presented my trapping permit.

John wasted no time in giving me some advice about the area. 'One of the best spots around here is about two miles down the south track,' he said. Pointing to a map, he suggested that we set up the new camp on the banks of the Noolteresh River, one of many clear, icy-cold mountain streams which sometimes flooded during the rainy season but, at the time, was little more than a few feet wide.

There, we went about selecting the best possible campsite. It had to be on a level surface to accommodate the tents and cages; reasonably close to fresh water with an abundance of available firewood nearby; and well away from dangerous-looking trees — for a falling heavy, dead branch could easily pierce a flimsy tent. Not to mention a leopard could leap, or a snake could drop down onto an unsuspecting victim.

After employing two new staff from the surrounding area, my new camp was complete. Moses was Kikuyu, Shadarak was Wakamba, and Elijah Maasai — all tribal enemies in the past, now working amicably together.

The daily adventures continued in my new location. I stepped out of my tent one morning and noticed huge, fresh footprints in the ground. Elephant. I glanced around nervously before calling Moses, who was busy stirring the posho. 'What happened?' I asked. Moses told me that an old bull had ambled into camp at around 2:00 a.m. and, despite its enormous size, made very little noise. Nonetheless, my assistants heard it; they quickly and stealthily made for the protection of the Land Rover which was close to their tent, where they all sat trembling until the elephant left. My annoyance at not being warned quickly dissolved when I realised that the slightest disturbance could easily have panicked the beast and caused destruction and havoc in the camp. We all laughed about the incident, but the idea of being crushed to death while asleep in my tent haunted me for some time.

While I was equally fascinated and frightened by all of the 'The Big Five' — elephant, rhino, leopard, lion and buffalo, so named by white hunters who considered them the most dangerous animals in Africa — it was elephants that intrigued me most. I could watch them for hours, especially the babies playing in the mud or slipping and sliding their way into and out of water holes. Elephants were so placid and tolerant for the most part, and yet could be menacing and violent at times.

Once, when I left the camp area to pick up firewood, I noticed what must have been a 20-foot python lying still in the tall grass. I couldn't believe my eyes. It was as thick as a man's torso. I had no way of measuring it, but it was a monster! I sprinted back to camp to get the Land Rover for a safer, closer look — I had no intentions of allowing it to squeeze me to death — but, by the time I got back, it had slithered away and disappeared into the thick undergrowth. Like the locals, I didn't like snakes at the best of times, but I realised that, as God's creatures, they too had a role to play in the greater scheme of things.

Being close to a town meant visitors. Among those who called were John and his German wife Annaliese; one or two local missionaries; and Robert, a hawk-nosed veterinary officer who, on seeing the camp for the first time, exclaimed to his African assistant, 'This chap's been doing it for years!' All were served tea, beer or cool drinks and shown around the camp.

There were other visitors — in the form of wildlife. Regularly at sunset for about a week, the same female giraffe accompanied by a baby, already six feet tall, would wander gracefully into camp. Gazing around at everything, after satisfying her curiosity and nibbling at the nearby acacia trees, she would then glance haughtily down at her youngster as if to say, 'There's nothing else here for us.' They would then amble off. *Yes, there could be few more satisfying and enjoyable moments than watching animals in their natural environment free to roam just as nature intended*, I told myself time and again.

After getting out of bed one morning, my big left toe started to itch like mad. I tried to ignore it, but it persisted and I finally got a good look at it. I now recognised the small pale bulge near the top as one of Africa's most loathsome pests — the infamous jigger-flea. This female parasite, less than a millimetre long, had burrowed beneath my skin. Now her abdomen was swelling with crops of eggs which she would drop into my toe. Left unattended, it would abscess. I was still peering intently at my foot when Moses noticed me. 'Do you want me to take them out?'

'No thanks,' I replied. 'I'd rather do it myself.'

Working very carefully with a long needle, I opened the surrounding skin in which the minute insects were embedded, pulled them out and threw them into the fire. I then washed my foot in disinfectant. It was a long time before the hole in my toe healed. Luckily, there was no recurrence or infection. *It was your own silly fault*, I told myself; I should have worn shoes and socks. Wearing only flip-flops with my feet exposed was asking for trouble.

Coming back from Loitokitok with Moses after doing some shopping one evening, we accidentally hit a dik-dik, a tiny, delicate animal and one of the smallest of the antelope family. Blinded by the lights of the Land Rover and frightened by the noise, it had sprung straight into our path. If it wasn't already dead, no doubt it was badly hurt, I reasoned as I got out of the vehicle to take a proper look.

I found the little creature huddled on the grass verge of the track, deathly scared and unable to stand or walk. 'Lette panga,' I called to Moses, intent on putting it out of its misery. But when I raised the weapon to do the job, the tiny animal looked at me mournfully with enormous, soft, sad eyes as if pleading for its life. I lowered the weapon, unable to deal the fatal blow.

'Maybe it is suffering from shock,' Moses suggested, for there was no gaping wound or blood. We decided to take it into camp. There, we laid it close to the fire and covered it with a blanket for the night. The next morning it was gone. No doubt it had been badly dazed and, when well enough, simply got up and went on its way.

As I sat in the Land Rover one morning waiting for Shadarak to finish checking some traps, several, tall, lean Maasai *moran* (young warriors) stepped out of the bush. For a second, I hoped they weren't hostile. After all, these were the men who, to prove their manhood, took on a lion single-handed with nothing but a shield and spear. (This practice was banned by the government in the early 1960s as it was deemed to be too dangerous, and too many lions were being killed.)

Their matted hair in long ringlets, flies clinging to their faces and eyes, they wore nothing but their everyday red-dyed togas which the breeze occasionally lifted up to their buttocks, revealing their full nakedness. Forming a semicircle around the vehicle, they casually adopted their favourite stance of leaning on their spears with one hand while standing on only one leg.

The most distinguished of the group, wearing a brightly coloured bead bracelet and necklace and corks in both ear lobes, stepped up to my open window. He saluted me Maasai fashion with open palm, greeting me with, 'Good morning, sir. How are you?' in perfect English. No sooner had I got over my surprise than began a brief but fascinating conversation.

In line with government policy, and in an attempt to bring the tribe into the 20th century, many chiefs' sons had been sent abroad to be educated. He was one of them. After spending three years at Oxford where he studied economics and became 'Westernised', the time came to return to his tribe. But in order to do so and exert some influence, he had to revert to his original way of life. Thus, he would be accepted instead of shunned. He seemed quite philosophical about what must have been an extraordinary transition on both occasions but admitted to preferring the free, nomadic Maasai existence over the Western way of life. After inviting me to visit his *manyatta* (village) 'anytime', we wished each other good luck and parted with a firm handshake, African style.

One long weekend, Anna-Maria took the long-distance bus from Nairobi to Kajiado, followed by a local bus crowded with people and chickens to Loitokitok, where I picked her up. We then drove to Kimana Swamp, soon to be designated a wildlife sanctuary, where both Ernest Hemingway and Robert Ruark had once hunted. There, after finding a suitable spot well away from the road, we stripped off to our underwear to sunbathe and watch game. Much to our pleasure, the stimulating setting proved to be too good an opportunity to miss and was the perfect place to make love.

We then drove to Amboseli Game Reserve with its wonderful views of snow-capped Kilimanjaro and a mecca for wildlife where for much of the day we excitedly watched lions snoozing, wildebeest and zebra grazing, and red-billed oxpeckers removing ticks from the hides, eyes and nostrils of buffalo and giraffe. We were also lucky to get quite close to Gertie, a black rhino famous for her three-foot horn. Sadly but not surprisingly, she would be killed by poachers in the not-too-distant future.

The next day, after overnighting at my camp, we drove to Tsavo West, where we stayed at the new and exquisitely located Kilaguni Lodge. It was the first lodge in Kenya to be built in a national park and would become one of my favourites. There, with the magnificent scenery of the vivid green rolling Chyulu Hills in the background, and the red earth against a backdrop of bright blue sky, we watched spellbound as a never-ending herd of buffalo and a variety of other wildlife came to the water hole to drink.

After two exciting days of game viewing with Anna-Maria, she returned to her job at the Italian embassy.

As much as my staff and I tried, there were simply too few vervets to be found in the area of Loitokitok. There were plenty of Sykes' and Patas monkeys, and black and white colobus species in the vicinity, but they weren't required. So the Nairobi office suggested I move and establish yet another camp. By this time, I'd spent just about six weeks in this magnificent area, and the excitement of living there had given me an even greater appreciation of the wonders of Africa.

I'll be back, I promised myself — the next time to stand on the summit of Kilimanjaro.

The Nairobi office recommended the best place to try next; Makindu, midway between Nairobi and Mombasa, where I presented myself to the district game warden, Barry Chappell. Aged 27, Barry was fair-haired, tall, broad-shouldered, and deeply tanned. Married, with a baby daughter, he was an extremely busy man. In addition to his regular duties, he was looking after a young, orphaned rhino and a variety of other animal strays. He was also building a new airstrip not far from his base of *rondavels* (circular, one-room mud dwellings with straw roofs). A year after our one and only meeting, his life was tragically cut short in a light plane accident when he crashed into some trees on his first qualified flight.

'That's your best bet,' said Barry, pointing to a spot on the map. He indicated a suitable camping site, again close to a river, the Kiboko. Although it was far less spectacular than Kilimanjaro district, there was an abundance of monkeys, Barry said. This area was also the home of the Wakamba tribe, Shadarak's people, who still hunted with bows and arrows.

Half a mile from camp was a *shamba*. It was owned by a member of the Wakamba tribe named Dixon. As a long-time resident in the area, he seemed the ideal person to turn to for assistance. With his enormous, rolling eyes and deep-throated, infectious laugh, he was popular and influential among his tribe. Thanks largely to him, the monkey business there flourished.

Another neighbour was Nick Carter, a celebrated rhino hunter who had introduced the anaesthetic dart; he spent most of his time capturing rhinos for zoos and game parks. A few weeks previously, Pete had made contact with him by letter and was invited to join him for a weekend. He was very excited to visit Nick and see his operation.

Once, en route to meet up with Dixon to pick up monkeys, we passed a native on an old bicycle with a tiny tame baboon squatting on the handlebars. The baboon was such an irresistible sight that I purchased him and named him Pepe on the spot. He provided no end of entertainment at the camp. (Ultimately, he became a star attraction at Nairobi's famous animal orphanage.) Fully grown, he was no more than a foot tall but boasted a voracious appetite. He soon had the complete run of the camp and, when he wasn't hunting for scraps in the kitchen, he was sliding down the guy ropes or pattering across the roofs of the tents. Whenever I sat down to write or eat, he crawled up my legs and curled into my lap.

Other pets included a bush baby and a semi-tame grey duiker that slipped forlornly into camp. The bush baby led us a merry chase before being caught. When I eventually managed to pounce and pick him up, he stared at me with his huge, bright eyes for a second then he sank his tiny, sharp teeth into my left index finger, penetrating my nail and making me shout out. That finger had to be bandaged and was painful for days. The last I saw of the little creature was a few weeks later, snuggling happily on the shoulder of Mr Noon's teenage daughter.

Time permitting, I occasionally drove to Hunters Lodge midway on the main Nairobi to Mombasa road. It was a hotel, petrol station and rest stop built by the late John Hunter, one of Kenya's famous white hunters. An idyllic spot, it was a little oasis surrounded by flat-topped acacia and palm trees, and tropical plants and flowers. There, in the late afternoon, I would relax over a cup of tea or a sundowner.

One day, I noticed a platoon of the General Service Unit (Kenya's crack paramilitary force, whose duties range from quelling riots to the protection of VIPs). They were casually standing around. I thought nothing of it until I stood admiring some freshly cut kudu horns lying on the grass to dry. 'Handsome, aren't they?' said the brown-skinned man in the dark green safari suit standing alongside.

'They certainly are,' I agreed as he walked away.

A big brawny white fellow who could only have been a hunter then approached me. 'Do you know who that is?'

I shrugged and replied, 'No idea.'

'That's Mahendra, the King of Nepal,' he told me. *Well, well, I'm certainly in good company*, I couldn't resist telling myself.

Checking the calendar, I had now been on safari for a little over five months. It was already August 1963, and about time to move on. I wrote to Pete. He replied that he too was ready to continue our trip.

Fortunately, Mr Noon understood the situation. He had explained when he took me on that the job was strictly seasonal. The rains would soon be coming, and trapping would have to be suspended for some months. Faced with the prospect of leaving, I felt quite depressed. The last few months had been an exciting and hugely rewarding experience; in fact, they were the richest and fullest days to that point in my life.

Besides feeling as free as a bird, there was no doubt that those wonderful, mostly carefree safari days had made a huge and unforgettable impression on me, and I would miss them dearly.

Mr Noon had set in motion arrangements to close down the camp and had given me an excellent reference, but even so, I was overwhelmed with the sadness of leaving on the day of my departure. It was with a huge lump in my throat and tears in my eyes that I shook hands and bid 'Kwaheri' (the Swahili farewell) to Elijah, Shadarak and Moses. And as I left camp for the last time, I couldn't for the life of me, look back.

I'd really been involved in some monkey business and had also been well and truly blessed. My love of the so-called Dark Continent knew no bounds and, thanks to the golden opportunity of living and working there, I'd come under its magic spell.

Author David Skillan, 22, outside parents' B&B, Hunstanton, England c. 1961

With Eve at Tyrella Beach, County Down, Northern Ireland

Pete, third from right, with members of his Signal Platoon, Ballykinlar, Northern Ireland

SCHWEIZERISCHE VEREINIGUNG
FÜR INTERNATIONALEN ZIVILDIENST

ASSOCIATION SUISSE
POUR LE SERVICE CIVIL INTERNATIONAL

Sekretariat:
Gartenhofstrasse 7, Zürich 4
Postcheckkonto: VIII 33387
Telephon: (051) 259705
Telegrammadresse:
Civilservice Zürich

S.C.I.
Le Bel Air,
5 Bd Clemenceau,
TLEMCEN,
Algeria.

To Whom it May Concern

    It is very pleasant to have the opportunity to say some good words about Edward Matthew, David Skillan, Peter Barrington and Michael Page, which may be of help to them on their world journey.

    For four weeks they have worked with us, becoming passing volunteers of our SCI team, at El Khemis in Algeria, where slowly we are building a village. It was our good fortune to encounter these travellers in Tlemcen and to succeed in luring them to our fort in the mountains. Their help has been great, their companionship worth as much, and they now depart to our loss but as friends. I hope SCI teams whose paths they cross will welcome them, and perhaps where possible benefit as we have done. And others as well. I think, anyway, they speak for themselves.

                                  Paul Rowland.
                                  Camp leader.

The Fort.
El Khemis.
Commune of Sebdou.

*Service Civil International – a Swiss-based
international peace organization established in 1920*

Our fully equipped Land Rover

Classic Nile River steamers

Four adventurers (L-R): Tony, David, Eddie and Mike in Kenya

Setting off to climb Mount Kenya

Kenya Regiment's final parade (Mike is second from right in front row)

Anna-Maria on beach at Malindi, Kenya

Eddie and Carol on their wedding day, Nairobi, Kenya

Author at Great Ruaha Camp, Kenya

On safari near Loitokitok, Kenya

PRIMATES (East Africa) LTD

Suppliers of Animals for Medical Research

P.O. BOX 2261 NAIROBI · TELEPHONE 23863 · CABLES "AIRBORNE"

30th September, 1963.

TO WHOM IT MAY CONCERN

    Mr. David William Skillan has been employed by this company from April 17th to September 30th, 1963, as a trapper of monkeys and baboons which are exported to laboratories in various parts of the world for Medical Research purposes.

    To mention a few essentials, the nature of this job calls for physically hard work, control of African labour, ingenuity, honesty and patience. David Skillan has proved to my entire satisfaction, his ability to conform with these requirements.

    David has always taken a great interest in the job, and has many times come forward with good ideas, which have been most beneficial to the running of the company. His clerical work has been meticulous, especially the keeping of imprest accounts.

    David Skillan leaves at his own request to continue on a world tour. I wish him the best of luck in the future, and can recommend him as being a man of good character, honest and conscientious.

A.M.T. NOON
DIRECTOR

Monkey business reference letter

Vervet monkey

Lion and lioness in Amboseli National Park, Kenya

Pete on truck near Loitokitok, Kenya

View from Kilaguni Lodge, Kenya

Naas, the rigger, at mast site

The riggers gang at mast site

Author at winch with mast in background

On the road in South Africa
Author (left) and Pete (right)

Veronica at family farm near Franschhoek, South Africa

Briefing of convoy drivers in Karoo, South Africa

New car roadside check

Waiting for 'all clear' to start engines in Cape mountains

Xhosa woman in Transkei, South Africa

Mounted policeman in Transkei, South Africa

Best ride ever – light plane in Tanzania

Okavango Delta, Botswana

Author

Author and Charlie

Pete

Pete and Charlie

At Victoria Falls

On summit of Mount Kilimanjaro

Coming down from Kili

CHAPTER 5:

Playing With Steel

'It isn't what we say or think that defines us, but what we do.'
— Jane Austen

Weeks earlier, as his relationship with Christiana became more intense, Mike had declared he was staying in Kenya. However, he somehow learned that Christiana had found a new boyfriend. So had Anna-Maria; she had met a Royal Navy Officer while attending a cocktail party at Nairobi's British High Commission. In short order, both Mike and I had been unceremoniously dumped. With that, Mike abruptly changed his mind and decided to come with us.

Interestingly, a few years later out of the blue, a letter postmarked 'Florence, Italy' caught up with me. It was from Anna-Maria. She explained that both she and Christiana had left Kenya in the mid-sixties to settle in the beautiful city of Florence and that she was happily married with two sons. And even though she had once ditched me for someone else, she still remembered me. She wished me well and thanked me for sharing so many 'happy times' as she put it 'on the safari'. I replied to her letter, telling her my latest news. That was the last time I heard from her. But I often thought of her as I did many of the new friends I made during my marathon journey.

While all this was happening, Eddie married Carol in Nairobi Cathedral in what was a shotgun wedding after Carol got pregnant. It was followed by a small reception which Mike, Pete and I all attended. In Eddie, Carol had what she wanted most, a husband and an affable and willing partner, and soon, judging by her actions, she was showing off her prize. Dear old Eddie didn't know what hit him! But he was so easy-going he didn't mind and seemed to relish married life. After being away from England less than 12 months, Eddie's travels had come to a halt, and he remained in Africa for the rest of his life.

With Eddie now married and out of the running, Gavin MacHutchin, our helpful friend, made overtures to take his place, while offering at the same time the use of his Land Rover. The offer was accepted with pleasure. Having another Land Rover at our disposal felt like old times. With that, our transport arrangements were complete. Our destination was South Africa. The 2,000-mile journey to Johannesburg would take us a little over two weeks. We would drive leisurely and divert at will, as the mood and circumstances took us.

Two days after leaving Nairobi and driving on mostly unpaved roads, we arrived in Dar Es Salaam, Tanganyika's sun-drenched capital on the Indian Ocean. There, on the street beside their battered old Volkswagen, we met two bearded young guitarists, an American named Steven Kroll and a Swiss fellow named Bruno Messmer. They called themselves The Wandering Minstrels. No mean musicians, they had already played before a considerable number of heads of state. By singing and strumming, they were earning money to travel around Africa.

Learning of our plans to visit the Island of Zanzibar, named 'the clove island' from the sweet scent that permeates the entire island, Steve and Bruno promptly joined us on board an Indian schooner for the overnight trip. Fortunately for me, the crossing was as calm as could be.

On arrival at Stone Town, the small capital of Zanzibar, we first visited the open-air market, which was full of newly caught red snapper and barracuda; fresh vegetables and fruit; and the heady scent of spices, nutmeg, cinnamon and ginger. We then hired well-used bikes to explore the exotic island known for its white, pristine beaches and tall palm trees. Later, after setting up camp, which merely involved throwing down our sleeping bags on a beautiful, empty beach, we played cricket using a hefty palm leaf as a bat and a dried-up coconut as a ball.

Our short stay wasn't all pleasant, though, for we saw our first terrible case of elephantiasis — a grotesque, severely deformed native, who shuffled pitifully along the street with the aid of a crutch and only one normal leg.

Returning to the mainland in an ocean-going dhow on a calm moonlit night provided a taste of the Arabian nights, until the captain, a hawk-nosed, djellaba-clad Arab appeared on deck, produced a tube of Colgate, brushed his teeth vigorously and then spat over the side.

When we clambered back into the Land Rover which we'd parked in what we thought was a secure spot in the city's docks, Mike immediately noticed it had been raided. A quick look confirmed the loss of Gavin's camera, together with a torch and panga. It was a bit of a blow, but we shrugged it off, considering ourselves lucky that was all the thieves took.

After leaving Dar, we headed south along the main murram highway, which was a continuation of the Cairo to Cape Town principal road, full of potholes and bum-numbing corrugations (washboards). Most nights we camped by the roadside in the bush, rising early when the sun woke us, anxious to make the most of the day. How many lions, hyenas and jackals sniffed around our sleeping bags during the night, we'd never know! Nor did we care as long as we weren't molested or attacked by either animals or humans. A foolhardy attitude, but typical of our devil-may-care approach to life at the time.

Once or twice when driving, we stopped and fell into conversation with passing motorists and were invited to stay at their homes, including once on a wattle estate. A friendly argument ensued between the manager and his assistant as to who should stay with whom. As in many isolated regions of Africa, they had few visitors and were only too happy to see us, so we compromised and split up. Mike and Gavin stayed with one couple while Pete and I stayed with the other. Without exception, these encounters were always lively. Our hosts would ask about 'the trip' while over a beer or two we asked questions about their home, work and lifestyle, with lots of laughter and good cheer.

One day while we were parked on the side of the road eating sandwiches, a cyclist pulled up to talk. His name was Heinz from small-town Germany, and he was also making a journey around the world. On a bike! By that time, he'd had been on the road about three years and had covered a considerable amount of ground as his bulging leg muscles revealed. His bicycle was so overloaded that only his head could be seen from the front. After sharing our sandwiches and swapping information, we parted, wishing him well but not envying him one bit.

Our entry into Northern Rhodesia — which long ago reverted to its original name of Zambia — presented no problems, and we stayed one night at Kariba in a *banda* (a circular thatched hut) at a tourist resort close to the mighty dam. On arrival in the small, modern capital of Lusaka, ablaze with purple jacaranda and deep red flame trees, we booked into a cheap hotel near the city centre, and then celebrated the occasion by attending a noisy, multiracial dance.

The next day we drove to the famous Victoria Falls. When we spotted the spray from a few miles away, we anticipated a marvellous sight — and it certainly was. Named by David Livingstone during his momentous African travels and known by the local tribes as Mosi-oa-Tunya (the smoke that thunders), the falls are one of the seven natural wonders of the world. And in no way did they disappoint us as we explored them from every vantage point and angle.

We stayed in the immediate area for two whole days, during which we chased two-foot-long monitor lizards; got soaking wet as we wandered through the rain forest; and gazed in wonder at the falls from Danger Point and the Eastern Cataract.

Later, in Hwange National Park, known for its abundance of wildlife, we pitched the tent at Robin Camp, then went off to spot game. In the space of four hours, we saw great numbers of eland, warthog, kudu, wildebeest, sable, waterbuck, giraffe and buffalo.

Next on our itinerary came Bulawayo, where we visited the Matopos and the grave of Cecil Rhodes, the controversial explorer, business tycoon, and millionaire entrepreneur who was equally admired and reviled. His dream had been to build a railway from Cairo to the Cape, which despite the best of intentions ultimately failed.

On arrival at the South African border at Beit Bridge, we encountered some slight difficulty, when an immigration official sternly demanded to see our onward tickets. Or at least a substantial amount of money. Unfortunately, none of us had saved very much during our stay in Kenya and our funds were fast diminishing. After discussing our dilemma, Gavin came to our aid and loaned us some money, which we thrust in front of the officer, in lieu of air tickets. Our passports were stamped, and we were allowed to proceed. That evening, after driving through the nation's capital of Pretoria with its tree-lined streets of white and purple jacarandas in full bloom, we arrived in Johannesburg, South Africa's largest city.

Apartheid was still going strong at that time. Introduced in 1948, it wouldn't end until 1994, nearly 30 years hence. Grossly unjust, it seldom affected us on a daily basis. But it certainly concerned us. It was simply wrong and unfair. Our British upbringing had taught us to treat everyone with respect, no matter what their origin or nationality.

We made the best of it by chatting to and befriending Africans and Cape Coloureds, the largest non-white populations, whenever we could. None of us liked discrimination of any kind and we hated the way buses, park benches and even public washrooms were segregated. Neither did we like the idea of being slapped in prison and serving the dreaded 90-day 'detention' for fraternising with black women. Upsetting as it was, we all agreed that apartheid was in a word 'horrendous' but realised we were in no position to change anything. So, like everyone else, we turned the other cheek and pretended not to notice. What's more, we told each other that as visitors we had no business interfering in the politics of any country. Even though we often discussed apartheid and all had our own strong opinions about it.

However, as we were soon to discover, despite the awful apartheid policy, South Africa was, by far, the most developed and progressive country in all of Africa. In June that year, 1965, Nelson Mandela's original five-year sentence for 'treason' was extended to life in prison. After 27 years of incarceration, most of which were in a small cell on Robben Island followed by house arrest on the mainland, he would eventually be released and become president of his magnificent country. He would also become one of my heroes because he had the courage of his convictions and stood up for what he believed in.

An earlier phone call to an uncle of mine resulted in an invitation to stay at his comfortable Bryanston suburban home. Stan Williams was a small, wiry, talkative man with a Roman nose, wavy dark hair and matching moustache who had immigrated to South Africa years ago. Soon we were relaxing over drinks with him, my aunt Francine from Mauritius, and their 12-year-old son, Paul. As we talked, my uncle admitted rather proudly to having 'knocked around a bit'. One of nine kids, and my mother's oldest brother, he'd left home at 15 to seek his fortune. He first went to New Zealand before landing up in South Africa where he became a company director of a successful, Johannesburg-based security alarm firm, a fast-growing business. He was obviously quite well off.

Impressed by our adventures and experiences, Stan set about arranging jobs for us. He had many friends and contacts in the city and wanted to help as much as he could. Thanks to him, we got jobs with SABC, the South African Broadcasting Corporation.

Television had not yet then come to South Africa, but giant radio and television masts were in the process of being constructed and made ready for broadcasting in strategic parts of the republic over the next few years. After Stan made contact with their senior management, and we had a brief interview, they decided to employ all four of us. To start, it meant driving the roughly 850 miles to Cape Town to be briefed by the chief engineer in charge of the Cape Province, the district where our services would best be utilised.

We camped along the way and drove through the endless, baking-hot Karoo semi-desert region. We also stopped at the large town of Kimberley where we browsed around the diamond museum and peered into the 'Big Hole' of the open pit diamond mine.

Some four days later, we arrived in Cape Town late at night and enquired at the main police station about a cheap central hotel. The sergeant on duty, a big Afrikaner, kindly invited us to stay. So, our first night in the 'Fairest Cape' was spent on mattresses on the floor of a cell! The next morning, another sergeant presented us with a freshly cooked breakfast of bacon and eggs.

At the seafront offices of SABC in Sea Point, we reported to Mr Wilkinson, the chief engineer. He welcomed us to his department, gave us a short briefing, and then assigned all four of us to Napier (pronounced Nă-pier'), a small Afrikaans-speaking *dorp*. The village was about 130 miles southeast of Cape Town, between the towns of Caledon and Bredasdorp. Fortunately, we still had Gavin's Land Rover at our disposal, which was to come in handy over the next few months on mostly local sightseeing excursions.

Shortly after, we were just leaving the city when we spotted Heinz, the German cyclist, outside the general post office in Adderley Street. He was selling two-page brochures about his exploits to passers-by. The money, he told us, would finance the next leg of his journey. By sheer coincidence, he and I would meet each other again in Mexico some four years hence.

On arrival in Napier, we drove up the steep hill where we would be working and introduced ourselves to the rigger, our boss for the next few months. Naas Kok was a dark-haired, 27-year-old, bearded Afrikaner, married with four young children. A strapping giant of a man and as strong as an ox; he was six feet, four inches tall, weighed 240 pounds and had a voice to match. His English was rather limited, consisting mostly of what he had learned over a Monopoly board! Nevertheless, he loved discoursing about every subject under the sun. Being so big and conspicuous, he was easy to spot on the construction site in his white overalls and red hard hat, with a pipe clenched between his teeth.

Affable and charming in his own way, he also possessed a keen sense of humour. One of his favourite jokes was about his home, the Orange Free State, which he described in rather earthy language, 'There are no … oranges, nothing is … free and it's in one hell of a … state.'

Under his supervision was a team of 16 happy-go-lucky Bantustans of mixed tribes who referred to him as *Baas* (sir). Without a doubt, they respected and liked him enormously. He, in turn, was extremely fond of them.

At first, we occupied tents in a field on the outskirts of the town where Naas's large caravan stood. He and his attractive wife, Hentia, lived there with their two daughters and two sons, all under the age of six.

Then, as it got colder, we moved into the town's one and only hotel. By now, we were fed up with sandwiches and eating out of cans, so we chose to eat breakfast and dinner at the small and only café across the street, which also served as a newspaper and confectionery shop. Mrs Van Tonder, the middle-aged proprietress, was delighted to meet us, especially as customers were few and far between. She took great care over the meals she prepared for us, many of which included thick, juicy steaks.

Napier, population 800, then stood virtually isolated within a valley surrounded by rolling sheep and wheat country. As Naas used to say, 'It's a pop and plumb town. By the time you pop your head out of your car window, you're plumb right out of town.'

Although the main road passed through the centre of town, most motorists rarely stopped except to refuel or to down an ice-cold Lion beer. The town itself had four shops; a hotel and bar, which in the evenings was always full; the one café where we ate most of our meals; two Dutch Reformed churches with adjacent houses for the pastors; two schools, senior and primary; and a doctor's surgery, while the nearest hospital was at Bredasdorp, 30 miles away.

It was outside Mrs Tonder's café on the morning of November 23, 1963, that we were horrified to see in the two English newspapers, *The Cape Times* and *The Argus*, the large black headlines: 'KENNEDY SHOT DEAD'. Greatly shocked and saddened, we learned the assassination had happened in Dallas, Texas, the day before, and now the whole world mourned. We, too, were terribly upset, for like most people, we greatly admired JFK and, for that matter, the United States. Collectively, our hearts went out to his family and the American people. We couldn't think or talk about anything else for days.

The entire population of the immediate area where we now lived consisted of Afrikaans-speaking people, few of whom knew much English. It wasn't long, however, before the small community knew of our existence and went out of their way to be friendly and helpful. For our part, we tried learning some Afrikaans but, despite our well-intentioned efforts, none of us progressed much further than 'Bei danke' and 'Tot siens' (thank you and goodbye).

For the first few weeks at work, we joined the small army of Bantustans, using acid and wire brushes to clean long sections of steel lying on the ground. It was hard, tedious work, and we quickly became bored. As Naas was the immediate boss of the project, he simply watched or occasionally shouted in Fanakolo, an African dialect used in the gold mines. He often jokingly reminded us, 'I am paid for what I know, not for what I do.' Highly qualified in his job as a rigger, he had served a five-year apprenticeship as a 'rope man' in the gold mines of the Transvaal.

Every day after a hearty breakfast in the little café — usually sausage, bacon, eggs and toast — Naas would pick us up in the five-ton International. We took it in turns to sit in the cab or to ride in the back with the Bantustans on the six-mile drive to the top of the nearby hill, the construction site from where we had a fine view of the Indian Ocean and surrounding countryside.

Disappointingly, there wasn't much wildlife in the region except for a few zebra and springbok that came and went, and birds of prey. We saw plenty of hyraxes scurrying around the rocks and marmots whose shrill cry could be heard for miles. A number of times, we spied different species of buzzards, kites, eagles and hawks circling in the sky. And once or twice, we spotted a large blue crane, South Africa's national bird, soon to become endangered, foraging in the fields lower down.

Summer was fast approaching, which meant the days were sunny and quite hot. The nature of the work, painting and re-painting all the steel on the ground before it was assembled into a 640-foot mast, meant that our backs and wrists soon ached from stooping and the exertion. It was also incredibly monotonous. So we were very glad to stop for the coffee breaks — one in the morning and one in the afternoon — and to consume our sandwich lunch. For the same reason, we were also relieved when it sometimes rained, making it impossible to work, and we had the day off.

Periodically, Naas took us to the neighbouring town of Bredasdorp to collect fresh batches of steel. There, we were only 20 miles from Cape Agulhas, the southernmost tip of Africa. These trips broke the monotony for all as the whole gang accompanied him to lift the extremely heavy, half-ton sections. After taking positions, we would move them gingerly from the railway bogies to the truck, accompanied by the chants of the Bantustans. Whenever any task required everyone's cooperation, one of the crew would begin to sing and everyone would join in.

Nearly all the Bantustans were of Zulu or Xhosa origin whose names ranged from Jackson, the boss boy, to Machete, Amos, Shorty, and Engineer. Their duties varied from painting and sorting out different sections to actually assembling the various parts of the mast. All were energetic and hard-working and as much at home hundreds of feet in the air as they were on the ground. For their efforts, they received a small weekly wage, food and accommodation in tents. All of them spoke no fewer than three languages — English, Afrikaans and their own African dialect.

Early during our stay, I expressed interest in horseback riding and asked Mrs Van Tonder where I could do this. She promised to make enquiries on our behalf and, a few days later, we were all invited to a nearby farm about four miles from town. It was owned and managed by Um (uncle) Jan and Tante (auntie) Helnie, as they were affectionately known. Arriving there for the first time, we sat down to tea. 'Ve know how all English people love tea,' said Tante Helnie in her Afrikaans accent as we ate freshly made biscuits, scones and cakes laced with cream; the first of many such pleasant visits.

Soon afterwards, we were escorted around the farm by Kuis, the eldest of their three sons.

From then on, we were frequent visitors to the farm, volunteering in our spare time to drive the tractor or to load and unload trucks, in return for which we were welcome to go horseback riding whenever we liked. As the farm covered a very large area, it proved ideal for this particularly enjoyable pastime, and I can recall many happy times galloping flat out across the open veldt. I'd ridden a few times as a teenager in Berlin but, with such constant practice then, my ability improved. Pete was also very eager to learn, so Kuis and I gave him a few riding lessons — much to his delight.

On one occasion, we were both asked to help muster some sheep, so we set off on horseback with Kuis, armed with great bullwhips. No sooner had we reached our quarry than it poured with rain. But it wasn't light rain; it was torrential, half blinding us at times. Drenched to the skin within seconds, we still managed to accomplish our task, laughing, shouting, and cracking the great whips. Back at the stables, we agreed that, despite the soaking, the day was one of the most memorable we had known.

Once I rode King, a spirited gelding, bareback, intent on taking him from the field to the stables to saddle up. He, however, had different ideas as I discovered after jumping onto his back. With only a rope halter around his neck, I was unable to control him when he cantered towards a six-foot barbed-wire fence. *This is it. I'll be torn to pieces*, I thought, but he stopped abruptly inches short of the fence, and I sailed over it, somersaulting once and landing heavily, with a dull thud, on my right hand and back! Staggering up slightly dazed, I was convinced, albeit briefly that I was badly hurt. But, incredibly, nothing was broken. My right hand, though, was badly sprained and had to be bandaged up for a couple of weeks.

Soon after my injury, the SABC Branch office decided that other riggers could use some additional help. As a result, Gavin and Mike were assigned to Piketberg, another small farming town within the Cape Province. Pete and I stayed at Napier for the time being.

Meanwhile, work on the steel was progressing very fast. After completing the scrubbing, Pete and I were assigned to the equally mundane task of painting what we had just cleaned. As we worked, we adopted the popular

South African habit of chewing biltong (dried shredded raw meat) known elsewhere as beef jerky.

By this time, Naas's annual holiday was due and being on light duties, as my hand remained sore, I was selected to run the site in his absence. It became my turn to drive the truck and do the necessary shopping.

Although single girls were relatively rare in the Napier District, I nevertheless managed to meet a 24-year-old teacher named Herina, who taught at the primary school and lived in the only residential hotel in town. She came from Walvis Bay, a coastal town in South West Africa, formerly a German colony, then administered by South Africa. After independence a few years hence, the country would become Namibia. Herina wasn't really my type, whatever that was, but she had a sense of humour, a beautiful body and she loved sex. Which suited us both. She had long brown flowing hair which she normally did up in a bun and which she let down when making love.

And as we both lived in the same two-storey hotel, it was easy for me to leave my room, which I shared with Pete, and go to hers. Pete also met and went out with a girl by the name of Cornelia, who worked as a telephone operator at the tiny post office. This was the time when people shared party lines, which led to some amusing stories as neighbours listened in and then brought friends up to date.

Having girlfriends meant we now had more to look forward to at the end of what had become mostly boring days. So, when we weren't visiting the farm, horseback riding or enjoying fine South African wine and chatting with the locals at the hotel bar, we met our girlfriends individually or went together for walks or an occasional game of tennis.

Over a period of several years Pete and I had only half a dozen girlfriends each. Most were casual, often innocent relationships, but we were not averse to occasional intimate female company. The infrequent more serious romances were fleeting and passionate, with neither partner expecting anything of each other except stimulating conversation and sexual gratification.

Christmas 1963 arrived, and we were all invited to spend it with the local physician, Dr Mangold, with whom we occasionally played tennis. He and his wife more or less adopted us and, like all the town folk, welcomed us with open arms. The family had a cottage in Hermanus, a well-known beach resort and whale-watching destination southeast of Cape Town. The days were hot and balmy as we celebrated the occasion in style, swimming, beachcombing, and eating our fill.

Naas returned from holiday with his family and, with the painting at last completed, work began on construction of the mast. Slowly but surely it grew. The work became more interesting as we learned something new. Occasionally, we joined the Bantustans in climbing the mast and assisted with the bolting of the massive pieces of steel. As Naas operated the heavy-duty winch, he gave instructions from the ground in his powerful voice. 'Jacksoooooooon,' he would shout to his boss boy. Jackson would respond with an equally loud cry of, 'Baaaaaaaaas?' which would echo through the nearby peaks. However, as the mast grew taller, even Naas's booming voice couldn't be heard, so he resorted to using a walkie-talkie.

At 300 feet tall, the steel structure was relatively easy to climb, for the narrow metal ladders were fitted with a safety backing. Higher up, on the FM section, the ladders became intensely narrow and completely exposed. Higher still, on the TV section, climbing was fairly safe, for the steelwork was such that it was like climbing up the inside of a chimney, with holds. Fortunately, neither Pete nor I suffered from vertigo, but there were several times when we latched onto the steel with bated breath and pounding hearts! Climbing was hard work at any time — one false step meant a fall and instant death, so we made sure to secure our safety belts once we had found a suitable perch. Sometimes a strong wind blew up, threatening to blow off anyone who was not properly fastened or hanging on tightly. When extremely windy, work was halted and we were called down.

As the mast grew higher, climbing up became harder, tiring our arms and legs. Often, we watched the Bantustans with envy as they jumped and walked as though they were on the ground. Naas, too, was a daring man who, when not supervising from his position at the winch, was up the mast checking to ensure that all bolts were properly tightened. Like the Bantustans, he walked, leapt and climbed at an alarming rate at the eventual frightening height of 650 feet or a fifty-storey building, making it look like the easiest thing in the world. Down on the ground, he would slap us playfully on the back, roaring 'Rooinek' (redneck), for fun. Unfortunately, he seldom realised his own strength, as we picked ourselves up from the ground!

Sunday was the big day of the week, when the once-weekly bus arrived and pulled up outside the hotel. As we sat on the *stoep* downing beer shandies, we looked at the passengers as they stared back at us! Saturday was Bioscope (movie) night and we often joined almost the entire population in the town hall, watching films that were usually as old as the hills.

One friend, called Hannes, frequently took us for rides in his brand-new Volvo. He liked nothing better than to race it and show off. One evening, he took me for a drive, and it soon began to rain hard. The road was very winding. Coupled with the rain and darkness, conditions were far from favourable. Hannes, however, remained unconcerned, intent on showing me what the Volvo could do. After taking several bends at 70 miles per hour, I'd had enough and told him so. After telling him at least three times, he eventually slowed down, and thankfully we got back to Napier safely.

Some two hours later, as I sat in the bar, there was the sound of a fearful crash from one end of town. Minutes later, someone rushed in to say that Hannes had overturned his car. Luckily, he had escaped serious injury, but his new vehicle was badly damaged and required substantial repair, while I felt exceedingly grateful at not being involved!

Occasionally we'd join friends to drive to nearby beaches, where we cooked *breifleis* (literally meaning burnt meat and equivalent to a barbecue) over open fires. While the food was prepared by a dedicated few, the rest would indulge in swimming or downing liberal quantities of Lieberstein wine. Then, after consuming large juicy barbecued sausages and huge steaks, we would descend on the enormous apples and pears. Dancing followed to the music of a record player. Then, one by one, couples would detach themselves from the party and stroll off into the night.

After four months, when the mast was nearing completion, Pete decided to hand in his resignation as he wanted to move on to see more of the country. After telling me his plans, he headed up the coast to the small resort town of Mossel Bay where he quickly found work in his profession. I stayed to see the mast completed.

Different employment had forced us to split up on previous occasions. In spite of having a great deal in common and being the best of friends, Pete and I were often interested in different things or sometimes got fed up with each other's company but any differences of opinions were short-lived and quickly forgotten. We would, however, correspond and meet often at prearranged rendezvous in different places, teaming up to travel and to exchange impressions and news.

A month later, the mast stood fully erected, and I was assigned to another site under a rigger called Ernie Webber, an English-speaking South African in his mid-forties who lived with his wife, Elizabeth, in a small caravan onsite. We worked in the mountains close to the town of Villiersdorp, a prosperous fruit-growing district of apples, grapes — and apricots and peaches as big as grapefruits. The campsite was a bleak, open area and the tent I was allocated stood under the only tree. Summer came to an end, and I had my first taste of the beginning of winter in the Cape Mountains. Like Naas, Ernie had a gang of 16 Bantustans who also answered to an assorted mixture of Christian nicknames. A bluff, hardy man like Naas, he led a tough, nomadic existence. Whereas being somewhat isolated apparently didn't affect Naas, it did affect Ernie, and he became cantankerous at times.

The actual mast was 12 miles from the camp on the top of a steep 5,000-foot mountain. To reach it involved a lengthy, tortuous drive up a precipitous winding 1-in-5 gravel road which was, for obvious reasons, restricted to the public. Whereas the countryside at Napier was soft and undulating, at Villiersdorp it was wild and rugged, and the mountain air was fresh and invigorating.

Fortunately, when I arrived, this mast was well under construction and the painting and scrubbing of the steel was completed, so my work was more varied. One was to contact other sites by radio every day. In place of an International, we had the use of a Unimog. With its mass of dials and wide windscreen, it felt like sitting in the cockpit of a small plane. At first, its six forward gears and two reverse gears made it a little difficult to master. When fully laden as it usually was, the steering was heavy and the slightest touch on the footbrake made a fierce impression. However, after driving it without mishap on fairly steep roads, I felt confident of driving anything.

At Napier, it had often been necessary to scale the mast, but at Villiersdorp, instead of climbing, I was raised by winch in the bosun's chair. This was a tremendously hair-raising experience. Seated on a flat, thick piece of wood attached to the half-inch steel cable and buckled to my safety belt, I was slowly winched up. Every few feet, the hawser would 'give' a few inches, causing the 'chair' also to slip slightly. As I hovered and swung slightly in mid-air, with nothing below me except solid ground, I had to fend myself off from the mast to avoid being crushed. Every time I was hauled up, I suffered from the same complaints — racing heart, sweaty palms and shaky knees — and it was with a great sigh of relief that I arrived at the top and reached out for the safety of the beckoning steel. It was like sitting on an open swing outside the window of a fifty-storey building. *God knows how high-rise window cleaners coped*, I asked myself more than once.

Safety precautions were strictly enforced, and work was called off in wet or very windy conditions. Most of the time, though, the weather at Villiersdorp was fine. As we worked, small herds of springbok, hartebeest and mountain zebra strayed into the vicinity, and hyraxes screeched and scurried around the rocks.

Sometimes, the wind dropped completely, making everything quite still, and the slightest sound of a falling rock would reverberate through the mountains. Those days Ernie didn't need the walkie-talkie to give instructions to the Bantustans on the mast. Like Naas, he used his powerful voice. 'Sooollloooomooon,' he would shout to Solomon, his boss boy who would respond with a loud cry of 'Baaaaaaaaaass'.

At night, the wind usually blew up, and it became intensely cold with occasional early morning frost. For the first time on the trip, I felt pangs of loneliness. I was no longer seeing Herina, as Napier was too far away. By the end of the day, with nowhere to go, I would turn in after supper at the ridiculously early hour of seven or eight o'clock.

Fortunately, there were a couple of times when I took the Unimog to the nearby towns of Paarl, Worcester, or Villiersdorp to collect supplies. Coming back, my route passed through the interesting university town of Stellenbosch and over the impressive Franschhoek pass. As night fell, it was particularly exhilarating driving beneath a full moon and twinkling stars, the road surrounded by high, rugged Cape Mountains silhouetted against the dark, velvety sky. No doubt about it, I was very much enjoying living and working in South Africa but, once again, it was time for a change of scenery. So I quit my job and headed for Cape Town.

CHAPTER 6:

The Open Road

'Here today, up and off to somewhere else tomorrow!
Travel, change, interest, excitement!
The whole world before you,
and a horizon that's always changing.'
— Kenneth Grahame

It was during those few trips in the Unimog that I'd seen convoys of new motor cars travelling to and fro along the main highways, and the thought occurred to me, *what an ideal way to see more of the country*. So, after hitch-hiking from Villiersdorp to Cape Town, I immediately checked the city telephone directory, located the office of Controlled Deliveries run by the Westcott brothers, and applied for a job as a convoy driver there and then.

Quite by chance, it was my good fortune to be interviewed on the spot by the managing director personally. Richard, better known as 'Dick' Westcott, was in his late thirties and a former cricketer who'd played for the South African team. His older brother, whose name I forget, was a partner and the company accountant. Impressed by my military background and adventurous spirit, Dick decided to employ me as his personal representative to be his 'eyes and ears on the road'. He concluded by offering me a reasonably attractive salary, although I would gladly have worked for almost nothing to see more of this vast, magnificent, diverse country. I couldn't wait to get started.

Travelling with convoys became one of my favourite jobs, for I liked nothing more than to be on the open road, no matter where. I was at my happiest when the road lay ahead of me like life, with all its ups and downs, twists and turns.

By this time, I'd spent a lot of time under canvas in Kenya and at Villiersdorp, and felt I wouldn't be sorry if I never slept in another tent again! I'd also had enough of playing around with steel. After checking out places for long-term stay, I rented an inexpensive room in a private home on the slopes of Table Mountain.

Meanwhile, Gavin sold his Land Rover and left for Southern Rhodesia, to be commissioned in the army (after which he would immigrate to the United States; marry the American Carole with whom we'd shared the house in Nairobi; and become a professor at an Oregon university).

Mike left the SABC too. He joined Pete in Mossel Bay for a while and subsequently drifted to Durban, where he found a job in insurance and a girlfriend, as he informed me in one of his once-in-a-while letters.

My new employer, Controlled Deliveries, was responsible for the delivery of new cars from assembly plants to auto dealers throughout the country. It was a simple but effective way of running in brand-new vehicles. The company employed about 400 Cape Coloured drivers and about 20 white supervisors. As brand-new cars rolled off the factory assembly line, they were put in a securely fenced outdoor holding area. The new cars would then join a convoy of 15 to 20 assorted vehicles. Once a convoy was assembled, the cars would proceed under strict supervision to their respective destinations. There, they would be thoroughly cleaned and placed in holding areas of auto dealerships until space became available in showrooms. The entire enterprise had to be run like a military operation to ensure that each new vehicle arrived at its destination in pristine condition. When the proud new owner picked it up, any problems had been fixed, and it was already run in. Otherwise, if there was the slightest fault, there would be questions from both the dealer and the new owner. Serious questions.

My job was to travel with any convoy I chose, stopping it at will, to make spot-checks for drugs and alcohol which were forbidden and to ensure that each driver was sober and capable. Due to the fairly exhausting circumstances, drivers sometimes secretly imbibed, usually when overnighting in motels. At times, I would change to another convoy travelling in the opposite direction in an attempt to keep the drivers on their toes.

The convoys operated seven days a week under all conditions. During the course of an eight-hour day or longer, depending on changing circumstances and weather, they travelled hundreds of miles. Each convoy had a senior driver who drove in the front car. The supervisor usually sat in the last car, with a mechanic driving beside him. With no cellular phones in those days, the convoy was halted by flashing lights or by walkie-talkie. In the case of a breakdown, the mechanic would spring into action to fix a flat or an overheated engine.

Before the supervisor allowed a convoy to get under way, the drivers were required to do a vehicle check. This meant checking tyres, window wipers, fuel gauge, radiator, and engine oil. Only when everything was in order was the convoy allowed to proceed to the sound of a whistle blown by the supervisor.

Over the next several months, I covered more than 80,000 miles, traversing the whole of South Africa in my exciting new job.

After a while, Dick Westcott asked me to take on additional duties and appointed me Training Officer. My job was to teach the ins and outs of convoy driving and enforce the rules. The technique is not as easy as it looks; for example there is a risk of accidents when all the cars are stopping or starting at the same time. After new drivers were taken on, I taught them everything I knew about convoy driving, which to begin with, was not a lot. Other times, I would sit in a car one-on-one with a driver in the compound where the vehicles were kept and explain the finer points of convoy driving. I would see the drivers again when riding in convoy and assess and report in writing on their skills.

When it became known that I had some expertise taking photos, I was also asked to be the company photographer, and I took many pictures of the convoys coming and going and stopped along the road. From these, I created a pictorial story of the company.

As each day involved hours of driving, frequent halts allowed the drivers to rest, stretch their legs, check over their vehicles, and compile progress reports. The company-regulated speed for the convoys was 45 miles per hour and, because of the huge distances, this demanded extremely careful and relentless driving. Delivery points included Johannesburg, Port Elizabeth, Durban, Pietermaritzburg, East London, Bloemfontein and Kimberley, the routes criss-crossing the entire country.

We drove — I often relieved drivers at the wheel — in all conditions, including mist, rain, and occasional snow and ice, but the compensations made up for any discomfort as we drove through interesting small towns and exceptionally beautiful scenery.

Over time, my limited knowledge of motor vehicles increased. I'd climb out of a Ford Escort or Vauxhall Victor and into an Austin Cambridge or Mini Minor and then return along the same route in a Hillman Minx or the latest model of Jaguar. Sometimes, it was a convoy of new trucks such as Leylands and Land Rovers. The job also provided the opportunity to stop in fairly remote villages and districts, allowing me to observe from a short distance the interesting South African peoples. And sometimes, discreetly, take photographs.

It was interesting to chat with the drivers, most of who were married with families. Before I joined the company, they came to work in their street clothes and, as many were quite scruffy, I proposed the idea of uniforms to Dick. I would emphasize the importance of being well turned out at all times. Now, dressed in their military-style uniforms of long khaki pants, short-sleeved navy blue shirts and maroon berets, they looked very smart.

I remember one driver, a Muslim named Ishmael, who often complained about his three wives, which no doubt accounted for the reason that he always looked tired! If a driver felt unwell or was overtired, that's when I would take over while he rested beside me. With different things happening every day, I enjoyed most of my time on the open road. However, it wasn't all a bed of roses, and it was extremely tiring sitting in a car for hours at a time. Like the drivers and supervisors, I occasionally suffered from cramps and sore eyes.

The Garden Route, which I must have driven over 50 times, never lost its appeal, with its spectacular coastal views and small, interesting towns of Knysna, George, Wilderness, Storms River and the gorgeous beaches of Plettenberg Bay. At Tsitsikamma Forest, we would always stop and troop into the woods some 150 yards to see its tallest and widest tree — an Outeniqua yellowwood said to be 800 years old.

One regular trip to Port Elizabeth, a distance of 498 miles along the famous Garden Route, sometimes took 18 to 20 hours due to adverse weather conditions and occasional breakdowns. Depending what time we arrived, we had an evening meal and after no more than a few hours of sleep in a motel, we were up with the birds, climbing into different cars for the return trip to keep to the tight schedule.

One place we always stopped when coming or going was Sir Lowry's Pass, on the N2 Highway 30 miles east of Cape Town, with a particularly steep incline. Because of the danger of engines overheating, every convoy stopped at the top to allow the vehicles to cool down and for the drivers to check their engines, stretch, admire the view and maybe relieve themselves in the nearby bushes.

Another route I particularly enjoyed was through the Karoo, the vast, arid semi-desert, especially in the spring with its carpet of yellow daisies. For the keen-eyed lovers of wildlife like me, there were countless interesting sightings, close up or in the distance, of springbok, eland, hartebeest, bat-eared foxes, black-backed jackals, ostrich and zebra, and once in a while, a solitary lion.

Seated comfortably in a brand-new car with lovely scenery passing by me and the open road ahead, I was once more in my element. Even though I wasn't earning a great deal, I was only too happy to do what I was doing. Often, all that lay ahead of us for miles upon miles was an almost empty road stretching to infinity with red earth on each side and a perpetually blue sky above.

But like most jobs, it wasn't perfect and, apart from the constant driving, which sometimes wore thin, I saw my share of accidents, sometimes serious ones. It was company policy to always stop to offer help. And once the company had a rare fatality. It occurred in a mountainous region in foggy conditions when one of our drivers at the wheel of a truck ran into another vehicle and was killed instantly. Fortunately, I wasn't in that particular convoy but quickly heard about it. My heart went out to his family, and the company took care of all the expenses and the funeral arrangements. It was also a lesson that everyone, including myself, took very seriously: driving could be dangerous at times.

For the most part, I got on with everybody, but there was a *rooinek* Afrikaner, whose name has disappeared in the mists of time. Aged about 40, red-faced and beefy, he was employed as the company policeman; he dressed like one, acted like one, and rode a Harley Davidson complete with police siren. His duties were similar to mine; stopping and starting any convoy at will, and reprimanding or cautioning both drivers and supervisors if they were not following strict protocol and procedure. I steered clear of him because I could tell he disliked me by his offhand manner and perpetual scowl. Although I'd joined the company after him, I outranked him, which put his nose out of joint. I thought Dick Westcott had made a mistake employing him, but I kept that to myself.

Dick was an exacting boss with very high standards. He worked hard himself and expected the same of others. I would learn a lot from him. At first, I reported to him verbally until one day, quite exasperated, he insisted that I 'get it down in writing'. I didn't like doing the written reports at first, but gradually got used to it. This made good sense as it meant proper records were kept. Soon, it became a regular habit of mine. And one I adopted in many future jobs. I'd written reports before in the army and during my sales rep days, but from then on and under Dick's constant prodding, I kept copious notes about everything under the sun. You never knew when you would need to refer to them.

Dick also drummed into me the importance of always responding respectfully to letters, whether business or personal 'If someone takes the time and effort to contact you, the least you can do is reply. Promptly!' he commented more than once. Good common sense advice, which I took to heart and applied throughout my life.

I enjoyed writing so it was no problem for me. Ever since setting off on the trip, I'd been keeping a daily diary, a travel journal about my new life and itinerant wanderings. I also tried to write home every three or four weeks to keep them informed of my whereabouts in the hope my mother wouldn't worry too much. She would keep all 97 letters from my marathon odyssey describing what I was doing and where I was. Amazingly, over the course of several years, only a few of the more than 100 letters, postcards and parcels that I sent home would go astray. Unfortunately, I would lose some of my valuable slides.

Having a regular job meant having a (semi-) permanent address where family and friends could reach me. In this case, I used the San Remo Hotel. Other times, while moving around, Pete and I mostly used Poste Restante c/o Main Post Office in the larger cities or sometimes British embassies or consulates. And if we didn't or couldn't pick up our post for some reason, we'd have it forwarded.

Most of our contacts were through correspondence; phone calls were expensive, so we seldom used the telephone except for rare emergencies or sometimes for brief local calls. To minimise costs, we mostly sent aerogrammes and postcards as they were cheaper and quicker than ordinary mail. Telegrams were still then much in use for important or urgent messages, but I can't recall ever sending more than one during those early world travels.

I fell in love with Cape Town and it quickly became one of my many favourite cities. A large port city situated at the foot of the African continent, I loved everything about it — the temperate climate; the exotic flora and fauna; its multiracial population; the cobble streets, quaint coffee shops and restaurants; the always-busy harbour; and the many interesting walks in all directions. There were dramatic views and vistas from almost every angle of the city, especially from the top of Table Mountain.

Whenever in town, I took most of my meals at the San Remo Hotel, a small, private, residential hotel near The Gardens at the foot of Table Mountain, and just around the corner from my rented room. The hotel was comfortable and quiet, and the majority of its residents were middle-aged or elderly. The guests reminded me of Terence Rattigan's *Separate Tables*. They included a retired British colonel, a judge, one or two divorcées, a couple of bachelors (one of whom wore an ominous black eye patch), a few elderly spinsters, and a sprinkling of young professional people, all with their regular tables in the dining room. Most of the older people were retired and spent their time reading on the veranda or sitting and chatting in the garden. The hotel was run by two white single sisters, with their obliging, all-coloured staff, all of whom spoke both English and Afrikaans.

It was at the San Remo that I met Veronica de Wet, whose family was of Afrikaans stock. An attractive, sensuous and willowy 20-year-old with blue eyes and shoulder-length auburn hair, her skin was as pale and as smooth as porcelain, unusual for someone who had lived all their life under the hot African sun. She was a legal secretary staying at the hotel. I'd noticed her in the dining room eating alone, made eye contact and one day, passing in the corridor, we stopped for a chat. She seemed quite lonely and eager to talk. My interest in her increased when she told me one day that her previous boyfriend had been a lot older than her. This led me to assume that young as she was, she may well have some expertise in the art of sexual relations. A couple of evenings later, after my return from my latest road trip and a nice dinner with Veronica over drinks, my hunch was confirmed. Much to my great pleasure, she not only knew but happily shared with me in the comfort of her bedroom some of her lovemaking secrets.

As Veronica and I continued dating, I learned about her family. Her father had died tragically in a bulldozer accident on their property. After taking some time to get over her husband's death, her widowed mother embraced the task at hand and now operated a highly successful farm and vineyard.

Veronica was an accomplished horsewoman, and a couple of times we drove to the family farm near the town of Franschhoek. With her two faithful dogs, a black lab and a fox terrier, trotting behind us, we'd go riding. Afterwards, we'd swim in the nearby stream. And then, after making sure nobody was around, we'd make love on the garden lawn surrounded by all kinds of exotic plants and flowers and the scent and blossoms of frangipani and magnolia. Later, we relaxed over a glass of fine South African wine in the kitchen or living room. Such happy, carefree times.

Once during my stay, snow fell on Table Mountain, the first in 40 years, which gave everyone something to talk about. The accompanying southeasterly winds were also an interesting experience. Commonly called 'The Cape Doctor' by Capetonians, these winds were quite violent, at times up to 50 miles per hour, and when they blew, everyone made an effort to stay indoors. Temporary hand ropes to hang onto were quickly rigged in the streets, and the girls exchanged their skirts for trousers or jeans.

Despite the considerable amount of time I spent on the road, I was able to get away from it all. Three times, I climbed Table Mountain with its magnificent flora and fauna, twice on very hot days. Each time, after negotiating a not-very-well-used trail through the magnificent foliage, I spotted a pair of graceful Verraux's eagles circling in the sky. At the top, I was met by many rock hyraxes, or dassies, as they are locally called. From a distance, I spotted mountain zebra, ostrich, bontebok, a few eland and once, lurking in the dense *fynbos* (Afrikaans for fine bush), I even saw a caracal, similar to a lynx.

I'd also met Laurie Warren at the San Remo. He was a thirtyish civil servant from Port Elizabeth who joined me for a few local excursions. On occasion, with either Veronica or Laurie, I visited the Cape of Good Hope, the famous peninsula jutting southward into the sea. There, we encountered the bold chacma baboons that snatched sandwiches from unwary visitors and dove into parked cars to see what else they could find. With their long fangs and razor-sharp claws, these primates could be vicious and dangerous as I had learned during my monkey catching days. We also went sunbathing and swimming at the magnificent beaches of Muizenberg and Fish Hoek where the locals flocked at weekends.

Once, as I waded into the ocean at Sea Point, I noticed a young boy struggling in the water, obviously out of his depth. Seeing his predicament, I waded slightly further out. As I did so, I felt a strong riptide trying to sweep me away. As I struggled to keep my balance, I leaned out as far as I could while shouting at the boy to grab my hand. With his mouth full of water and panicking and spluttering, he grabbed it. With all my strength I pulled him to shore, where he stood shivering for a while and then ran off to play on the beach. His mother had seen the incident and thanked me profusely for rescuing her seven-year-old son. I was only too glad to help. My efforts that day were nothing compared to Pete's. He would later be involved in a far more dramatic life-and-death rescue which I would personally witness.

Before long, Christmas 1964 arrived, a circus came to town, and festive lights were strung up along Adderley Street. I celebrated with friends, swimming at Muizenberg with its famous long and mostly deserted white sandy beach, followed by a sumptuous turkey dinner with all the trimmings at the San Remo Hotel.

By then, I'd decided it was time to hit the road again and, besides, Veronica already had someone else. So I submitted my resignation to Dick, who understood my reasons for leaving. A week or so later, I resumed my South African journey.

My first destination on leaving Cape Town was Oudtshoorn to visit the famous ostrich farms. Um Elias, a jovial, middle-aged fruit farmer, typically dressed in only shorts and vest, was one of several drivers who gave me a lift. He picked me up in his ancient Ford truck, on his way home from delivering a load of fruit to the Mossel Bay docks.

As we drove, he kindly suggested I break my journey and spend the night with his family at his home. With nothing better to do, I happily agreed.

The spectacular Swartberg Pass was then still a gravel road and known for its very steep incline and sharp bends. After a somewhat frightening drive over the pass which had me gulping and holding my breath, we arrived at his farm where I was introduced to his plump wife and five of his six grown-up daughters. All of whom were most attractive and friendly. More and more, I realised things were happening to me that one only hears or reads about!

The next morning, after picking and eating as many juicy and succulent peaches as I could stomach and accepting two heavily filled bags 'for your journey' from Um Elias, I continued on my way. An hour or two later, after another couple of rides, I arrived at Highgate Ostrich Farm, where I was one of a group of tourists to volunteer for an ostrich ride.

Riding an ostrich was a peculiar sensation, rather like sitting on a tall barstool. I sped around and around in circles on the back of a male ostrich, hanging onto its wings for dear life, while two well-known ostriches, Jack the Ripper and Jean the Stripper, stared at me steadily with their beady eyes from their respective fields.

A resident guide explained the purpose of breeding these huge, flightless birds, which stand roughly eight feet tall and weigh on average 250 pounds. Every part of them is put to good use. Their meat is lean and delicious; an ostrich egg when scrambled or made into an omelette can feed 20 people; their feathers are used as dusters and boas; and their skin when processed is ideal for the manufacture of fashionable leather shoes, belts, and handbags. In addition, their eggshells are painted and sold as handsome souvenirs.

Next on my itinerary was the spectacular Cango Caves, a vast labyrinth of limestone caves where I joined a group of tourists for the one-hour tour.

Meanwhile, Pete had left Mossel Bay for Port Elizabeth, where he teamed up with a group of other young men he'd met at work. Together they made an exciting journey by road to South West Africa. The trip ended in Durban, where he met up with Mike outside a recruiting office for mercenaries — to fight in the Congo. They considered it but quickly dismissed the idea. There were nicer things to do in the world. Besides, we were soldiers of fortune in our own way, always on the lookout for new experiences. As the weather was fine and to conserve funds, they resorted to sleeping on the beach until one night some of their clothes were stolen from under their heads!

Shortly after, I too found myself in Durban, South Africa's most popular beach resort. After a relaxing few days at an inexpensive beachfront hotel and swapping news with Mike and Pete, I took a bus to Johannesburg, curious about the gold mines. A helpful receptionist at the department of mines informed me that it was, in fact, possible to visit a gold mine and issued me with two complimentary tickets on the spot. Pete joined me a few days later after I contacted him. Together with some 30 others, mostly foreign tourists, we were flown in a chartered South African Airways DC-3 to Hartebeestfontein. There, we first sat down to a delicious buffet lunch and were welcomed by the mine manager before slipping into protective clothing and helmets.

After descending the 6,000-foot-deep mine by swift, silent lift, we stepped out at the bottom into an unexpected and almost overwhelming dry heat. An interesting hour underground followed as a mining engineer pointed out reefs of gold and how the operation worked. Interesting though it all was I felt distinctly uncomfortable that far underground, and I couldn't wait to get back on the surface to sunshine and fresh air. It was my first ever bout of claustrophobia, and an experience I was in no hurry to repeat. After our visit down the mine shaft, we were invited to try to lift a gold ingot with one hand. Despite everyone's supreme efforts, nobody succeeded, but, had anyone done so, the gold bar would have been theirs.

Pete and I had now both been in South Africa some 14 months, the longest time we would spend anywhere. We had each managed to save a considerable amount of money, the equivalent of several hundred pounds, far more than we had originally set out with. Despite the strong feeling of financial independence at the time, we knew only too well we would have to be reasonably careful spending our hard-earned cash if we wanted to do everything we intended.

During some of our many conversations, we decided to travel back through Africa, hoping to accomplish two objectives: climb Mount Kilimanjaro and visit Ethiopia. Although part of the intended route meant retracing some of our steps, we decided to divert whenever practical and possible. As much of our route did not include regular bus or train services, we would hitch-hike where necessary. It would be cheap and, as usual, prove to be an excellent way of meeting people.

Mike was quite happy to remain in South Africa for a while longer before heading to Southern Rhodesia and promised he would catch up with us sooner or later. But both Pete and I felt it was a half-hearted gesture, and we somehow knew that he'd never join us again. He'd considered the trip a 'great' idea, but perhaps because he hadn't taken part in the original preparations, it didn't seem to mean as much to him as it did to us. So now we were two again.

Back in Jo'burg, we called on my Uncle Stan to thank him for his earlier help. He had been a valuable contact who had been of great help. We had no other future contacts apart from two of my distant relatives. As usual, we would have to fend completely for ourselves.

Armed now with more money than we'd ever possessed, we were determined to travel as far as we could before having to seek employment again. After not carrying our backpacks for some time, they seemed heavier than ever. With our cameras loaded with Kodachrome film and our water bottles filled, we got back on the road and headed north. Or as professional tramp, Jim Phelan, would say, we 'hit the grit', eager for more adventures. Plenty would come our way.

CHAPTER 7:

Wandering through Africa

> *'The real voyage of discovery*
> *consists not in seeing new landscapes,*
> *but in having new eyes.'*
> — Marcel Proust

From Johannesburg, we hitched a ride by truck to Mafeking. The town is known for the 1899 siege when the Boers tried to take the British garrison commanded by Colonel Robert Baden-Powell. Next, we caught a local bus to Barolong where we crossed into the British Protectorate of Bechuanaland (renamed Botswana in September 1966). This time, with 'sufficient funds', we had no problem with immigration officials.

We spent the night in a semi-completed house in Gabarone, the new capital then under construction. The next morning we waited an hour for a ride. 'Hop in the back,' said Tim, a white Rhodesian businessman and driver of the truck. We travelled with him for a few hundred miles, one of several similar long-distance rides we had travelling through Africa. With such vast distances and hardly used, desolate roads, most drivers enjoyed the company of friendly strangers.

For nine hours, we endured a bone-shaking journey with Tim, cringing at times as we travelled over deep corrugations along the wide, red-earth highway. There was virtually nothing to be seen except endless thorn scrub as we skirted the famous Kalahari Desert. The monotony was occasionally broken by the sight of a herd of springbok, zebra or impala running and leaping across the road. Soon, we were covered in thick red dust, something we had become used to but never enjoyed when driving on Africa's mostly dirt and murram roads.

At Francistown, the largest town in the country, we made our way to one of the town's two stores. There, we fell into conversation with the two Greek proprietors who also ran the small hotel, one of many we discovered in isolated parts of Africa, where we stayed. That evening, over sundowners, they told us something about the vast Okavango Swamp, nowadays known as the Okavango Delta, which reputedly has the freshest and cleanest water on earth. The river originates in the Angolan Highlands and ends in the Kalahari Desert, in the northwest of Botswana.

The delta supports large concentrations of birds and common African mammals like elephant, lion, hippo, and leopard as well as wild dog, roan and sable antelope. This habitat is home to specialised wetland antelope such as lechwe, puku and the shy sitatunga, a small bushbuck-like creature which lives among the reeds. It was also a great place for fishing with apparently large numbers of tilapia, bass, bream, catfish and even snapping turtles. A place we couldn't wait to see for ourselves.

To get there, we had to go to Maun, a small, remote township and safari base only a stone's throw from the delta, but the road was a hardly used track, traffic was exceptionally infrequent, and we were advised that we might have to wait several days for a lift. Fortunately, we were in luck. After enquiring at various offices, we were told at the veterinary department that a vehicle would be leaving for Maun the next day. And yes, we could get a ride in it.

It was already hot and sunny when we met Tom, a white veterinarian, and Isaac, his African assistant, early the next morning. After tossing our packs into their Chevrolet van, we set off due west. Tom was a lively and flamboyant character who loved the African bush. Because the 302-mile track was so rutted and sandy, we were sometimes forced to travel at little more than a snail's pace. Twice, we had to fix punctures.

We took hours skirting the vast, unpronounceable Makgadikgadi salt pans, eventually pulling up at dusk to make camp. After dinner, we fell asleep on camp beds supplied by Tom. Twice during the night, we were woken by a lion growling in the vicinity, and each time Isaac tossed another log on the fire to keep it at bay.

Breakfast was tinned bully beef, beans and fresh fruit —our fairly regular diet while travelling through Africa. As we continued on our way, we passed great plains dotted with game and tall, weirdly shaped termite mounds, including one with a monitor lizard sunning itself on top. Several hours later, weary and dusty, we bumped and rattled into Maun where we parked outside the local police station.

A young British police inspector named Ian came out to greet us, promptly inviting us to stay at his bungalow home. 'I'll be glad of the company,' said Ian. After listening to our intentions to explore the swamp, he suggested that we meet the local chief, as a courtesy measure. 'He'll give you advice and probably help, too. You'll find him in the bar at sundown.'

That evening, we joined some of the few local whites over glasses of beer in Maun's one and only small hotel. Sure enough, around seven o'clock, a youngish, slightly built, African smartly attired in khakis walked in, and everyone respectfully stood. As Chief of the Tswana, the dominant tribe, he'd been educated in England but was clearly very shy. We were introduced and, after a short and pleasant conversation, he promised to see what he could do about hiring a dugout canoe and guide for us. 'Meanwhile,' he said, in impeccable English, 'I shall arrange for you to explore the area on horseback, if you like.'

We passed him the next morning as we made a tour of the township on borrowed horses. Immaculate in his tropical whites, he sat in the shade of a giant bottle tree, presiding over a meeting of tribal elders and headmen.

We met again in the hotel garden where he introduced us to an elderly, grey-haired African who owned a dugout canoe. Barefoot and wearing tattered khaki shorts and a dirty vest, the old man, whom we promptly nicknamed Charlie, spoke no English but smiled constantly. The chief negotiated and, shortly after, Charlie agreed to act as our guide in exchange for a basic wage and rations of tobacco and maize. Soon after, stocked up with supplies for several days, we stepped gingerly into Charlie's *mokoro*, which we subsequently learned had taken him two years of intermittent labour to hollow out.

As usual, we were dressed in short-sleeved bush shirts, shorts, and sandals. We were also armed with a powerful .365 Mauser Magnum rifle, which had kindly been loaned to us by a local white hunter whose name I forget. But it could have been Harry Selby, the most famous of white hunters in the region. One well-aimed shot from this deadly weapon would stop an elephant in its tracks in the unlikely event we were attacked. We borrowed it somewhat reluctantly but knew only too well there was a first time for everything, especially in Africa.

At first, at the slightest movement in the dugout, we almost capsized, but we soon developed the art of moving without overturning the flimsy craft or falling overboard. As Charlie punted with a long pole, Pete and I paddled with crude handmade paddles. Gradually, we made headway along the river, a tributary of the Okavango. Then, a while later, we entered the swamp.

As we penetrated deeper into the swamp, Charlie produced a panga to cut a channel through the tall, dense papyrus reeds. As we went further, we were dazzled by the huge purple and white water lilies. Then, as the sun set, we searched for a suitable camping spot on firm, dry land. As darkness descended, we stoked the fire, and Charlie made a meal of *seswaa*, a traditional meat dish made of beef, goat or lamb boiled with onions and pepper. As we ate, Pete and I discussed the day's events and then slipped under our mosquito nets and attempted to sleep. Eventually, under the watchful eye of Charlie, who catnapped by the fire, we dozed off. This became our routine as we explored just a tiny part of this stunningly beautiful, remote, and relatively unknown area.

In complete contrast to the days which were mostly quiet except for birdsong and the occasional grunt or splashing of hippos, nights in the swamp were amazingly noisy. Owls hooted, monkeys screeched, and baboons barked while the surrounding bush came alive with the buzzing of thousands of crickets.

For three days, we journeyed into the heart of the swamp, blazing our own trail. The birdlife was fascinating, but the only game we saw was an occasional waterbuck, some reedbuck, one magnificent male sable, and a few small herds of puku and red lechwe which inhabit the wetlands all year. Because it was so hot, every so often we'd jump into the fresh, clear water to cool off and once we spotted a snake swimming among the lily pads. Luckily, we saw no sign of crocodile. The further we went, the less water there was. Before long, we were manhandling the dugout through waist-high mud from one pool of water to another. Finally, we could go no further. Ahead lay only more dense reeds and mud. As it was the dry season, the water had all dried up. So we retraced our steps.

Despite not going as far as we had hoped, we were ecstatically happy. We had been somewhere that few people had been before. We returned to Maun none the worse for wear, covered in mosquito bites and wearing our deepest-ever tans. With its natural beauty, remoteness and abundance of wildlife, the Okavango Delta would eventually become a magnet for the well-heeled tourist.

Back at Ian's house, after checking our map, we announced our intention to travel to Kazungula on the Caprivi Strip, where Angola, Botswana, Zambia and southwest Africa all meet. He advised us to contact the local tsetse fly control officer who occasionally travelled that way. Again, more luck. A young field officer named Jonathon was leaving almost immediately for Kazungula and agreed to give us a lift.

Our route along yet another seldom-used track yielded marvellous scenery, the tall elephant grass gradually giving way to vast plains and dense woodland. Soon, we had made up for what we had missed in the swamp when we encountered great herds of blue wildebeest, zebra and buffalo. As we neared the Chobe River, a wide tributary of the Zambezi, we halted at a distance to give a massive herd of elephant the right-of-way.

Passing close to a fenced mud-hut village, we stopped and were immediately surrounded by small, semi-naked women and children, all chanting in unison, 'tobacco', the limit of their English vocabulary. I distributed all that I carried while Jonathon informed us they were members of the Bushman tribe who still hunted with bows and arrows. In time, it would be scientifically proven that all mankind originated from these people of the Kalahari Desert. As we posed for photographs with them, Pete stood next to a smiling, pregnant woman but made it abundantly clear that he wasn't responsible! As we proceeded further, we came across members of the Herero tribe, whose women had adopted colourful long German-style dresses and headwear, whom we couldn't resist photographing.

At the tiny township of Kazungula, we bade farewell to Jonathon, heaved our packs onto our backs and walked a few yards to the south bank of the Zambezi. As we gazed across the turbulent, fast-moving river, waiting for the little ferry to transport us to the other side, we wondered what else lay in store. Africa was full of surprises, and thank God, so far mostly very interesting and enjoyable ones.

Having been wildly excited on seeing Victoria Falls the first time, we couldn't resist returning for another look. So, for two days, we viewed them again from every conceivable vantage point. At night, we sneaked into the garden of the swanky colonial-style Victoria Falls Hotel and slept under a huge baobab tree. When we awoke, we went inside to use the washrooms and clean up, and then indulged in a sumptuous buffet breakfast.

With more money at our disposal now, we occasionally stayed in hotels, but, as budget travellers, we considered the tourist ones too expensive, and unfortunately, the African-run places all too often were rundown. At times, we simply threw down our sleeping bags wherever we found ourselves.

On the outskirts of the neatly laid-out town of Livingstone, we hitched a ride with a talkative white South African; then at the small town of Kapiri M'poshi at the junction of the Great North Road, we transferred to a Volkswagen minibus. It was driven by a young Rhodesian man accompanied by two female teachers all going away for the weekend. We'd been in Africa long enough to know that the roads can be extremely dangerous any time due to hazards such as extreme weather, reckless driving and poorly maintained vehicles. Even knowing that, we certainly didn't expect what happened next.

After a brief tropical downpour, the road had become rather slippery, to say the least. Despite the slick conditions, our young driver, smoking a pipe, insisted on driving at 50 miles per hour and also kept turning around to talk. A recipe for disaster. Sensing trouble, Pete and I kept glancing at each other. Suddenly, and as we anticipated, the vehicle skidded out of control and, before we knew what was happening, it careered from one side of the road to the other. Then abruptly overturned.

After landing upside down and extricating ourselves from the vehicle as quickly as we could, we checked each other out. Everyone was shaken up but, surprisingly, no one was badly hurt. I'd grazed a leg, and one of the teachers had cut her ear, but that's about all, which was remarkable considering the speed we had been going and the way the minibus had rolled over. Almost immediately, some Africans came running to our aid. With their assistance, we righted the vehicle and carried on our way. Stupidly, our youthful driver apparently hadn't learned anything from his grievous mistake and continued at his original speed. By this time, Pete and I had had more than enough of this maniac's driving and my comment, 'Slow down for God's sake!' obviously met its mark, for he finally reduced his speed. But enough was enough, and we still insisted on getting out! He may have wanted to die young, but we certainly didn't.

This incident reminded us that there were other risks and dangers inherent in hitchhiking. For example, sitting by the roadside, we too often narrowly escaped injury from empty beer cans and bottles casually thrown from passing cars. We also learned over time of several hitchhiker fatalities, including one fellow who was killed when the car he was travelling in was hit by a train. Another by a flying stone. And two others who died in head-on collisions.

Still, we willingly stuck out our thumbs to complete strangers.

An Indian businessman took us to M'pulungu at the southern tip of Lake Tanganyika, on the condition that we talk nonstop to keep him awake while he drove. Twice, however, despite Pete's incessant chatter and my various questions, we nearly veered off the road.

We'd heard about steamers heading up and down Lake Tanganyika. At 420 miles long and 30 miles wide, it is one of Africa's largest lakes and so vast that sailing on it was like being at sea. We purchased second-class tickets and made ourselves at home on the *SS Liemba*, formerly the *Goetzen*, a First World War German gunboat which had been converted into a passenger and cargo boat.

After weighing anchor, our boat headed for the port of Kigoma. Relaxing on deck, we looked across the great lake in the direction of the Congo, then experiencing revolution and war. For most of our three-day boat trip, it was calm — until late on the third night. We'd just climbed into our bunks when a violent thunderstorm struck and lightning lit up the sky. Gale force winds descended on us, creating huge waves and making me groan and moan for the next few hours, miserably sick. On a lake, of all places! 'Who else but you could be seasick on a lake?' Pete laughed. And despite feeling sorry for myself, I had to laugh too.

When we docked at Ujuji for a few hours, despite the debilitating heat, we set off at a brisk pace to visit the scene of Stanley's famous meeting with Dr Livingstone. At the tiny village by the lakeside, a large concrete memorial had been erected with the words, 'Under the mango tree which then stood here, Henry M. Stanley met David Livingstone, 10th November, 1872.' It was here too that Stanley recorded his feelings when they met, writing in his journal:

> *I pushed back the crowd and passing from the back, walked down between the lines of people until I came in front of the group of Africans where the white man with the grey beard stood. I wanted to run to him but I was a coward in the presence of such a crowd; I wanted to put my arms around him, only as he was an Englishman I did not know how he would receive me, so I did what cowardice and foolish pride suggested. I walked up to him and said, 'Doctor Livingstone, I presume.'*

There, we asked a fellow standing nearby to take a photo of the two of us. Then Pete and I shook hands for the picture and paused, momentarily lost in thought about the famous meeting between the explorer journalist and the missionary, before wandering down a narrow, overgrown path through the trees to the lakeside. It was a tremendous thrill walking in the footsteps of those great men in such a beautiful, peaceful setting. The more we travelled and saw of Africa, the greater our respect and admiration for such men grew (and indeed for all the early explorers and pioneers, not only in Africa but all over the world.)

On arrival at Kigoma, close to the borders of Burundi and Zaire (later the Democratic Republic of Congo), we were subjected to questioning for a period of two hours by Tanzanian immigration and police officials. They initially considered us mercenaries infiltrating from the Congo. Their suspicions weren't so far-fetched as we'd all considered it at one time or another.

We travelled to Dodoma by train in a crowded third-class compartment and spent the night in our sleeping bags on the lawn of the big railway hotel. We rose early, walked inside to use the washroom, and, as at Victoria Falls Hotel, ordered a hearty English breakfast.

Back on the road, a Peugeot stopped. The driver, Derek, was a missionary as well as a pilot. He was heading to a nearby airstrip to fly his parked aircraft to Nairobi. Although the plane was already fairly full, after learning where we were headed, he offered to take us with him. Ten minutes later, after excitedly loading our packs behind the rear two seats and assembling around his Cessna 175 Skylark and bowing our heads to pray, we were airborne. Two and a half hours later, we landed at Wilson Airport on the outskirts of Nairobi. That totally unexpected experience proved to be our best ride ever!

It had been nearly 16 months since we were last in Kenya's capital. Eddie and Carol who now had a second child, a daughter, were delighted to see us and insisted we stay with them for a couple of days. They were preparing to move to Cape Town, where Eddie was transferring to the South African office of Singer. This suited Carol. She'd be back in her home country with her husband, toddler son and baby daughter.

It was now March, and we knew we'd have to hurry if we wanted to climb 'Kili' before the rains. We also knew the weather on such a high mountain could be unpredictable. To accomplish the task we'd set for ourselves, we took a long-distance bus to Namanga, followed by another to the town of Moshi in Tanzania, known for its coffee farms and as the gateway to Kilimanjaro National Park. Then we caught a minibus to the Marangu Hotel, the main base for would-be climbers at roughly 4,500 feet above sea level. There, after sitting down to a nice sausage and mash dinner, our last good meal for a while, we sorted our stuff, and decided what to take on this next adventure and what to leave behind.

In those days, there were few people attempting the climb, a round-trip hike of about 40 miles — the tourist boom was years away — and most of those were accompanied by guides and porters. Scorning such aid, we carried our own packs, and, for most of the climb, I wore my safari suedes and Pete, his sandals!

It was blazing hot on the first day. Using stout sticks left behind by hikers who had no further use for them, we left Marangu and skirted the verdant coffee and banana plantations and *shambas* of the local Chagga people. As we hiked through the tropical rain forest with its thick and lichen-covered trees, we caught occasional glimpses of both colobus and blue monkeys. But little else. We plodded steadily uphill for about eight miles until we both felt we could go no further. The heat and our 50-pound packs, loaded mostly with tinned food, were exhausting us, so we stopped for a break. When we felt better, we continued a little further. Rounding a bend in the track, we spied

Mandara Hut, at 9,000 feet, the first of the three overnight huts. 'Sweet relief,' Pete commented as we set our heavy packs down. It had been a gruelling day.

Once inside the rustic building with its basic kitchen and shared sleeping quarters, we rolled out our sleeping bags and fell into a deep sleep in two of the unoccupied bunks. Upon waking, we prepared a meal, wolfed it down, and then huddled back into our sleeping bags in an attempt to keep warm. When we awoke after another deep sleep, we chatted to other climbers, including a Royal Air Force team of eight men on a training exercise. Four of them would go on to reach the summit while the other four went no further due to lack of proper equipment, poor-fitting boots or acute mountain sickness (pulmonary oedema). This common high-altitude affliction can be fatal without immediate medical attention or descending to a lower elevation as soon as possible.

After washing in icy-cold water piped from a nearby stream, and waiting for the usual early morning mountain mist to clear, the second day seemed easier. After bypassing the route to Mawenzi, Kili's other, more rugged peak, our route was relatively flat and didn't require too much effort. Just a steady plod. Nevertheless, the extreme exercise, the weight of our packs, and the altitude soon got to us. The thinner air left us panting for breath. Both of us were just about dead with fatigue when we reached Horombo Hut and the next cluster of wooden buildings at about 12,335 feet. We endured another freezing night despite being in a building heated by a wood stove.

The next day several hours of heavy plodding across the alpine moorlands, barren and desolate except for giant lobelias, brought us to Kibo Hut at 15,520 feet. The highest we had ever climbed.

There, we spent a restless night. I was suffering from a raging headache, presumably from oxygen deprivation, and Pete had a nasty stomach ache. We also both experienced hallucinations and, even with a fire burning in the hut, we couldn't escape the cold. I had a fierce craving for orange juice and bacon and eggs, but we had to make do with heated tinned beans. More than once, as we got dressed, I glanced with envy at those who had a hot meal prepared for them by their porters. By now, of course, we'd both changed our totally inadequate footwear for more sensible boots.

The next morning brought heavy cloud and bone-chilling cold. It was dark when we set off at 5 a.m. on the final and most challenging climb to the top. We hadn't gone far when we both developed splitting headaches again. But we carried on. The higher we went, the more difficult our breathing and movements became. As we picked, panted and perspired our way above 16,000 feet, it seemed that our lungs would burst and our legs would give way. Our hearts pounded, and we gasped for air, but there was no respite as the hours passed interminably slowly. But we kept on, determined to reach the summit. After a while, our headaches cleared, to be replaced by a terrific thirst. By this time, our water bottles were empty. We continued struggling up the lava scree and, at the snow line, we satisfied our fierce craving for water by sucking pieces of snow and ice. We were now no longer cold, but very warm as the sun rose and beat down upon us. We took off our anoraks and sweaters. This helped but, despite being fighting fit, we were now feeling well and truly knackered.

After a brutally hard slog and what seemed an eternity, we gradually neared the summit. As the snow deepened, we inched forward cautiously. Suddenly, the crater rim loomed ahead. A last determined effort and a final scramble over half-hidden rocks brought us onto the ridge. As we stood on 'The Roof of Africa', fighting for breath, a tremendous feeling of relief and exhilaration enveloped us. 'Done it!' exclaimed Pete jubilantly between heavy breaths, leaving nothing else to be said.

After so much effort, I told myself, *there is nothing quite like standing on the top of a high mountain, particularly Africa's highest.* Doing it on our own without the aid of guides and porters had been a significant accomplishment, and we both felt pretty chuffed. It had been quite a test of endurance. Now we could cross something else off our long wish list.

As we stood there, we looked around to admire the view. Small, puffy, close-knit clouds sailed below us, stretching to the horizon. An icy wind penetrated our clothing, so we put our anoraks back on. There, at the stone cairn marking Gilman's Point and surrounded by the snows of Kilimanjaro, we tossed a few snowballs at one another, remaining only long enough to shoot some photographs and peer into the huge crater full of massive boulders and enormous icicles.

Then we turned around and began the descent.

Going down was much easier, and we soon slipped into a casual clip-clop pace, stopping only to chat to a young doctor and two nurses, all from the

UK. Well-equipped and fit, they were on their way up. Lucky for us, we were to meet again.

On our fifth and last day on Kili, we met two English women, teachers from Nairobi. Both were dressed in flimsy blouses and shorts, and one carried a biscuit tin, inside of which were three or four sandwiches.

'How far are you going?' I asked. 'As far as we can,' one replied.

Pete and I grinned at each other, wished them 'Good luck,' and continued on our way down.

No sooner had we reached the Mandara Hut, the camp where we had spent our first night than torrential rains began, pelting down as only the rains in Africa can. Our thoughts turned to the two teachers. Definitely unprepared for a lengthy stay on the mountain, they might need some assistance of some kind. So we decided to go back and look for them.

It was still pouring when we discovered them about to enter the rain forest in a cold, bedraggled state, slipping in the mud and falling down. Assisting one each, Pete and I escorted them back down to the shelter and barren comfort of the closest hut where we lent them our towels to dry off and our sleeping bags for warmth. By the time we got a fire burning in the stove, they were still trembling from the rain and cold, greatly relieved that we had come along. They would have given us a ride in their rented car, they told us, but were headed in the opposite direction. They thanked us profusely as we exchanged hugs and said our goodbyes.

As we were not far from Loitokitok, my second monkey business location, I insisted that Pete see my old campsite for himself. We got there after hitching a ride on top of a Fanta Orange truck, which we frequently had to help push out of the mud. On seeing the lush, spectacular district for himself, Pete's hands went up in exaltation, a habit of his when something impressed him a lot.

The tiny one-street township had grown since I was last there. After the death of Jomo Kenyatta, his vice president and successor, President Daniel arap Moi, would invite people to move to the area to farm and take advantage of the fine black soil found on the lower slopes of the mountain. Over a short period of time, Loitokitok went from a small, sleepy place into a large, sprawling town. At the colourful, local outdoor market, we purchased a couple of pineapples and were given three Aspirin for change.

An African police superintendent gave us a lift back to Nairobi, stopping several times en route to shoot yellow-neck and quail. A pleasant, friendly man, he insisted that we join him for dinner at his headquarters at Kajiado, a small township where, after his cook prepared it, he served us the day's shoot.

A day or two later, Eddie and Carol saw us off as we left Nairobi. We would keep in touch through occasional letters. About ten years after their wedding, I was leading a tour in South Africa and invited them to dinner at the posh Mount Nelson Hotel, where I was staying in Cape Town. We had a lovely evening reminiscing. That happy get-together turned out to be our last meeting. Eddie's life was cut short by illness at the much-too-young age of 57. We would miss him dearly. He'd been a fine friend as well as a wonderful travel companion.

Next on our itinerary was Murchison Falls in Uganda, relying on local buses and sleeping in inexpensive government rest houses to get there. Along the way, we relaxed and discussed our route over tea and cakes on the lawns of a plush hotel, with a spectacular view of the snow-capped Ruwenzoris, otherwise known as the Mountains of the Moon.

Arriving in Murchison Falls National Park, we were just in time to catch the tourist launch. With an East African pound note in each of our hands, the price of the trip, we jumped on board. Soon, we were excitedly taking photographs of huge crocodiles, elephant and hippo within the vicinity of the thundering falls.

We were obliged to spend overnight at Paraa Lodge, a luxurious tourist establishment, as it was the only place to stay within the park. There, we watched elephants indulge in their evening habit of eating the lodge's monumental leftovers.

In the morning, after a hearty breakfast, we were just about to leave the lodge precinct when we noticed an elephant bull chewing the leaves of a huge fig tree at the foot of a nearby hill. Hoping for some good close-up pictures, we cautiously approached him on foot. We actually got to within 30 yards when he caught our scent. Turning our way, he suddenly trumpeted, flapped his giant ears, lifted his front legs, and started lumbering towards us!

Simultaneously, Pete and I reacted the best way we knew how — scrambling quickly back up the hill we had just come down, slipping and stumbling as we went. Halfway up, Pete lost a sandal in his haste to escape as, with pounding hearts, we found a safer spot and stopped to look back. Fortunately, the elephant had stopped at the foot of the hill where he glared at us for a while before ambling back to his tree.

'Phew. That was a close one!' I gasped to Pete. Then we both burst into laughter. On reflection, we were thankful that we had not been in a flat, open area where he could have built up speed and we'd have been at Jumbo's mercy. As far as I was concerned, encounters with elephant were becoming too much of a regular occurrence!

Faced with returning to Juba in the Sudan to take another Nile trip, this time going north and downstream, we pondered how we'd get there. As we sat by the gates of the lodge, a Land Rover drove up. To our surprise, inside were the young doctor and nurses we had previously met on Kili and, although we had only exchanged a few words at the time, we now greeted each other like old friends. They were heading the same way as us and immediately suggested that we join them. Hugh, Anne and Debbie had met in Cape Town, working at Groote Schuur Hospital, soon to be made famous by Dr Christian Barnard's first-ever heart transplants. They were now travelling overland back to England.

We had made ourselves comfortable in their Land Rover by the time we reached the town of Nimile, a small, dusty town on the Sudanese border. There, we decided to drive through the night to ensure catching the next steamer the following day. We hadn't been driving long when our headlights picked out two armed soldiers, one on each side of the track, their rifles aimed straight at us.

This is it, flashed through my mind. Simultaneously, Anne uttered a scream then broke into a sob. As we slowed down, expecting bullets to smash the windscreen any second, one of the soldiers signalled us to stop.

'Better do as he says,' suggested a tight-lipped Pete to Hugh.

Anne, utterly panic-stricken, then began to cry hysterically, declaring, 'It was just as everyone predicted. Friends told me not to come this way' while the rest of us looked at each other uneasily and wondered what would happen next.

Fortunately, our fears were unfounded. One of the soldiers informed us that it was merely a 'security check'. He asked to see our passports, and then signalled us to drive on. We all then breathed a collective sigh of relief, but Anne was still so petrified that Hugh had to give her a sedative to calm her down. We hadn't gone much further when the same thing happened again. This time we were prepared for it. A Sudanese army officer told us that we should have joined a convoy for our own safety.

Upon arrival at Juba, we had no sooner been told that the steamer was leaving the next day when we learned that it was leaving within a few hours. Time had erased the miserable memories of our previous Nile trip. This time, instead of travelling deck class, Pete and I could afford second-class cabins. The smell was appalling when we went on board, picking our way over sheep, goats and semi-naked natives. But all in all, that trip was far more pleasant than the first one.

As we were travelling with the current this time, the journey only took five days instead of ten. As before, every time we hove to near a village, the entire population turned out, the women with breasts exposed and the men as naked as the day they were born.

Hugh, the doctor, was very reluctant to discuss his medical background, especially with strangers, knowing that if he did so, he would be besieged by 'patients'. As far as he was concerned, he was travelling to get away from it all. However, as a doctor, he couldn't resist revealing his profession when an injured Dinka tribesman was brought aboard on a litter. The man had been badly gored by a buffalo and was obviously in great pain; so Hugh examined him, diagnosed a fractured pelvis, and gave him some pain killers. Enough to last until he got to a hospital.

Thanks also to Hugh, our knowledge of birdlife increased. Both Pete and I were by now very keen birdwatchers, but we couldn't keep up with Hugh. He knew them all. As we made our way downriver, we happily whiled away many interesting hours spotting and identifying numerous species.

At Kosti, we disembarked and, with so few trains, drove to Khartoum along a railway track — the so-called road being little more than rocks and sand and impassable in places. Unlike our previous stay in Khartoum, this time, Pete and I could completely relax. We visited the palace where General Gordon had made his fateful last stand, and we stood on the bridge at the confluence of the Blue and White Niles. As we all sat on the lawn outside the British embassy licking ice creams, the military attaché, a colonel, stopped to chat. At his invitation, we wound up the day sipping iced tea at the exclusive Sudan Club.

The temperature in Khartoum was 108 when the time came to go our separate ways. Hugh and the girls headed north to Egypt, while Pete and I trudged east to the outskirts of the city. As we admired some camels attached to the Sudanese desert police patrol, a truck laden with crates of Pepsi-Cola picked us up, and we joined some Arabs on top. For roughly eight or nine hours, we experienced another bone-rattling journey across the flat, barren Nubian desert in the insufferable heat. We had been hot and sticky before, but never quite to that extent. Little did we know worse was to come as we came to a halt and night fell.

After sleeping under the truck for only about three hours, we were shaken awake by the driver in order to be on our way. Back on top of the truck, it was now bitterly cold, and we were exposed to the freezing wind. With not enough room to slip into our sleeping bags, we snuggled up close to our fellow passengers, who looked miserable and equally uncomfortable.

Finally, we lurched into the market town of Kassala near the Ethiopian border and the high Taka Mountains in the distance. With a population of roughly 300,000, it was full of people, camels and donkeys. Beneath the onslaught of the midday sun, we checked into a spotlessly clean Greek-owned hotel, where we immediately enjoyed much-needed showers, a good meal, and the luxury of air conditioning.

While enquiring about bus schedules and acquiring visas for Ethiopia, we continually bought cool drinks to ease our perpetual thirst. 'Lemoon' was our favourite. Made with fresh lemons and sugar in water, we considered it then to be the greatest thirst-quencher in the world.

The evening before leaving Kassala, we visited the local open-air cinema. The film was an old Indian comedy, and the audience responded wildly by clapping, whistling and throwing stones at the tattered screen.

A dilapidated local bus, half-filled with scruffy, unshaven Ethiopian army deserters handcuffed to a corporal and each other, took us to Tessenei on the Sudan-Ethiopian border. As we stepped off the bus, we were frisked by smartly uniformed, tough-looking border troops. As a sentry covered us with a Bren gun, we wondered what we'd let ourselves in for while our passports were stamped by an immigration officer for entry into 'the Christian island in a Muslim sea', as Ethiopia is sometimes known. And as both Pete and I were soon to agree, it was the most extraordinary country we had so far visited.

Like so many African countries, Ethiopia — or Abyssinia as it was once called — was fascinating, mysterious and forbidding. It has its own share of natural and man-made wonders, including the Tissisat Falls, the Simien Mountains, rock-hewn churches and ancient manuscripts. We were determined to see some of them for ourselves.

We had heard from fellow travellers that the best way to describe Ethiopia was 'fantastic'. Now, we would be able to judge for ourselves. We had learned that Emperor Haile Selassie, the diminutive emperor and self-proclaimed Lion of Judah, King of Kings, ruled the country with an iron hand. Eventually he would be violently deposed. It was he, when the Italians invaded Ethiopia during the Second World War, who issued the following declaration, 'All men from the ages of 14 will answer the call to arms. Women, children and men too old to fight will seek refuge in the mountains. Those failing to comply with this order will be hanged.'

En route to Asmara, soon to be the capital of the breakaway country of Eritrea, the roads in the spectacular Simien Mountains were so steep that our bus could only move in the lowest gear. And the bends were so sharp that we had to manoeuvre backwards and forwards to proceed — a particularly disturbing experience, for there were no safety barriers, just sheer drops of hundreds of feet. On this route and others we travelled in Ethiopia, most of the women passengers hung out of the bus windows dizzy and vomiting while the local men closed their eyes and clutched tightly to anything they could lay their hands on. No wonder nearly everyone prayed! As for us, although greatly impressed by the jaw-dropping scenery, we were at times frightened to death and much relieved when each journey ended.

It cost exactly two Ethiopian dollars (about 25 cents) to stay in a clean, comfortable, little pension in the heart of Asmara, a beautiful, Italian-influenced town. After showering and changing, we strolled around the centre of town to find ourselves confirming at least one of the reports we had heard: the women were very beautiful. Coffee-coloured with slender figures and long wavy black hair, their Caucasian features were enhanced by rich, sensual mouths and dark, laughing eyes. As Coptic Christians, many had crosses tattooed on their foreheads, which in our eyes seemed to make them more attractive. We were fascinated and enchanted by them. When they saw us looking, they returned our gaze steadily and smiled.

That evening, impatient to enjoy ourselves, and in search of fun, we hired a 'garry' (pony cart). Using sign language, we told the driver to take us where we could find some girls. Minutes later, we pulled up outside a rather shabby house in the seediest and darkest part of town. Our driver rapped on the door with his whip. An old, ugly, hunched woman cautiously opened up and beckoned us inside. Pete and I looked at each other, shrugged, and walked in.

Inside was a dimly lit hall, with doors on either side. The old woman invited us to open one, and we peered in. For a second, we could scarcely believe our eyes. The room was richly furnished, lined with rugs, and in the centre stood a large double bed. On it, reclining sensuously and casually reading a book, was a gorgeous young woman dressed completely in white. Immediately she saw us she sat up, smiled, beckoned to us and patted the bed. 'Just looking,' whispered Pete as we stepped back into the hall. We opened the opposite door to discover an almost identical scene and another raven-haired beauty, also dressed in white. She too gestured for us to sit with her on the bed. It then became obvious. We were in a brothel, a whore house, a den of iniquity! Two minutes later, we were back outside and, between spasms of laughter, sank back into the garry and told the driver, who no doubt considered us the fussiest men alive, to take us back to our hotel.

We took the 'Litterina' (mountain train) to Massawa, Ethiopia's largest port, where we swam in the Red Sea, which was like taking a warm bath. The town itself was dirty and depressing, and we were pleased to leave the stench and heat and return to the coolness of Asmara.

Travelling by bus and occasionally hitchhiking, we took to the road, marvelling at the high, rugged mountain peaks and lush, green valleys. Often, we noticed monasteries accessible only by rope ladder, perched precariously on the side of sheer mountains or jagged cliffs. We learned these were inhabited by monks who seldom entered the outside world.

The more we travelled through this incredible country, the more our respect increased for the Italian engineers who had constructed most of the mountain roads. However, despite almost four weeks of criss-crossing the country, we never got used to the hair-raising bus rides. As for the many cheap hotels in which we stayed, we seldom escaped being bitten by bedbugs and fleas. Nevertheless, we happily survived on the country's national dish, wat and injera, a thick chicken stew served atop sourdough bread, accompanied by a cup or two of the famed Ethiopian coffee.

Every time that we found ourselves in a town or village, we were quickly surrounded, especially by children, who cried out in Amharic, 'Ten-as-ta-lign' (Hello), followed by, in English, 'From where do you come?' or 'Where do you go?' Then, as we departed, they would sing out, 'Go in peace.' It was delightful to be greeted that way.

The Coptic priests of Ethiopia enjoy enormous prestige and respect as we discovered when visiting a monastery. Approaching the massive, ancient building followed by a crowd of about 20 excited, shouting boys, a priest suddenly appeared, and the boys immediately fell silent and bowed their heads. One by one, they kissed the Holy Cross hanging from his neck. Like so many people we met, the priest asked, 'From where do you come?'

At Axum, another ancient and fascinating former capital, a 12-year-old boy acted as our guide as we wandered around the giant obelisks, the Tomb of the Kings, and the pool where the Queen of Sheba supposedly bathed. We also met a couple of American Peace Corps workers keen to have company. They invited us to stay with them when we reached Addis Ababa.

Something else high on our list that we both wanted to witness was the hyena man of Harar. After catching a bus to the old walled city of Harar with its maze-like alleys, we checked into another cheap hotel and then, as it grew dark, made our way to the outskirts of town. There, we found an elderly Ethiopian man seated on the ground beside a paraffin lantern surrounded by spotted hyenas eating great chunks of meat out of his hands. This nightly ritual had begun to prevent hyenas sneaking into the town to scavenge. He made his living getting tips from travellers like us.

Later, we relaxed over an Ethiopian beer and talked about what we'd seen. We fell asleep, as we had in many African towns and villages, to the sound of barking dogs and the now-familiar whoop, whoop of hyenas.

After visiting Gondar (another former capital), we sat by the roadside for two days waiting for a lift. What little traffic that passed us was already full. But our perseverance eventually paid off and a middle-aged Ethiopian businessman drove us in his old Citroën sedan over a very rough and tortuous road to our next destination. We reached the spectacular Tissisat Falls, also known as the Blue Nile Falls, located not far from the pretty town of Bahir Dar and Lake Tana where the Blue Nile originates. There, we gazed in awe at the hardly known but once-seen-never-to-be-forgotten falls, which during the rainy season thunder over the rocks and into the gorge below. At 1,300 feet across and about 140 feet high, they were not as impressive as Victoria Falls but still well worth the visit.

As it got dark and after exploring much of the surrounding area, we threw down our sleeping bags quite close to the falls, hoping to get a good night's rest. Alas, what little sleep we got was disturbed when the wind changed direction, and the intense spray quickly drenched us and soaked our sleeping bags. We spent a couple of hours viewing the falls from different angles as the sun rose and while our sleeping bags dried out. Throughout our brief stay, there was not another soul to be seen. We had the lovely place entirely to ourselves.

Not long after viewing the falls we arrived in Addis Ababa, the teeming capital of Ethiopia. At more than 8,000 feet above sea level, 'Addis' is the highest city in Africa, a bustling, colourful city full of dirty, old buses, and taxis that carried no meters and free-wheeled down the many hills.

It was a public holiday when we arrived and just about everywhere was closed, but we managed to find an Italian restaurant open where we ordered spaghetti bolognese to celebrate our arrival. We decided to call on our Peace Corps friends, whom we had recently met.

One of them, Mike, about our age and with a great sense of humour, was a teacher at one of the large secondary schools. During our brief stay, he invited us to speak about our travels to his class of 18-year-old girls and I took on the task. Most of the girls in the class were quite beautiful, which I found slightly disconcerting at first but, as I relived some of our adventures, I soon forgot their physical attractions! The 'lesson' concluded with questions and such comments as, 'We would like you to stay in Ethiopia' and 'You are very pretty.' When the bell sounded, I left blushing but feeling very pleased with myself.

That night, we went with Mike to visit what the local expats knew as 'the desert', an area not far from the city centre consisting of masses of shack-like bars. Here, we ventured into Teeny Weeny's, where we met the proprietress of the same name. Disarmingly attractive, she wore a tight-fitting dress and boasted a figure comparable to the most fabulous of Hollywood stars. Think Jayne Mansfield. When we 'farang' (foreigners) entered the bar, two of the girls left their stools, extended a warm, somewhat limp hand and led us to a table. Seated next to us, they requested drinks for themselves in a naïve, innocent way, then teased and flirted with us unashamedly.

While in the main post office, Pete and I fell into conversation with a young VSO worker who suggested that we could work at the Garfasa Orphanage if we had the time. Having worked as volunteers in Algeria, this idea appealed to us. When we got there, we liked the location and the idea of helping out so much we stayed two weeks. The orphanage itself was situated in rolling hills 20 miles east of Addis at Wyechacha and, at that time, was still in its early stages of construction.

Two VSO workers, named John and Alan, together with half a dozen Ethiopian workers, were employed at the site while Her Royal Highness, Princess Sybil — a slim attractive woman and a granddaughter of the Emperor, Haile Selassie — was president of the scheme.

Once again, as in Algeria, Pete and I rolled up our sleeves, set to with enthusiasm and were soon laying piping, mixing cement, fixing fences and a multitude of other things which we were happy to do for nothing. Our food and accommodation were provided, and we slept in one of the newly completed dormitories.

There was a mongrel dog living at the site which I accidentally annoyed one day. He retaliated by biting my foot. It remained painfully sore for a number of days and made me wary of all dogs for a considerable time. As rabies was prevalent in the area, the others joked about the prospect of me foaming at the mouth and going mad. This was not as far-fetched as one may think; in Africa, it could well have happened.

After shopping by myself in Addis one afternoon, I noticed a naked man walking nonchalantly along the street. He must have been out of his mind, I thought. As everyone else in the city was reasonably clothed, I expected this strange fellow to be picked up by the police. However, my assumption was not correct as I was soon to find out.

As it was intended that the orphanage would have horses for the youngsters to ride, Princess Sybil, or 'Her Highness' as we referred to her, suggested that I accompany her by car to some stables in the city and ride a horse back. As she drove her sleek Mercedes chatting to an assistant, presumably a lady-in-waiting, I sat quietly in the back. No sooner were we in busy traffic when, lo and behold, the same naked man I'd spotted earlier appeared in front of us. Her Highness tut-tutted and said something in Amharic like 'disgusting' as the fellow dodged the traffic.

At the stables, a small white pony — Abyssinian ponies are famous for their strength and endurance — stood saddled and ready for me, and I immediately set off to ride the 18 miles back to the orphanage. I was apprehensive about riding a horse through the busy city centre, and had no sooner ridden half a mile from the stables when who should I see but the same naked man! This time, he was carrying a rock and looking my way. As we drew closer, he started running and shouting, definitely intent on throwing the rock at me! My immediate reaction was to run him into the ground or simply gallop away but, unfortunately, my sturdy steed was very obstinate, preferring to go its own way.

Using all my strength, I managed to turn the pony's head and, digging my heels in for all I was worth, we trotted slowly back in the direction we had come. Lucky for me, the fellow, obviously mentally unstable, suddenly stopped, stood still and quietly watched. Back at the stables, annoyed and frustrated, I demanded a change of horse and set off again, this time on a more spirited and responsive animal. By now, the madman had disappeared and, after negotiating the city uneventfully, we arrived safely at the orphanage just before dark. Pete laughed his head off when I related the incident and was convinced I was exaggerating, especially as I'd seen the same naked individual three times, and he hadn't seen him once.

'Thank you so much,' said Princess Sybil, thanking us personally in her finishing school English as we wound up our brief stay at the orphanage and said goodbye to the other workers. Shortly afterwards, we boarded a train for the Ethiopia-French Somaliland border. As in Egypt, we were forced to share our carriage with chickens and goats and, yet again, we were bombarded with food from a motley array of travelling companions, a common occurrence in so-called undeveloped countries where people with so little still like to share.

A long, hot, tiring train ride across barren desert took us through the tiny country of French Somaliland to Djibouti, its capital on the Red Sea, where we decided to try to work our passage on a boat to Aden. The docks were crowded with ships of all nations but after two days of sweatily climbing up and down numerous ships' gangways in 120-degree heat, we gave up. Either the captains weren't available to grant permission or the ships weren't going our way. But mostly, we were dismissed and told 'no chance' for not being in possession of a seaman's ticket, the official document required to sail as a member of the crew.

Disappointed, Pete and I changed our plans and decided to fly to Aden, where we would look for ships sailing to India. As we sat somewhat apprehensively in the Aden Airways DC-3 droning over the Red Sea, I felt a twinge of regret on leaving the African continent. It had been good to me and given me some of the happiest and most exciting times of my life. In return, I had loved it with all my heart and soul, and would in the future promote and publicise it as much as anyone could. I'd come to realise it had absolutely everything — from an agreeable climate and incomparable landscapes to its extraordinary wildlife and the warmest, friendliest people you could hope to meet. I comforted myself with the thought that one day I'd be back.

Years later, much to my great pleasure and delight, I was to visit many of East Africa's national parks and game reserves during my long career in the travel industry as a professional tour leader and safari operator, and I would share the wonders of Africa with many people on my personally escorted tours.

CHAPTER 8:

The Bum's Way

'There are stranger things to be seen in this world than what lies between London and Staines.'
— *Sir Walter Raleigh*

In the then British Protectorate of Aden, we checked into the Hotel Marina in Steamer Point in quite the hottest temperatures and stickiest heat we had ever encountered. Before long, a Marxist regime would overturn the government, and the country would be known as Yemen. We set about making enquiries for ships bound for India or Ceylon (now Sri Lanka). After calling at the offices of just about every travel agency, looking for the best deal, we settled for the French liner SS *Vietnam*, leaving in ten days for Japan via India and, our destination, Ceylon.

As Aden consists of little but rock, sand and mosquitoes, our movements were considerably restricted. The 'emergency' — the fight for independence — limited our movements even more. At first, stepping out for a cool drink after dark was eerie and frightening but, after doing it a few times without being held up or gunned down, our fears diminished.

There was little to interest us except shops and cheap duty-free goods, so we were quickly bored. Bartering was difficult at first, but we soon acquired the technique of starting at some ridiculously low price and gradually compromising. Both Pete and I went on a spending spree and bought some new clothes. I purchased a small Panasonic transistor radio in a nice leather case, which would come in handy for listening to the news and music to cheer me up.

After boarding the ship, we made ourselves at home in *classe économique*, with four bunks to a cabin which we shared with a fellow traveller, an Englishman. Douglas was a professional smuggler who spent most of his time travelling back and forth to India from various free ports. 'I don't work. It's a mug's game,' he explained. So, on the very persuasive advice of this 'gentleman of leisure', both Pete and I each bought another camera and watch, hoping that we'd have no difficulty in disposing of them.

Most of the other passengers were young Europeans heading to Japan to travel around plus a few Japanese people going home. At the time, I couldn't have guessed that I would travel again on this same vessel and also on her two sister ships, the *Cambodge* and *Laos*, all three owned by Messageries Maritimes (MM), a French merchant shipping company. These ships plied back and forth from Marseilles to Yokohama via many of the world's most exotic and colourful ports. As the line was the most reasonably priced in the Orient, it was very popular with people of all nationalities, travelling on the cheap.

The crossing of the Arabian Sea to Ceylon via Bombay took eight days, and for most of that time, I was again desperately seasick. My *mal de mer*, as the French call it, was persistent and debilitating. Prostrate on my bunk, I existed on only tea and orange juice and couldn't wait to reach dry land. Well-intentioned advice led me to try all types of pills, medicines, beer, cheese, but nothing worked. My fellow passengers assured me that if they reached such a pathetic state, they'd throw in the towel and save up to fly. I felt that sea travel was an essential part of the travel experience, not to be missed. Besides, air travel was too expensive and saving enough money would take too long. I still hoped against hope that in time I'd find a cure.

Suffice it to say, most of my boat trips were thoroughly spoilt. Despite often feeling I was on death's doorstep, I did reach the stage of laughing at my ridiculous situation, even joking to whoever was in earshot, 'Here we go again!' I would say with a groan. When I did feel more like myself, I ventured up on deck for air, usually in a dead calm sea.

After docking at the port of Karachi, one of the most densely populated cities on earth, we all swarmed ashore for the day. Not long after, we dropped anchor in Bombay, later to revert to its original name of Mumbai. There, in the midst of all the hustle and bustle of this dynamic and dirty city, we sold our contraband for a substantial profit and regretted not buying more. Douglas left the ship just before it set sail, armed with two bulging suitcases full of cameras and radios. We heard about him again when he was apprehended by Indian police on another trip and was deported for his illegal activities.

A few days later we berthed in Colombo, capital of Ceylon, where we quickly discovered that almost everyone was openly involved in black marketeering of some kind! Everywhere we went businessmen or street hawkers approached us with, 'Good rate for dollar or English pound' or 'You got watch, radio, camera to sell?' Pete's new watch drew attention and admiring glances from everyone, as did my bright red towel. Such items were either unobtainable or expensive in Ceylon, we learned, and we were constantly asked to sell. Indeed, the black market was so prevalent and widespread that even had we been the richest tourist, it would have been difficult to resist the temptation of some of the offers. So, after being beckoned by various shopkeepers with the words, 'Come inside and we talk,' we changed our money with whoever offered the best price — usually at twice the bank rate!

It didn't take long to find out that Ceylon lives up to the name of 'Island Paradise' with its all-pervading scent of the tropics: incense, spices, frangipani and magnolia. Internationally renowned for its precious stones — rubies, sapphires, garnets and cat's eyes — it is further blessed with friendly people, lush, beautiful green countryside, and magnificent sandy beaches. We made the most of our stay, often tossing and turning throughout the night excited at what the next day would bring.

At Mount Lavinia, we swam in the crystal-clear waters of the Indian Ocean, and then relaxed on the veranda of the luxurious old colonial hotel, sipping iced tea. We hired flippers and goggles at Tinamaharama and discovered some of the delights of life under the sea.

Ceylon boasted some extremely good rest houses, constructed and maintained by the government. We frequently stayed in these accommodations and always found them reasonably priced and the service unobtrusive. It was like a dream, we thought as we thoroughly enjoyed freshly caught red snapper and ripe mangoes.

As Pete and I continued our peripatetic lives, roaming the world, back home in the UK, Dusty Springfield, Cliff Richardson, Tom Jones and Cilla Black were singing their hearts out, becoming more popular and richer by the day. How did we know? Because we heard their music here and there, and we kept up with the news and current events by listening to the radio, watching television, and reading English-language newspapers whenever we could. Also, like so many of the places we visited, much of our information came from fellow travellers we met in cafés, restaurants and hotels.

As we travelled, we resorted to local buses and third-class trains, and it always amused us as we read our fellow passengers' minds. 'They must be mad!' But whatever their thoughts, we found kindness and goodwill wherever we went. We never did get used to the locals' peculiar habit of chewing betel nut, and were often confused when they nodded their head, meaning 'no' and shook it in assent!

Once, on a bus, I jumped out when it stopped to purchase some biscuits. When I returned, the Ceylonese fellow sitting next to me casually asked how much I'd paid. When I told him, he exclaimed in his singsong accent, 'My God, you've been robbed.' Without further ado, he jumped up and, accompanied by two or three passengers descended on the shop proprietor who cowered under the sudden onslaught, returned my money and insisted that I keep the biscuits, free of charge!

Pete and I then split up for a while with plans to reunite in a few days. He visited the tea district of Newara Eliya, while I decided to visit the Veddah people, an aboriginal tribe. Ever since living in Africa, where I acquired a great liking and respect for so-called primitive peoples, one of my great interests was to see them in their natural environment.

On arrival at the tiny town of Maha Oya, named after the nearby river, I presented myself to the district officer for permission to visit the nearest Veddah tribe. Soon after, I set off to walk along the eight-mile jungle track to the nearest aboriginal village, accompanied by a young clerk named Gomez, who had been instructed to act as my guide. Twice en route, despite the heat, we stopped to bathe and refresh ourselves in hot springs.

The Veddah village was a quiet, peaceful place, where birdsong was the only sound. We met two or three families who were very shy at first and a little afraid of me. But within minutes, when they realised they had nothing to fear, they smiled and laughed, and I took photographs very discreetly while watching them go about their daily tasks.

Ancient shotguns had replaced their blowpipes, bows and poison arrows, and they wore a minimum of clothing — a sarong for the women and a loincloth for the men. One old, bearded man still hunted wild pig with his trusty blowpipe, and he showed us how to make poison from the gum and juice of a peculiar-looking tree.

It was some time after I'd trekked to the Veddah village that I learned Ceylon has a huge number of poisonous snakes, most of which live in the jungle and countryside. Not to mention the largest leopards in the world. Luckily, I didn't come face to face with any.

Pete and I teamed up again in Kandy, a former capital, where we went down to the river to watch the elephants bathing. Well trained and docile, they carried out their mahouts' instructions, given with a shout or gentle prod of a sturdy stick. After paying one of the mahouts a couple of rupees each, we took it in turns to photograph each other taking a short elephant ride in the river.

As Pete headed east to see the ruins of the famous city of Anuradhapura, I went north to Batticaloa to swim and relax. Afterwards, I decided to go to Trincomalee (a former British naval base), which boasted an interesting fort and a glorious, sun-kissed stretch of beach. After reading the railway timetable for trains to Trincomalee and plodding and perspiring through the streets of the coastal town of Kalkudah, it dawned on me that I might miss the infrequent train, so I quickened my pace. As I did so, a young student on a bicycle drew alongside and asked me where I was going. When I told him, he invited me to jump on the back. Pedalling as if his life depended on it, we reached the station with minutes to spare.

United again in the town of Jaffna, as Pete and I changed money in the local Standard Bank, we fell into conversation with the young British bank manager named Chris who invited us to stay at his home prior to our flight to India. At his comfortable house with wraparound veranda, and electric fans, we enjoyed a delicious meal of curry and string hoppers.

Soon after, I developed a particularly nasty toothache and was obliged to call on a local dentist. His equipment was quite obsolete — an ancient wooden chair and pedal-driven drill — but I was prepared to face anything for relief. While the dentist poked around inside my mouth, mosquitoes hovered around in the heat and humidity. Half an hour later, the treatment completed, I walked away, frantically scratching my ankles and wrists.

As the plane fare cost virtually the same as travelling by sea, we flew to India with Air Ceylon in a DC-3. Ever since my safari days in East Africa, I loved the thrill of flying in light aircraft, but both Pete and I were nervous about flying in larger aircraft, preferring our feet on solid ground. We accepted it as an occupational hazard, but while in the air, we kept glancing at each other for moral support. It would be a long time before we both relaxed during flights. And there would be times when, after a particularly bumpy flight or scary landing, I was so relieved I could have kissed the ground!

Landing at Trichy, we took the night train to Madras, where we shared a dormitory at the YMCA with several other travellers. They passed on some tips gleaned from their own experiences. One Frenchman summed up India with the expression, 'In zis country, all is possible,' as we were soon to find out for ourselves. Having just travelled in a third-class compartment accommodating 16 people but built for eight, we already understood what he meant.

Pete and I spent several weeks travelling through much of India. Most of the time, it was not a very pleasant experience. But, I had to admit, in its own bizarre way, it was well worth doing, if only to open one's eyes.

Many other travellers confirmed our observations and conclusions about travelling around the subcontinent. It was a different matter, of course, for wealthy tourists, isolated and cocooned in their first-class railway compartments and luxury hotels.

Apart from the obvious dreadful poverty, disease and countless other problems, there seemed to be a total lack of organisation. As one government official put it, 'People want to be paid just to come to their offices. If they work, they expect to be paid overtime!'

Whenever we arrived in a town and were looking for a place to stay, we asked the nearest person, 'Hotel?' Many times we were given the wrong directions, we assumed because people were too proud to admit they didn't know. Once we questioned two dhoti-clad men sitting on a bench. Simultaneously, they pointed in opposite directions. At the same time, we noticed a hotel right behind! Another time, we asked no less than five people the departure time of the next train, and although all were waiting for it, they all quoted different times. This is funny to hear or read about now, but was incredibly frustrating when it occurred.

Every time we stepped out of the relative quiet and privacy of our guest houses and hotels, we would be swept up in a frenzied mass of humanity. A chaotic scene of noise and confusion with men, women and children heading in every direction along with sacred cows, decorated elephants and so many vehicles — taxis, scooters, buses, trishaws, cars, buses and trucks simultaneously sounding their horns!

Train travel was another experience. Every time we boarded a train and fought our way to a compartment, we seldom found a seat despite having made a reservation. But nobody cared — except for themselves. Once, we counted no less than 22 people in our compartment meant for eight. To say we were crushed and lacked fresh air was a huge understatement. This miracle of contortionism was accomplished by occupying the luggage racks, windowsills and leg space — at our expense! We wouldn't have minded so much except for the fact that we were invariably the only genuine ticket holders, the other occupants not bothering or being able to afford them. It was a case of 'every man for himself'; the whole fiasco made worse by wandering, pushy beggars on crutches with hands outstretched, pleading for annas, the smallest of Indian coins.

Whenever the train stopped, somebody hurled their baggage from the platform in through the window, only to find when he'd pushed and kicked his way into the compartment, that someone else had thrown his baggage out of the other side! And several times, upon reaching our destination, the only way out was to literally climb through the windows.

Apart from the rampant black market — we couldn't walk more than ten yards down any main street without being accosted to change money — we learned more about the widespread corruption. At the post office, we were advised that, after sticking stamps on letters, we should actually watch them being franked. Apparently, if we failed to do so, the stamps would be quickly and deftly removed — a common practice in the Indian post offices.

By this time, we were mostly using Poste Restante at major post offices for collecting our mail. It was easier than using embassies. We followed the same procedure when leaving a place — completing the appropriate forms with our forwarding address. Alas, this didn't work in India, for it seems nothing was ever re-addressed.

Travelling on the subcontinent, there was seldom any escape from the crowds. No matter where we went, we were constantly surrounded by a heaving, pulling, pushing mob of people. There was little or no privacy and people relieved themselves anywhere they could find, which amused us somewhat. When travelling by train in such cramped conditions, sleep was almost impossible, so we often saw dawn break. As the sun rose, we noticed the fields dotted with squatting figures of men and women doing their business while holding black umbrellas, which they used as screens as the train passed.

In New Delhi, we saw our first snake charmer seated on the ground playing his peculiar-shaped flute while three cobras hovered in the air. Cries of 'Baksheesh, baksheesh' accompanied his act. Then we saw something that could only happen in India — two men working with the same shovel! After the one holding it had scooped up some sand, the other pulled it with a short rope to another pile not three feet away, where the load was dropped.

After Delhi, we visited Benares, since renamed Varanasi, India's most sacred city, located on the banks of the holy Ganges and quite the smelliest, dirtiest city we'd ever visited. There, we stood transfixed as we watched dead bodies being cremated before our eyes. Hovering above were black kites while Indian vultures waited patiently like statues for an opportunity to grab a piece of human remains. The stench of burning flesh was overpowering, and we had to hold handkerchiefs to our faces as we took in the grim ritual. The friends and relatives of the dearly departed were so engrossed in cremating the bodies, they didn't seem to notice it. And were oblivious to us.

In such circumstances, it wasn't surprising that both Pete and I were sick several times, usually from the water and occasionally from the food. In an attempt to safeguard ourselves, we made a point of eating in the restaurants of the reasonably priced and relatively comfortable railway hotels but, even in those establishments, the butter was often rancid and the curry and rice so hot that it required gallons of suspect water to swill down.

The water throughout India was nearly always polluted, hardly surprising when people clamber over fences to wash and swim in the reservoirs. Many travellers carried their own water bottles and purification tablets, as we did. But we couldn't always wait long enough for them to work — and paid the price. I'd coped with the occasional Tanganyika trots when travelling in Africa, but now, much to my great discomfort and horror, I learned the true meaning of Delhi belly!

One day while travelling by train across the great Indian plains I ran out of water, and my throat became so parched that to relieve my thirst I joined the queue at the water tap at the next station platform. Another big mistake! An hour later, my stomach wracked in pain, I disappeared into the train toilet and more or less remained there for the next four hours. No sooner did I reach my destination than I ran as fast as my bouncing pack and queasy stomach would allow, to the nearest building, which sported the magic sign: Hotel. Turns out I had a full-blown case of dysentery and was in great discomfort for a few days. It left me deathly ill, dehydrated, and terrified of drinking any more water.

Eventually, Pete and I learned the best way to avoid getting ill from the water was to drink soft drinks and beer. We were assured these were processed with the finest, cleanest water, which we assumed meant from the lower Himalayas. We also made sure before taking a swig that the bottle tops had not been tampered with, and re-used. Often they were.

One of the most depressing aspects of our stay in India was seeing the huge number of beggars. We frequently dug into our pockets for these miserably poor people. Most of them wore little more than rags blackened from dirt and sweat, and many, we learned, were purposely blinded or maimed in childhood in an effort to supplement the family income. There were so many, and after seeing signs on every railway platform stating simply, 'Do not encourage beggars', we became hardened to the situation. Much as it distressed us, there was only so much one could do.

Despite the contradictory nature of many things, we retained our sense of humour, essential in that part of the world. Once, trudging through the town of Poona, hot, tired and weary, we had to laugh at the notice outside a barbershop. Written in English, it said, 'Hair cutter and clean shaver. Gentlemen's throats cut with very sharp razors, with great care and skill. No irritating feelings afterwards. A trial solicited.'

In Agra, after some haggling, we hired bicycle rickshaws to visit the Taj Mahal. Seeing it for the first time, we were spellbound like so many travellers before us. As we admired it, I remembered reading somewhere that a trip to India just to see the Taj would be worthwhile. I fully agreed. It is widely considered the most beautiful building in the world. Built as a mausoleum by Emperor Shah Jahan to commemorate his beloved wife, Mumtaz Mahal, it is made entirely of ivory-white marble and semi-precious stones.

It was at the Taj that I took one of my favourite photos of which my old boss, Geoff Cardew of the Nestlé Company would have been proud. As always, I had only the film in my camera plus a spare, so I rationed every shot very carefully, making each one count. After standing against the wall of the massive arched entrance for ten minutes, three women dressed in colourful saris suddenly walked in front of me and paused for just a few seconds before descending the few steps to take the long walk to the monument. As they lingered, *click*, I snapped the shot.

As India then was well known as one of the cheapest and most corrupt countries in the world, it wasn't surprising to find it a haven for undesirables, some of whom had entered the country illegally by inserting money bills into their passports when presenting themselves to immigration officials. We encountered numerous 'overlanders', who had hitch-hiked across the already well-worn route via Greece, Turkey and the Middle East, many intent on reaching Australia.

The overland trail from Europe to India was rife with stories of both male and female travellers who went missing in the mountains of Afghanistan and Pakistan, lured by drug dealers who literally and figuratively took them for a ride. Young and restless like Pete and me and loved by a family waiting for their return home, they disappeared in a strange foreign country never to be seen again.

In addition, we met numerous 'aimless wanderers', usually young people from Europe, now living in the era of hippies and free love, who suddenly decided to travel purely on a whim — for 'kicks', and to 'find' themselves. Setting off with incredibly little money, even less than we had, they would take the increasingly popular overland trail to India, and see how far they could get before running out of cash. Then, rather than use any initiative, they would present themselves at their respective embassies and ask for a plane ticket home. No wonder embassy staff didn't welcome those people with open arms!

There were others, too, who came to realise that travel is not always as romantic as it's made out to be, and that getting around and seeing the sights can be hard work.

Many of the backpackers we met boasted how they had travelled so far and spent so little money. Practically all spent the nights in Sikh temples, Sikhs being very hospitable and never turning anyone away. Many wore their scruffy appearance as a badge of honour. One day, we met a young American fellow who passed his days begging on the streets. Clothed in rags and wearing a long red beard, he apparently did quite well. Another fellow of indeterminate origin churned out International Student Cards for the equivalent of 50 cents each to make money. As students received a multitude of travel reductions and other privileges, they were in great demand.

After spending a day in the ancient city of Patna, we headed to Kashmir. It was a tiring mountain journey by decrepit bus, including an overnight at a cheap roadside hotel. Travel on the tortuous Jammu-Srinagar National Highway, then undergoing widening and repairs, was at times frightening and consisted of one-way traffic, regular mandatory halts and the constant danger of falling rocks. Eventually, we arrived in Srinagar, the picturesque capital.

Kashmir, the land of celestial charm, claimed by both India and Pakistan, is often referred to as 'the heaven of the world'. Consisting of canals and lakes, it is surrounded by alpine meadows, forests and the magnificent mountain scenery of the lower Himalayas. As the cost of living was cheap, we lingered in the area for about ten days, most of them happily spent hiring *shikaras* (an Asian kind of gondola) to explore the numerous waterways, and swimming and learning to water-ski on beautiful Lake Dal.

On the bus ride to get there, we had met two American couples who suggested sharing a houseboat to minimise expenses. Upon stepping out of the bus, we were surrounded by no less than 20 grown men, all shouting and beckoning us to follow them. They were houseboat-owners who hoped, as the season was not then in full swing, to receive some early business. Not content with shouting, they started to pull and push us until we became annoyed and raised our voices.

We decided to go with a little man who had remained quietly in the background and walked with him to his boat. After inspecting it and bartering over the price, we agreed to hire it for several days. Comfortably furnished with three twin bedrooms, a lounge and dining room, and Kashmiri made rugs and carpets throughout, it was quite luxurious. Rashir, our middle-aged bearded boatman, did all the cooking and, when he wasn't available, his son Ali took over the chores.

While on the houseboat, we lived like royalty with our new-found friends, enjoying three excellent meals a day and the customary English-style tea breaks. Frequently, we decided to take a picnic lunch, in which case Ali would accompany us, smiling shyly all the time. Beneath the shade of a tree, he would lay out the food. There, we'd enjoy roti, pappadums, dahl (lentil soup), and fish curry with rice. No effort was spared, and everything was complete: tablecloth, cutlery, crockery, food and drink.

During this brief but ecstatic interlude in what was to become another of my favourite parts of the world, Pete and I took a ride in a rickety bus to Gulmarg, a nearby mountain resort in the Vale of Kashmir, reminiscent of a Swiss alpine village. Amid lovely green meadows, we hired a 'pony-wallah' and two of his ponies for just a handful of rupees. As we rode on horseback up the mountainside, he trailed behind on foot!

After reaching the highest valley, we hired a 'sledge-wallah' by the name of Subzali to act as our guide, and set off on foot to reach a frozen lake at an altitude of nearly 15,000 feet. It was a glorious, warm sunny day, visibility was perfect, and we could see for miles. After a fair amount of hiking and exertion, we stopped to admire the view. It was then that Nanga Parbat — the third highest mountain in the world — in all its snowbound glory came into sight, so clear and so close we felt we could almost touch it from where we stood. The brisk mountain air and magnificent scenery were intoxicating, and we could have stood there forever lapping it all up.

We were now at snow-level and, after a brief snowball fight, all three of us climbed onto the sledge. Subzali got in front; Pete was behind him with his legs over Subzali's shoulders; and I had my legs over Pete's shoulders. No sooner had we performed this miraculous feat, than we were off hurtling down the mountain, gathering speed and experiencing one of the most frightening but exhilarating rides of our lives!

As it was midsummer, there was little snow at that altitude at that time of the year, but rocks and boulders in great abundance. Partly exposed, they loomed before us as we careered along at breakneck speed with our hearts in our mouths. Thanks purely to Subzali's masterly steering with the use of a thick stick, we missed them by inches. We'll never know how fast we travelled, for we were much too busy hanging on to one another for dear life!

We spent that night in one of the old-fashioned, chalet-type hotels. Despite being huddled in our sleeping bags covered by two or three blankets and with a small, inadequate electric heater, our teeth chattered throughout the night.

Back in Srinagar, our houseboat was besieged daily by merchants intent on selling their wares. No sooner had we shooed them away than their shikaras were again alongside, and almost before we could say 'Jack Robinson', their goods were displayed on the carpeted floor. I bought a swordstick with a nicely carved head. Ostensibly a walking stick, I thought it might come in handy both for walking and as a deterrent. Like my rucksack and camera, it would accompany me everywhere throughout the rest of my worldwide journey.

As we relaxed and enjoyed ourselves, we noticed army trucks and tanks rolling into the area — the build-up prior to the Kashmir dispute. Two days after we left, the first of what would be two or three wars between India and Pakistan erupted when the predominantly Muslim Pakistan invaded Hindu India. After the ceasefire, the Tashkent Agreement was signed in 1966 by the diminutive Lal Shastri, Prime Minister of India, and President Ayub Khan of Pakistan.

On our return to New Delhi, we chatted to a group of travellers from Europe. After only a few months, they were already disillusioned. All had suffered from stomach problems and bemoaned their travelling days. Although hardly ecstatic about India, Pete and I remained as cheerful and excited as we had ever been. The difference, of course, was that we were now seasoned travellers and had faced all kinds of difficulties travelling mostly 'off the beaten track' while the others had covered the well-established route, which by all accounts in the words of one fellow had been 'thoroughly spoilt'.

Another traveller we met, Allen from Manchester, had recently arrived in India and was impressed by everything he saw. 'It's so fascinating,' he crooned.

'Don't worry,' I said. 'You'll soon change your mind.'

Allen insisted he wouldn't.

We left him with my rather cynical parting remark of 'Give it a little time.'

One day, we fell into conversation with a young Indian student named Raju. Dressed in dirty shorts and shirt and sandals, his one ambition was to be rich. 'With money, you have anything. Nice life,' he remarked rather sadly. We didn't want to disillusion him, thinking of some of the wonderful times we'd had which hadn't cost us a penny, but later when we discussed it, we concluded that, living in India, he definitely had a point.

En route to Kathmandu, Nepal's capital, we held our breath as the DC-3 flew so low we thought we might crash into the surrounding hills. We checked into the newly completed youth hostel and then explored the city on hired bicycles. Having heard so much about the well-known temples, we made a beeline to the most famous ones where we stared at what most Westerners would consider obscenely carved figures and phallic symbols.

At the Globe restaurant in Kathmandu, a popular rendezvous for travellers, we enjoyed our first water buffalo meat, which tasted just like beef. Our intentions to make a trekking trip deep into the Himalayas were quickly forgotten when we learned that the monsoons were imminent.

At that time, the local expatriate community were concerned that the remote and peaceful life of Kathmandu would soon be spoilt with the gradual influx of travellers. Their prediction proved correct, for before long it became a drug haven for hundreds of young people from all over the world. Although we personally made a point of trying just about everything, we had no desire for hashish or any other drugs. Life, we often reminded each other, was quite exciting enough!

In Africa, we had learned from other travellers about teaching English in Japan. Much as the idea rather alarmed Pete, he decided to give it a try. He returned to New Delhi and took a ship to Yokohama — experiencing a typhoon en route!

While Pete flew from Kathmandu, I rode back by bus to the Indian border. It was easy to get a ride as all the trucks and Sikh drivers assembled in the main square each morning. Most of the six-and-a-half-hour journey was in low gear through thick mist. We climbed, descended, twisted and turned, occasionally catching glimpses of the mighty snowbound Himalayas.

At the town of Raxaul, I walked across the border and reported to the immigration and customs office, where I found an overweight middle-aged customs official reclining on a sofa. Too lazy to stand, he merely looked up, uttered, 'Anything to declare?' then, with a casual wave of his hand, motioned me to leave.

It was at Raxaul station that I bumped into Allen again. Now that he'd spent longer in India, his attitude had so completely changed that he could hardly wait to leave. His stay was made even more depressing when his train was late due to an accident on the line. A woman had been run over and decapitated and Allen, on stepping from the train at the platform, had accidentally trod on the severed head — which had lodged beneath the steps. We met outside the station toilet where he'd just been violently sick!

Later that same evening, arriving at another station in the dead of night, and deciding against a hotel, I threw my sleeping bag down on the empty platform. I lay down on top of it as it was too hot to get inside. 'Ah, peace at last,' I told myself. That proved to be one of my worst nights ever, as I was constantly pestered by bugs, beetles and mosquitoes. I didn't sleep a wink. Travelling rough and cheap certainly had its drawbacks!

On arrival in Calcutta, a general strike was taking place, so the city was exceptionally quiet. The previous day, students supported by thousands of workers had staged a mass demonstration, burning trams and creating havoc when tram fares were increased by the equivalent of a farthing! Nothing to a Westerner, but a lot to poverty-stricken Indians, I realised.

When I strolled down Park Road in the early evening, I was accosted by several pimps and rickshaw boys, all of whom offered me 'Nice English lady,' or 'Chinese or Indian girl'. This, in fact, reminded me of one of the few consolations of travelling through India; the beautiful women. Clad in the most exquisitely coloured saris and brocades, covered in gold bangles and precious jewels with their wrists, fingers and necks tattooed in henna, they walked and carried themselves with a certain dignity, a special grace, and great pride.

One day as I walked along a busy street, a man in dhoti and sandals walking towards me suddenly stopped, lifted one leg and then let out a resounding fart. He then carried on walking as though his lewd display was the most natural thing in the world. And it was, in India!

It was now my intention to travel through East Pakistan — which a few years later would change its name to Bangladesh. From there, I would make my way overland to Rangoon, the capital of Burma, to catch a flight to Bangkok. A visa for Burma was only good for 24-hours transit, the stipulated length of time for all visitors then entering what was known as 'the people's road to socialism'.

I left India on the Ganges River steamer, which was chock-a-block. I was just contemplating where I could best lay my head for the overnight trip when the chief steward interrupted my thoughts. Noticing my dilemma, he said, 'Sleep there' nodding to a small, roped-off vacant deck where I could escape and be by myself. I could have kissed him with relief, for I was dead tired at the time.

I was still fast asleep when we entered Bangladesh, the land of catastrophic floods, cyclones and numerous other natural calamities. A train took me to Dacca, the capital, since renamed Dhaka, a city of mosques and rickshaws where everyone moves at a frenzied pace, and I checked into a small, centrally located hotel. Stopping to consult my street map for the airline office, some 20 adult Pakistanis immediately surrounded me, all standing and staring at me as if I had two heads. Children I could understand being curious about a foreigner, but grown men! I'd been stared at often before, but not quite to that extent, and I resented it after a few minutes had elapsed. When two of them stood so close that their heads obliterated my map, I moved away. Even then, I was followed until I nipped into a nearby office to escape them.

Neither Pete nor I ever got used to those exceptionally inquisitive people of the Indian subcontinent who, if they spoke English, repeated the same questions time and again. And whenever we chose to eat in public, on a railway platform or in a train, we were stared at and scrutinised as though we were birds in a cage.

Once or twice I asked some Pakistanis what line of work they were in, and as in India, nearly all professed to being 'businessmen' of some kind. Usually 'Import, export' — which I soon realised covered a multitude of sins. However, they were not the only ones to engage in bribery and corruption. On securing my visa from the Burmese consulate, I paid 30 rupees to the consular official, who issued me a receipt for ten!

The monsoons suddenly began — five inches of rain in eight hours — and as much as I'd hoped to visit Cox's Bazaar (supposedly the world's longest unbroken beach), it was impractical to go at that time. So I boarded the

Green Arrow Express for Chittagong, from where I would leave for Burma, years later to be renamed Myanmar.

On reflection, it was just as well that the rains had started, for it gave me an excuse to leave. I'd just about had enough of the poverty, squalor and frustration of the subcontinent. On my day of departure, I summed up my feelings about the last few weeks, writing in my diary, 'If you want to learn about life, go to India or Pakistan.' It was also the place, I had decided if I had my way, where I would send wayward youth and spoilt brat Western kids for a month to see and learn about the realities of life!

Despite the ups and downs of my recent challenging travels, I learned to treasure them, for they taught me not to take anything for granted and made me appreciative of many things.

In due course, India would change for the better and be almost unrecognisable, becoming an international powerhouse and highly respected nation on the world stage.

Years later, I would return there several times as a tourist and tour leader and get a totally different perspective — thanks in part to good food and first-class hotels.

CHAPTER 9:

Relief at Last!

'The soul of a journey is liberty, perfect liberty,
to think, feel, do just as one pleases.'
— William Hazlitt

Two days later found me in Rangoon, after flying in a Pakistan International Airways Fokker Friendship plane, one of only two passengers on board. As accommodation was included in the price of my air ticket, I was lodged at the elderly Strand Hotel which, like the rest of the city, was in dire need of renovation.

After checking in, I took an interesting walk around the town. My impressions were fleeting due to my short stay. Despite the fact that the president, General Winn, wanted 'no outside interference', I was struck by Romanian, Yugoslav, Czech, and Soviet trade exhibitions all taking place at the same time!

Twenty-four hours after landing in Rangoon en route to Thailand, I was strapped into the seat of a Burma Airways jet. I was surprised to find a middle-aged, Burmese woman taking the customary place of a young, glamorous hostess and doing the work exceptionally well. After three hours flying over hundreds of miles of dense green jungle, we touched down at Bangkok's Don Muang airport.

It was with a feeling of enormous relief that I arrived in Thailand. After the subcontinent, it was a refreshing and welcome change. Everyone seemed incredibly happy and friendly, confirming why it is so often called 'the land of smiles'. It was also well known for its very long and sometimes impossible to spell and pronounce names!

A genial taxi driver drove me the 18 miles to the city centre and throughout the 45-minute ride, tried his utmost to persuade me to 'take massage — special service for you.' But I was in no hurry. That could wait, at least for a while.

After discovering a small, cheap hotel near the main railway station, I called at the government tourist office, which was staffed by the most exquisitely beautiful Siamese girls. They furnished me with maps and brochures, smiling all the time. Yet again, I realised I was in a strange, but this time, enchanting new world.

Like most visitors, I was immediately impressed by Thailand and in particular the people, known to be gentle, and carefree. As exemplified by Queen Sirikit, the Thai women are world-renowned for their elegance, charm and disarming beauty. Indeed, as I sat by the democracy monument in Rajdamnern Avenue, one of Bangkok's main streets, watching them pass by, I winced with pleasure and admiration. In fact, I couldn't take my eyes off them.

Bangkok, with its magical, mystical, colourful temples and winding, narrow *klongs* (canals) proved fascinating. It being a sprawling city, I was usually footsore and weary whenever I returned to my hotel. It was August and extremely hot and humid. I must have walked miles in every direction, trying to see as much as possible, never dreaming that one day, not far into the future, I would be living and working there. Several times, I walked along Sukhumvit Road, the main street of this massive city, which seemed to go for miles in each direction. Every so often I would spy another traveller, laden down with a backpack, making his or her way along Khao San Road, a favourite area for globetrotters where they could find a suitable place to stay as well as enjoy cheap, tasty meals at one of the many street stalls.

There were very few guidebooks in those days, particularly for those of us who were travelling on a tight budget. But I clearly recall the many times I referred to someone else's copy of John Gunther's *Inside Asia*. Gunther was an American newspaperman, a foreign correspondent turned author, who had lived and worked all over the world and wrote what then was considered essential reading about Europe, Asia, and Latin America, including, as mentioned earlier, *Inside Africa*. Whenever I came across one of his books, I'd thumb through it with great interest. However, my primary source of information remained fellow travellers, people who had been where I was going.

My first few days in Thailand presented some problems such as the language barrier. Initially, not knowing what to eat, I simply ordered what others were eating and spent money very fast. Through necessity, I was compelled to use chopsticks, a technique which seemed awkward at first until I got the hang of it. The rice seemed tasteless, and I felt hungry for an hour or two after, but gradually I developed a liking for the food, and 'flied lice' known locally as *nasi goreng*, with shrimp and topped with a fried egg became my favourite, everyday dish. Other delicacies I occasionally sampled included *pla priew waan* (sweet and sour fish) and *tom tung kang* (a famous soup, spicy and strongly flavoured with lemon grass, lime juice and shrimp, chicken or pork). I also enjoyed a variety of noodle dishes as well as different *pad thai*.

The Thai desserts, all very sweet, were a treat. The most common were banana fritters and sticky rice with mango or coconut. In fact, the types and number of dishes available in Thailand were quite overwhelming at times whether eating at a hole in the wall, a street stall, or a fine dining establishment. Being on the go constantly throughout my travels, I always had a hearty appetite. The tantalising aroma of freshly cooked prawns and vegetables only added to it. However, much as I liked and was daring enough to eat all kinds of food, the plainest, budget-friendly of dishes such as fried rice, papaya salad, and stir-fried noodles suited me just fine.

Very early in the morning on my third day in the city, I took a ride in a samlor or 'tuk tuk' as they are locally known. These three-wheel covered scooters are Thailand's cheapest form of transportation and are found everywhere. They became my principal mode of transport around this large, congested city. That day found me at the fancy Oriental Hotel on the south bank of the Chao Phraya River, Thailand's main river. There, I joined a group of tourists and transferred from the landing stage into an open, long-tailed boat to begin our journey to the famous floating market. Travelling via a series of canals, it was all too obvious why Bangkok is sometimes referred to as 'the Venice of the East'. A veritable maze of waterways stretched for miles in and around the city.

Our boat, powered by a noisy auto engine, emitted thick black smoke from the exhaust and sped along the canals, slowing down only to avoid other small craft laden with flowers and fresh produce. Sophon, our always-smiling, cheerful guide, pointed out monkeys (macaques) and wild orchids. He introduced us to the mouth-watering tropical fruits of pomolo, custard apples, lychees, rambutan and mangosteens, as well as durian, the huge green fruit resembling a rugby ball and emitting an unpleasant odour, which he laughingly described 'tastes like heaven, but stinks like hell'. As we gaped at the children swimming happily and cleaning their teeth in the filthy water, a young Swiss woman summed up everyone's feelings about Thailand. 'So much happiness,' she sighed. Very soon, we were surrounded on all sides by small boats, mostly paddled by women wearing wide straw sunhats and sarongs, and laden with fresh fruit, vegetables, noodles and spices. This was the famous Floating Market I'd heard and read about.

Bangkok, like most Asian cities, never seems to sleep and was just as lively and busy at night as it was by day, a seething mass of people and traffic all jostling to get somewhere — to a restaurant, a movie or a bar.

One evening, as I sat by myself in a small restaurant, a Thai businessman named Pithak introduced himself with the universal Thai greeting of 'Sawatdee,' with hands clasped prayer fashion under the chin, and asked me to join him at his table. Most Thai men practise what is loosely called *sanuk*, the pursuit of pleasure, and Pithak was no exception. An hour later, after a delicious meal of mixed vegetables and fried rice, he refused to allow me to pay the bill, saying, 'You happen to be a stranger in Bangkok. Please, I pay.' He then suggested showing me some of the 'city of angels' by night and, after attending a cabaret show, which consisted of Western-style go-go and striptease, we arrived at a massage parlour.

After entering a plush, dimly lit lobby to the sound of soft piped music, we faced a large room fitted with a large one-way mirror window, impossible for those inside to look out but possible for those outside to look in. Inside the room, seated on chairs, casually combing their hair, manicuring their nails or watching TV, were some 50 lovely young women of all shapes and sizes aged, from late teens to early thirties. All wore hairdresser's smocks with a number pinned to their lapel. A 'pimp' stood in the lobby extolling the virtues of numbers 28 or 43 to newly arrived customers which, Pithak informed me, probably meant those particular girls were having a lean night with few customers and therefore needed some promotion.

Relief at Last!

I gulped when Pithak urged me to 'choose a girl'. After somewhat timidly choosing number 36 and paying the price of baht 40 (about $2) to the cashier, I was met by the pretty young woman of about 21 and led upstairs to a small room containing a bath and massage table. There, she slipped off her shoes and turned the bath taps on. For the next hour, I was subjected to one of the most pleasurable and erotic experiences I had ever known.

Feeling very self-conscious and with my heart thumping madly in my chest, number 36 slowly proceeded to undress me, giggling shyly and speaking softly in Thai. After gently and expertly removing every stitch of my clothing, she then assisted me into the warm bath where she scrubbed, lathered and skilfully washed every square inch of my body.

I was by now fully aroused, but she pretended not to notice. She had probably seen many naked men — or was unimpressed — and simply got on with her job. The bathing completed, I was stood up, gently but firmly towelled down, and asked to lie face down on the table where I was massaged, pummelled and caressed in no uncertain manner. Among other things, my fingers and toes were almost pulled from their sockets; my legs were pinned back almost to my ears; and, for the first time in my life, I felt the weight of a woman walking up and down my spine. So thorough was she that by the time she had finished, I was convinced that I must have been the cleanest man in town. As I lay completely relaxed and content, I couldn't help sighing to myself, 'Ah. Relief at last,' as I thought of the Indian subcontinent and all its frustrations. What a pleasure it was to be in Thailand!

It was all over when the buzzer sounded, and after helping me dress, number 36 escorted me arm in arm back to the lobby where I tipped her a few baht. With a deep bow, she made the parting remark, 'Come again, please,' more or less the only English she knew. I made a mental note that when next I was tired, I knew where to go for an experience I'd happily repeat any time.

Around 1 a.m. over another meal with Pithak, I pointed out some insects in my fish soup but he seemed quite unperturbed. 'These insects like this kind of soup,' he replied. 'No harm.' An hour later, we rode together in a tuk tuk to my hotel, where he dropped me off. With head bowed and hands tightly clasped to his chest, he wished me goodnight — and goodbye. That was the last I saw of him as he disappeared into the night. My introduction to the Orient was already proving to be interesting, to say the least.

The next day, after visiting Bangkok's most well-known and ornate temples, including the veritable fairyland of Wat Phra Kaew (aka the Emerald Buddha temple, although actually jade), my tour concluded at Wat Benchamabophit, the marble temple, where I watched a young artist at work. Intent on practising his English, we fell into a halting conversation during which he insisted on presenting me with his newly completed picture. 'Please have,' he said. Then, as an afterthought, he quickly added what I thought to be his signature. When I looked closely, he had written '007'. Thanks to books and films, Ian Fleming's James Bond had become well known all over the world!

By this time, I was keen to see more of this exotic country, so I set off by train for Chiang Mai, northern Thailand's largest city. No sooner had we pulled out of the station than smiling, friendly Thais milled around me, offering me fruit and garlands of jasmine. After a brief but happy exchange, a young schoolteacher named Vichai seated in the same compartment insisted that I break my journey and stay at his home. So, at the small town of Uteridit, we got off and made our way to his teak wood house on stilts which stood among tall palm trees and paddy fields. Amid lots of laughter, his family and I did our best to communicate during a dinner of rice, pork and freshly cooked vegetables. That night, I slept in a hammock beneath a mosquito net.

The next morning, I was introduced to the family water buffalo, which just snorted when I went to pat it. Interestingly, unlike the Cape buffalo which is considered one of Africa's most dangerous animals, the water buffaloes of Asia are relatively docile, timid creatures used as beasts of burden to carry heavy loads, work in the paddy fields and take children on their backs to and from school.

Upon resuming my journey, I was invited three more times by fellow passengers to interrupt my rail trip, but I reluctantly declined, feeling at that rate, I'd never reach my destination! Making friends in Thailand, as I discovered, is one of the easiest things on earth and, as a *farang*, I had only to walk down a street and someone would sing out, 'I love you' — the one English expression everyone knew. So wide and genuine were their smiles that, after days of earnestly reciprocating, I literally suffered from an aching jaw.

It was late at night when my train pulled into Chiang Mai. Feeling tired and perspiring from the humidity, I hired a trishaw to take me to a massage parlour. I was so tired that I scarcely noticed when number 15 undressed me but, like my previous experience in Bangkok, it wasn't long before I felt like a new man! A short while later, I checked into a small hotel.

During my month-long stay in Thailand, I mostly stayed in cheap, but spotlessly clean, Chinese-owned hotels. Within minutes of signing in, a girl would bring a pot of cold tea to my room, then linger and smile from the doorway. And invariably, whenever I passed the reception desk, I was politely asked, 'You want girl?' Or, 'You want to see live show?' And I couldn't help thinking Pete was certainly missing something!

Having heard of some of the hill tribes in the area, I took a bus to Chiang Rai, a large town on the Burmese border, where I sought permission from the district governor to visit them. He not only approved my request but, owing to sporadic guerrilla activity, insisted that I be accompanied by a bodyguard for my safety and protection. Shortly after, I was introduced to two young border police officers who had been assigned to escort me. Dressed in jungle-green fatigues, both were armed to the teeth with submachine guns, a bayonet, and a revolver each. Neither spoke any English except for 'Yeh' and 'OK.' Nevertheless, we managed to communicate over the next few days when we stayed with members of the Ekone tribe.

It was baking hot as we walked several miles along a jungle track and up into the lush, green hills but noticeably cooler as we arrived at the first village, where, at first sight of me, the villagers scattered. Patience and a friendly approach soon won the day, however, and within a short while, I was poking inquisitively into bamboo huts and playing with the children.

The Ekone people, one of many different hill tribes that inhabit the so-called Golden Triangle, appeared to lead contented lives tending their small holdings and existing on rice, pork, chicken and bamboo shoots. I stayed with them for three days and nights, during which I slept in the tiny village school, donated by King Bhumibol, Thailand's popular monarch. While the women kept house, worked on the land, and fed the pigs and chickens, the men passed their time sitting around and smoking opium. It was also interesting to note that all the womenfolk wore identical headdresses, adorned with coins minted during the reign of Queen Victoria, yet no one could tell me how, where or when they had been acquired.

Upon returning to Chiang Rai, a young man named Chanoi fell into step with me, asking the usual questions, 'Where are you from?' 'Where are you going?' As nothing seems to give Thais more pleasure than showing a *farang* around, he took it upon himself to show me the local sights. At his suggestion, we ended up outside a private house. Seated cross-legged on the ground repairing fishing nets (almost everyone fishes in the rivers and canals of Thailand) were two most attractive young sisters, aged 18 and 19, dressed in blue jeans and colourful blouses. They looked at me with wide, flirtatious, almond-eyed smiles. After I happily joined them for an impromptu snack and refreshing ice-cold Singha beer, Chanoi smiled at me and nodding towards the girls said, 'Please choose.' For a moment I was speechless and couldn't believe my luck. Both young women were equally attractive. For a minute, I didn't know what to do. It was a difficult decision. But I didn't need any persuasion and quickly made up my mind. I'd always had a strong sex drive. It wasn't surprising that here, with the constant stimulation of travelling and surrounded by so many ravishing and willing young women, my libido increased.

During my stay in the land of smiles, I became acutely aware that Thailand had a huge sex industry comprised mostly of countless massage parlours incorporating everyone's wishes and desires. At times, it seemed like one big knocking shop. A gigantic brothel! Certainly, many women did it for a living but, as far as I could make out, the majority of them did it because they loved sex. And, as I discovered more than once, gave it willingly and freely.

Instead of returning to Bangkok by train, I decided to try hitchhiking. It wasn't long before a truck stopped, headed for the capital. The 300-mile trip took two days and a night during which I spent most of the time lying on the open rack above the driver's cab. Boongas was the driver's name, and throughout the trip, despite my strong protests, he insisted on buying most of my meals. Every time I offered to pay, he silenced me with what had become a familiar expression and used frequently by every Thai, 'Mai-pen rai' (never mind). Indeed, the world could be falling apart and still the Thais would nonchalantly and philosophically shrug it off with this same universal phrase.

We stopped in several towns, including Passang, where Boongas arranged for his dirty laundry to be washed by a girlfriend. Passang is renowned for its particularly beautiful women, but I couldn't help wondering how anyone could reach such a conclusion for gorgeous, smiling women were everywhere — in shops, offices, bikes and on Yamaha, Suzuki, and Kawasaki scooters and motorcycles.

Back in the vast metropolis of Bangkok and wandering along one of the busy streets, I was befriended by some saffron-robed monks who invited me to their temple. When they were not walking the streets, with woven shan bags over their shoulders and clutching food bowls, they spent their time praying. It amused me that the walls of their temple dormitories were covered in pin-ups; radios played Beatles and Beach Boys records; and the monks themselves smoked and chewed gum. In other words, they were normal young men.

Another side-trip from Bangkok took me 80 miles to Pattaya, Thailand's premier beach resort on the Gulf of Siam. Soon to become a major tourist resort, it was also a favourite spot for American GIs on R&R from Vietnam. There, I spent a relaxing weekend swimming and waterskiing. I also chatted to a British consular official on the beach who told me the following story.

Two young Canadian backpackers were travelling somewhere in the interior by bus when it was held up by bandits. With guns drawn, they ordered everyone to hand over their valuables. Everyone did as they were told except one of the Canadians, who defied the bandits by shouting and demanding his rights. One of the bandits then calmly shot him at close range in the stomach. The bandits then fled loaded with the stolen loot. Immediately they were gone, the bus raced the injured man to the nearest hospital where sadly he was pronounced dead.

I'd noticed that some of the buses carried armed guards in the more remote parts of the country. Now I realised how lucky I'd been. As in many countries, travelling in Thailand did not come without risk. There were always potential dangers threatening to spoil a trip if you didn't have your wits about you and failed to use common sense. Or were just plain unlucky!

My next stop was the bridge at Kanchanaburi over the River Kwai. Built by Japanese prisoners of war during the Second World War, destroyed by Allied bombs and rebuilt, it stands as a permanent memorial to the thousands of

Allied Forces men who built the 'death railway'. Twenty-five years later, the surrounding jungle was peaceful and quiet, but nearby, reminding visitors of the horrors of war, is a war cemetery. At the entrance, a plaque reads, 'In honoured remembrance of the fortitude and sacrifice of that gallant company who perished while building the railway from Thailand to Burma during their long captivity.' Seeing the hundreds of crosses and reading some of the epitaphs was a very emotional experience, and I couldn't hold back the tears as I thought of what those POWs had endured. It also reminded me of my uncle Frank, a former prisoner of war, whom I would visit in a few months' time at his home near Adelaide, Australia.

It was now my intention to travel south along the Thai peninsula and then take an overnight boat trip to one of the many coconut islands clustered in the Gulf of Siam. After arriving by train at the little town of Surat Thani, I boarded a boat for the trip to Ko Samui, then an unspoilt and little-known island in the South China Sea. Many years later, it would become a hugely popular place of houses, shops, restaurants, with an international airport. Casting off at midnight to coincide with the tide, the boat crossing took five hours and, to my dismay, was fairly rough. Eventually, we dropped anchor in the darkness a quarter of a mile offshore. Due to low tide, the boat could go no further, so the passengers had to disembark in knee-high water and wade ashore.

There, when the sun rose, I was confronted by a paradise, the perfect place to idle away a few days. Tall palm trees stretched as far as the eye could see along a sun-kissed, deserted beach of soft, white sand. At the time, the island was sparsely populated and rarely visited by outsiders. After checking into a ridiculously cheap bed and breakfast, I did little more for two days than relax, swim, and wander happily along the beach. I also watched trained monkeys run agilely up the trees upon commands from their owners to toss down the ripe coconuts while rejecting the others. My meals consisted of fresh seafood and papayas and mangoes bought at one of the few beachside covered stalls. Here, I was given cigarettes for change and a free glass of refreshing coconut juice.

During my brief stay on Ko Samui, I heard that the only other foreigner on the island was a Peace Corps worker. Inevitably, in such a relatively remote place, we ran into one another. Jim Adams was a fair-haired, suntanned fellow from San Diego, California. When we met at one of the few tiny cafés, he asked me to speak at the one and only primary school where he was teaching English. After telling the sarong-clad students a little about my travels and answering their questions, they all clapped enthusiastically.

After this brief and most pleasant interlude, I left on a small freighter bound for Songkla on the southernmost tip of Thailand. The weather was hot and sunny, and the sea was as calm as could be. I had all the deck space I could wish for and as much fresh seafood, 'flied lice' and fruit that I could eat. 'This is the life,' I wrote in my diary as I took it easy with nothing but blue sky overhead and a gentle swell beneath me.

When my boat arrived in the small port of Songkla, I refreshed myself with iced tea, then boarded a train for Malaysia. After successfully crossing the border where strict immigration officials maintained a wary vigil for communist guerrillas, smugglers and young, destitute travellers, I caught the ferry to Georgetown on the exotic island of Penang. It was September 5, 1965. I'd been travelling the world and away from home for precisely three years.

At that time, the land of rubber and tin had enjoyed only five years of independence when President Sukarno of Indonesia delivered a threat to 'smash Malaysia' and the 'confrontation' as it was known was taking place, and the British Army was involved.

Having heard that our old regiment, the Second Battalion, Green Jackets, was stationed on Penang, I decided to visit their barracks in the hope of meeting up with some people I knew. I presented myself at the main gate and introduced myself to the guard commander. When I asked about certain people I'd known, the corporal told me they had either left the army or were jungle-bashing in Sarawak. I left a little disappointed.

Malaysia was one of the countries I'd always longed to see. Misty green mountains, dense impenetrable jungle, picturesque *kampangs* (villages), and friendly people were but some of the attractions in this fascinating country flanked by the Straits of Malacca and South China Sea. The roads were exceptionally good, well surfaced and signposted, and fairly empty of traffic. As in Thailand, I stayed in cheap, immaculately clean Chinese hotels and once or twice in government rest houses. Like those on the subcontinent, the rest houses were a throwback to British colonial times with their own unique atmosphere.

Since Malaysia consists of a multiracial population, I never knew when hitchhiking whether it would be a Malay, Chinese or Indian driver I'd be meeting next. Whoever it was, it invariably meant a cheerful greeting, a lift to the next town, and sometimes a tour of a Sultan's palace, a tin mine or a rubber plantation. All of which I enjoyed to the maximum.

Once, while sitting on the roadside outside a small Malay town, an elephant with mahout came plodding towards me. I stuck out my thumb, and they stopped. The mahout then helped me up, and I travelled with them for a mile or so until they turned off into a jungle clearing to start the day's work hauling and stacking heavy teak logs. My most original ride yet!

By this time, my money was running low. After booking a berth at a travel agency for a ship that would take me from Singapore to Japan, I had two weeks to pass before it sailed. So I left the hot and humid capital of Kuala Lumpur and headed for the East Coast and the town of Kuantan, where I hoped to see the giant turtles wading ashore to lay their eggs on the beach. Unfortunately, I was a week too soon.

The East Coast was unspoilt in every sense of the word. Deserted beaches were dotted with swaying coconut palms. Fishing boats glided slowly in rhythm with the gentle ocean swells. Barefoot Malays pedalled their trishaws which carried not only passengers but frequently, chickens, ducks, and the occasional small goat or pig along the quiet coast road. Women dressed in *sarong kebaya* — the national costume consisting of ankle-length batik sarong and colourful blouse — sat outside their houses built on stilts, patiently weaving while small children played or kept watch on the family's only cow or water buffalo. Time meant nothing, and for the village folk, life hadn't changed for generations.

I don't recall where and when but, sometime during my stay, I heard of a nearby lake called Lake Chini, well off the beaten track shrouded in mist and mystery which, according to local legend, was home to a dragon, their version of the Loch Ness monster. So I decided to explore it. Surrounded by verdant tropical rainforest deep in the Malaysian jungle, to get there first required a 15-mile drive along a roughly hewn track into the interior to an iron ore mine. There, I lunched with the Japanese mining engineers and arranged for a small, elderly Chinese man who lived on the edge of the lake to be my guide. An exhausting three hours followed as we walked and sometimes hacked our way along a seven-mile overgrown track through what seemed like never-ending eerie jungle.

It transpired that my Chinese guide had fled into the jungle some 20 years previously to escape the Japanese army and, safely in hiding, he married a local aboriginal. It was with them and their young grandson that I stayed when we reached the village on the shores of the lake at sunset.

After a restless night in their primitive attap house due to the mosquitoes and following a breakfast of rice, bamboo shoots and fish caught in the lake, we went to the lakeshore where I climbed gingerly into their dugout. With their grandson in the bow, myself midship and the old man in the stern, we paddled off in search of the monster. After paddling around for three or four hours in search of the so-called dragon and coming up empty-handed, thunder and lightning suddenly rent the air, and a tropical storm began. Rain started pelting down and very soon we were soaked to the skin. So we turned about and paddled furiously for the shore. The rain put a halt to any further search, and I spent the rest of the day attempting to learn how to handle a seven-foot blowpipe, used for hunting by the Jakun, the tribe who lived in the area.

The next day, we tried again but, after investigating almost every part of the lake, including the tall reeds at the sides, we gave up. The old man and his wife then took me via a series of narrow water channels to another small village close to the main east-west highway where I paid them for their services, and we parted ways.

Next, I visited the historic town of Malacca, where I clambered around St. John's Fort and took a short ride upriver in a small motorboat. Finally, I headed for Singapore and was soon picked up by a young English woman and her boyfriend in an old Austin 7 who were headed my way. Rowena and Allan were their names. On arrival in Singapore, I utilised the good services of the YMCA while they stayed with friends.

For the next couple of days, we hired trishaws to visit various places of interest including Change Alley, Chinatown, the Tiger Balm Gardens, and House of Jade. Come evening, we made our way to the infamous Bugis Street, then a haven for transvestites, prostitutes, homosexuals, and seamen. There, we joined the tourists who had assembled to 'see the sights.' Seated at a table of a tiny open-air restaurant under the star-filled sky, we had a meal of crab and corn soup, sweet and sour pork, spare ribs, and fried rice. None of us felt like sampling snake or iguana despite the solemn assurance of 'well cooked', but we eagerly indulged in piping hot satay, a Malay specialty — skewers of chicken, beef and mutton soaked in spices, grilled over charcoal fire and eaten with spicy peanut sauce.

Singapore is a tiny tropical island city-state with a kaleidoscope of people of all nations (but predominantly Chinese, Indian and Malay). Interestingly, nearly everyone spoke no fewer than three languages and, among other observations, I was convinced that every Chinese woman owned a sewing machine, and all the men wore striped pyjamas on Sundays. Owing to the general squalor, poverty and high humidity, my initial impressions of delight quickly faded, and I left feeling that the place was in bad need of a facelift.

Five years later Prime Minister Lee Kuan Yew, an Oxford-educated authoritarian who clearly believed in the adage 'Cleanliness is next to godliness', imposed exceptionally strict fines for littering. He also had a no-nonsense approach to crime and law enforcement. Chewing gum was forbidden, long hair was banned, and anyone caught selling drugs faced the death sentence. Thanks to him, Singapore underwent a massive transformation and quickly went from being one of the dirtiest, smelliest, crime-ridden places in Asia to become the cleanest and safest place in the Orient. It served as an example to other countries that saw the advantages of quite literally cleaning up their own act.

On my last day in Singapore, Rowena and Allan, kindly offered to see my ship off. But first, they took me to the famous Raffles Hotel, once a favourite haunt of Somerset Maugham and other well-known personalities of the day. There, on the garden lawn surrounded by a variety of tropical flowers and plants and beneath the shade of tall traveller's palms, we casually sipped Singapore slings and talked about our future plans. Mine included the next stop on my itinerary, Japan, 'land of the rising sun'.

That evening, after saying our goodbyes, we exchanged waves for a long time as the SS *Cambodge* pulled away from the quayside. Soon Rowena and Allan, my new-found friends, now names in my address book, were no more than tiny specks.

Yet again, like nearly all my boat trips, I spent most of the ten-day voyage in my cabin, desperately seasick. *Will I ever overcome this peculiar affliction?*

I wondered. Nevertheless, the sun shone as we steamed through the South China Sea and a holiday mood prevailed throughout the ship.

Our first port of call was Saigon. With the Vietnam War in progress, it was with considerable trepidation that we sailed up the Saigon River. However, when no mortar bombs or rockets hurtled our way, we relaxed a little. Our stay in port was all too brief, a mere 24 hours, but long enough to be impressed by the French-influenced architecture and ambience. We noticed the numerous roadblocks manned by white-uniformed Vietnamese police; pretty girls in jeans seated side-saddle on the backs of Vespa scooters; and masses of crew-cut American servicemen, all of whom looked like giants compared to the diminutive Vietnamese population.

Our next port of call was the British Crown Colony of Hong Kong, where we anchored for 36 hours, during which I went on another shopping spree and purchased a few new clothes. Hong Kong was exciting and spectacular, and I made a mental note to return as soon as I had a chance. Two days later we berthed in Kobe, one of Japan's major ports. It would be the scene of a massive 7.2 earthquake 30 years later. But then, it was another exotic place waiting to be explored. It was early October 1965. Since leaving South Africa, I'd been travelling virtually nonstop for a period of ten months, the longest I would travel at any one time without stopping to top up my funds. I'd also fallen madly in love with the Orient.

Sri Lanka

Pete astride working elephant

Father and son of the Veddah tribe

Running on deserted beach

Indian peasants lining up to view the Taj Mahal

The Taj Mahal, Agra, India

Relaxing during bus break en route to Kashmir

Pete astride pony in the Vale of Kashmir

Pete and author sunbathing on glacier in Kashmir

Shikaras and houseboats on Lake Dal, Kashmir

Arriving via shikara at our houseboat in Kashmir

Girls of Ekone tribe gathering firewood, Thailand

Artist painting the marble temple, Wat Benchamabophit, Bangkok, Thailand

Deserted beach, Ko Samui, Gulf of Thailand

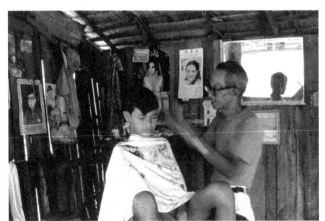
Rural barbershop in Malaysia

Cycling in Kathmandu, Nepal

Kids posing in Macao (façade of St. Paul's in background)

Street festival in small town, Japan

Student, Shigeko, in front of Osaka Castle, Japan

Taiwan

Travelling over the mountains with an army lieutenant

With Chen in Alishan National Park

Tattooed aboriginal women

CHAPTER 10:

Fuji, Sake and Donburi

> 'The true fruit of travel is perhaps the feeling
> of being nearly everywhere at home.'
> — Freya Stark

It was autumn, and there was already a chill in the air when I stepped ashore in Kobe, and I felt apprehensive about experiencing my first real winter in four years. On board the *SS Cambodge*, I had shared a cabin with a New Zealander named Bob, who had been to Japan before and had a Japanese girlfriend named Matsuko.

After the two met at the gangplank, they invited me to stay with them in Matsuko's apartment in Osaka, Japan's second-largest city, a short train ride from Kobe. This proved to be an excellent introduction to Japan, as everything seemed particularly foreign to me with few English signs.

During the next couple of days, I became accustomed to sitting and sleeping on the floor in Matsuko's small three-room wood, paper and bamboo 'aparto', and I also enjoyed my first taste of sukiyaki.

From Osaka, I travelled the 310 miles to Tokyo on the new Tokaido line, better known as 'the bullet train'. The super express left exactly to the second, and as it whisked along at 130 miles per hour, it was as steady as a rock and extremely comfortable. Once in Tokyo, often referred to as the world's greatest metropolis, I checked into Suidobashi youth hostel then telephoned Pete to announce my arrival. In one of his letters addressed to me care of various post offices, he had told me that he was staying with a Japanese family in the Tokyo district of Shin-Nakano.

Imagine my surprise and concern when I called Tadashi Mitsui, his host, and he told me in halting English that Pete had suffered an 'accident' the previous weekend. Stunned by this news, dreadful thoughts swept through my mind until Pete came on the line to elaborate. Yes, he admitted in a subdued voice, he'd been involved in a 'nasty incident' on Mount Fuji. I went immediately to see him and to hear more about his frightening, painful and unforgettable ordeal.

One of the things both of us had wanted to do in Japan was to climb Mount Fuji but, as Pete had no idea when exactly I would arrive, he'd impatiently decided to climb it alone.

Towering over the surrounding countryside at 12,350 feet above sea level, Japan's sacred and most impressive mountain is not considered particularly difficult to climb but, like all high mountains, it can sometimes be dangerous. Buses go halfway up the mountain from where it takes several hours of stiff hiking and scrambling to reach the top.

Having set off fairly late in the day, Pete had made good progress and had just about reached the snowline when darkness fell. Determined to reach the summit, he decided to spend the night on the mountain. Fortunately, he had taken his sleeping bag and climbed into it, attempting to sleep, but soon it became very cold. So he made his way to one of the overnight huts, only to find it locked as the climbing season had officially ended. It was now too dark to see or do anything, so he spent a restless, freezing night outside huddled in his sleeping bag. Throughout the night, he continually rubbed his feet to keep the circulation going. But he was fighting a losing battle. As the night progressed, his feet became steadily worse and when dawn arrived, they were so painfully frostbitten he was unable to stand. To this day he doesn't know how, he managed to slip, slide and stumble his way down to catch a train back to Tokyo.

Once there, Mr Matsui took one look at Pete's swollen feet and anguished expression and immediately took him to his family doctor. Upon examination by the doctor, Pete was rushed to hospital, where more doctors gently and expertly examined him. At first, the frostbite was considered so severe they considered amputating, and when Pete got a hint of this, he broke down in tears.

Fortunately, it never came to that. Thanks to skilled Japanese medical attention, the frostbite was caught in time. But had he been any later getting medical help, he may well have lost both his feet. He was restricted to the Mitsui residence for a week while a nurse came every day to massage and bandage his feet. After that, he had to report to the hospital every day and could only wear slippers for the next six weeks. Sheepishly, he admitted his impatience and foolishness in breaking one of mountaineering's golden rules — never to climb alone! But as I had already seen, it wasn't the first time his daring, devil-may-care approach to life got him into trouble.

Pete brought me up to date about what else had happened to him since we had parted in Nepal. On board his ship, the SS *Laos*, he had been befriended by an American who lived in Japan. This was a lucky break for, upon arrival in Yokohama, Pete had precisely $10 to his name. His new friend, Andie, was an English teacher at one of Tokyo's many English conversation schools. He gave Pete some teaching tips and also introduced him to Tadashi Mitsui and his family. With characteristic hospitality, the Mitsuis invited him to stay with them at their home behind their shop, a small camera and photo store. Pete graciously accepted the offer and very quickly became one of the family.

He then set about finding a teaching job. Before and after the 1964 Tokyo Olympics, a number of English conversation schools had sprung up in both Tokyo and Osaka, Japan's two largest cities. It was an opportune time for the Japanese to seriously study English as international trade and travel were increasing at a phenomenal rate. After the Olympics which were attended by thousands of foreigners, many of them liked Japan so much they decided to stay, and teaching English provided a decent way to make a living.

As the Japanese were taught basic English grammar, reading and writing in high school, they simply wanted to speak to a native English speaker. They were prepared to pay, and pay reasonably well. Although Pete absolutely dreaded the prospect of public speaking, he nevertheless decided it would be a good way of making money. Fortunately, thanks to our travels, he found that meeting new people and exchanging conversation came naturally; he enjoyed it immensely, and his fears quickly evaporated. It wasn't long before he became a successful and very popular teacher.

Prior to starting to teach at the conversation school, Pete got a break when he met some students in the street who invited him to give a talk about his travels the next day at their school. Willing, but somewhat petrified at the thought, he was accompanied to the school where a giant banner welcomed him in English and where a crowd of about a hundred people waited. Feeling quite overwhelmed and nervous, as he'd never done any public speaking, he was unsure how it would go. But once on stage, he forgot his self-consciousness as he warmed to both the audience and the subject matter — our world trip. He received a standing ovation for his talk — and a respectable fee. His confidence was immediately bolstered! In one fell swoop, that impromptu speech and positive audience reaction had made it much easier when it came to speaking to a group of people.

My money was now depleting fast, so I made my own enquiries about teaching. Meanwhile, I'd remained in touch with some fellow travellers from the ship, and one day I was invited to join them at a Japanese film studio. There we were fitted in rather baggy suits and spent the rest of the day in front of a movie camera as extras.

The Japanese film was a James Bond-type movie starring the famous Koyobashi, screen idol of millions. The scene was supposed to be a gambling casino in Monte Carlo. But it was boring work, mostly just hanging around waiting our turn to go on camera. The action started when a screen fight broke out and we all had to pretend to exchange blows. At the end of the day, we left a couple of thousand yen better off and with a tiny knowledge of the Japanese film industry! And I'd learned enough to know that I'd be bored stiff if I were a movie star.

After replying to an ad requesting English teachers in the *Asahi Shimbun*, one of Tokyo's large English dailies, I was asked to report for an interview. At the Foreign Language Centre in Iidabashi, one of three conversation schools under the same management, I met the managing director, André Roy, a 32-year-old French-Canadian from Montreal. A former Air Canada purser, he had travelled widely before deciding to settle in Japan and open his own business — a conversation school. Soon, business boomed, and he established two more schools in various parts of Tokyo — one at Roppongi, the other in Shibuya, both popular entertainment districts.

Mr Roy asked me a few questions, such as how long I intended staying in Japan — I didn't know but assumed it would be several months — then concluded by offering me a teaching position on a part-time basis. He also emphasized the importance of punctuality and 'no dating' of students for obvious reasons. If a student could have free lessons, why go to school and pay for them? Depending on my ability and performance, Roy added, I would be allocated additional classes later. This arrangement suited me perfectly, for I only wanted part-time employment, thinking I could do quite well teaching privately.

Like Pete, I, too, was a little apprehensive about teaching but soon discovered it only required common sense and a slow, clear, and precise way of speaking. Classes were small with a maximum of six students and lasted for an hour. The students were diligent, extremely polite and all eager to learn. In their day jobs, they were secretaries, doctors, nurses, shop assistants, engineers, and businessmen. In short, every age and occupation. Lessons consisted of basic, simple English conversation which became more advanced as the students progressed, with an emphasis on practice and pronunciation. The teachers were British, Irish, American, Canadian, Australian, and New Zealander — all native English speakers.

I needed a more permanent address than the youth hostel where I was staying. Around then, Pete heard of a vacant apartment in Shibuya, and together we went to inspect it. A couple of days later, we moved in.

Although very small and ideal for one person, we'd make do, and the location suited us. It consisted of six 'tatami' (reed and bamboo matting, each approximately 3 feet by 6 feet), a small kitchenette curtained off from the main room, a tiny toilet and a shower outside the front door. The place was furnished with a small foot-high table, one upright wooden chair, and a telephone, which we shared with our next-door neighbour, a Mexican girl with a hysterical laugh which could often be heard through the flimsy walls. It also had a large closet with *shoji* (sliding doors) in which we used to sleep, Pete on the lower level and me on the one and only shelf, when we couldn't be bothered to haul out the mattresses. Which was often.

Our immediate neighbours living in the same building were an American teacher named John; two male French students studying Japanese; Dorothy, the fun-loving Mexican girl who was also studying Japanese; and a young Japanese couple (to our consternation, the husband practised his golf strokes regularly outside our window in the dead of night, his only free time).

Now with a permanent home address, I set about advertising for private students to teach at the apartment while Pete was working full time in

Roppongi. I compiled an advertisement, along the lines of 'British teacher available to give private English lessons' and took it to the offices of the *Yomiuri Shimbun*, one of Tokyo's large Japanese dailies, where a charming young lady translated it into Japanese and arranged for it to be inserted. I then made some small posters and, with the assistance of a young Japanese friend named Masami, pinned them up in nearby colleges and universities.

The very next day, the telephone hardly stopped ringing as I was deluged with calls from prospective students. By this time, my knowledge of the Japanese language was limited to 'Hai' (Yes), 'ee yeh' (no), 'koo da sai' (please), 'arigato' (thank you), 'ogenki desu ka' (how are you?), and 'gomen nasai' (excuse me). Fortunately, most of those who called knew a little English, and somehow I managed to convey how to get to our apartment. Over the next few days, I acquired about a dozen people who wanted to improve their English language skills.

Most were university students with males and females equally represented. Coming in twos or alone, they all arrived exactly on schedule, some full of confidence and others a little shy. I invited them to be seated on cushions on the floor, which surprised and amused them, for most Occidentals living in Japan invariably equipped their 'apartos' with fitted carpet and Western-style furniture. Pete and I preferred the austere, simple Japanese way. Not to mention we couldn't afford furniture. I then compiled a file on each student, listing their names, address, hobbies, and interests, which would assist me during lessons. Once classes began, we would chat about a variety of subjects for an hour, after which I would be presented with my fee, in a tiny, colourful envelope. The students would leave bowing politely and deferentially to 'teacher' David-san after arranging for their next lesson.

It wasn't long before both Pete and I were totally smitten with Japan and even more so with its people, whom we found to be exceptionally polite and hard-working. Despite the difficult Japanese language, we found getting around was not so hard. Just about everyone seemed to know a few words of English and most made no bones about wanting to practise. At stations, for example, when we were unsure which platform we wanted, we simply quoted our destination to the ticket collector. Up would go three fingers as he uttered, 'Track number three'. Standing in the street, we had merely to pull out a map and scratch our head, and within seconds there were friendly, charming people, usually women, beside us smiling and enquiring in English, 'May I help you?'

As I discovered, Japanese women may appear to be demure and self-controlled, but there was no question who was the boss when they were married and at home! Every month their husbands handed over their paycheques. They also seemed quite attracted to foreigners, and it soon became obvious that one or two of my private students were looking for more than English lessons. They wanted a more personal, intimate experience.

When I started teaching at the Iidabashi School, it meant a 30-minute commute by bus and train. At first, I sat in for three lessons understudying a young British fellow called Shaun McCabe, a well-respected, popular, and capable teacher. Shaun had previously travelled overland and by boat to Australia, where he had worked in the mines of Mount Isa before coming to Japan. After 18 months teaching English, he decided to return to the British Isles for one reason or another. But he missed his girlfriend and the lure of Japan so much that he soon returned and married Michiko, his Japanese sweetheart.

Shaun was not only helpful in teaching me the tricks of the trade, but he also took me around parts of Tokyo, pointing out the most reasonably priced restaurants and coffee shops in the area.

It was also Shaun, on a trip through the countryside, who stopped one night at a small town and went for a bath at the communal *ofuru* (Japanese bath). He'd no sooner immersed himself in the piping hot water than in came a group of high school girls. All were naked! Upon seeing him, they all shrieked with surprise and rushed back to the changing room. Minutes later, when they'd overcome their fright, they shyly returned and joined Shaun in the bath where he proceeded to give them an English lesson!

All *guijuin* (foreigners) and, especially teachers, enjoy enormous respect in Japanese society and nowhere was this more obvious than in the classroom. The men sat upright, listening intently, and the women scribbled everything down in small notebooks.

The student names — Kawakami, Tsumura, Toyokishi and Uchinami — were difficult to grasp at first, but after a while they came easily. While I prefixed their names with Mr, Mrs or Miss, they referred to me as David-san. At home, in a more relaxed atmosphere, I called them by their first names of Koji, Masayo, Kenji, Toshia and Satchiko. As a foreigner with fair hair and blue eyes, I was sometimes told by the younger women that I was 'handsome, like film star' — flattering, but hardly true! All lessons served as a cultural exchange and, while they asked me questions about England and places I'd travelled, I picked their brains about Japan.

Sometimes Pete and I cooked meals in the apartment, but we mostly ate out, usually living on noodles, Japanese curry and *donburi* — a Japanese rice bowl topped with vegetables and chicken, beef, or fish — washed down with green tea or Japanese beer. Pete's favourite was donburi, while I liked Japanese curry. I could have eaten it every day!

Many restaurants displayed plastic samples in the windows, which provided a great deal of amusement to us and those in the restaurants. After deciding what we wanted, we would beckon a waitress outside and point to our choice. Giggles, accompanied by 'Ah so desku' (Oh, really) would then precede our order. No sooner were we seated — be it in a restaurant, coffee shop or bar — than we were handed hot *oshiburi* (scented towels) to clean our hands.

Now that we had money, we could afford to treat ourselves. Some of our favourite dishes were *yakitori* (small chicken shish kebabs dipped in a soya sauce), *tempura* (chopped vegetables and various seafood fried in batter), *shabu-shabu* (wafer-thin slices of beef dipped in boiling water) and *tonkatsu* (pork cutlets fried in bread crumbs). We also ate ramen (different types of noodles in broth) and, of course, sushi, which would take the Western world by storm many years later. All were served with salad and steaming hot rice and eaten with chopsticks, which by now we handled as well as the locals, much to their surprise and amusement. As we enjoyed the many different forms of sushi, there was one dish that neither of us could quite get used to. That was *sashimi* (raw fish by itself), which was definitely an acquired taste. We both liked our meat — be it beef, chicken or fish — cooked!

How well I remember going alone or with Pete, ducking under a *noren* (a traditional Japanese fabric divider) and entering a small, cosy local restaurant. There, we enjoyed any one of the above-mentioned delicious dishes together with a refreshing Kirin or Sapporo beer, all for a modest price.

At that time, most places in Japan had small convenience stores and vending machines on just about every corner where you could buy almost anything. Among my favourite places were the Almond coffee shops, which sold tea, coffee and pastries to the working folk.

Before winter began, I too wanted to climb Mount Fuji. So I invited John, our American neighbour, and a couple of Japanese friends to join me. We caught the late-night Friday train to Gotemba located at the foot of Fuji, in Hakone-Izu National Park, to enable us to get an early start the next morning. It was 2 a.m. by the time we arrived. Intent on saving money, we decided to sleep in the station waiting room. The night turned out to be freezing cold and, despite wearing a couple of sweaters and lying in our sleeping bags for extra warmth, there was little relief. It was then I realised what Pete must have gone through.

It was so cold that after an hour we got up and wandered around the darkened town looking for a place where we could warm up. Thankfully, we discovered a small café where we spent the next few hours huddled around a small electric fire, eating curry and rice and sipping gallons of hot green tea.

Fortunately, daylight brought sunny weather, and the mountain air was crisp and clear. After taking the bus to the last stop and halfway point, we set off through a magnificent forest ablaze with autumn colours and a thick carpet of leaves. Mount Fuji is a strenuous hike at the best of times, and we sweated as we plodded upwards for a couple of hours. When we stopped for a breather, I left the others in order to tackle a small rock face. No sooner had I reached it, than I noticed some yards away what looked like another climber lying in a small dip. I continued up the scree until I got close to him. He was lying very still in a peculiar position, as though he had fallen. As I drew closer to the crumpled figure, I shouted to attract the attention of my friends, 'I think this man is badly hurt.' I touched him, but he was stiff and lifeless and inert as only a corpse could be. 'It's no good,' I said to John and the others. 'I think he's dead.'

Feeling slightly nauseous, I gingerly searched the body for identification. While one friend sped off down the slopes to notify the police and get help, the rest of us sat guard over the dead man, speculating over his fate. When help arrived in the form of stretcher-bearers, it was getting too late to continue the climb, and none of us felt like continuing to the top. Slowly and forlornly, we made our way back down. *So much for 'beautiful, serene Mount Fuji'*, I thought. Like all high mountains, she had a deadly side.

We learned later from a news report that the Japanese climber was in his mid-twenties. He was well-equipped with climbing boots and winter clothing and showed no obvious signs of any external injuries except for a few scratches to his expressionless face. Judging from his crumpled position, it seemed he had slipped, knocked himself out and rolled some distance to where we found him. He may well have died from internal injuries, but most likely, he had failed to regain consciousness and died of exposure. We also heard that he was last seen climbing alone the previous day. That mistake had cost him his life.

Soon, winter arrived with strong gales and six inches of snow. So we bought thick top coats, the first time we had been out of summer clothes for about three years. Despite purchasing an electric heater, our tiny, flimsy apartment was still draughty, and our sleeping bags weren't effective enough, so we purchased *futons* (Japanese quilts) to be warm enough at night.

The weather didn't deter my students unless the buses came to a standstill. Often, at school or at the apartment, they presented both me and Pete with presents in the form of key rings, delicately painted pictures, or fresh fruit, a Japanese custom which impressed us no end. The men were always immaculately dressed, usually in dark-coloured suits. Only the older women regularly wore kimono, the becoming national dress, because it was so time-consuming to put on. But they were worn for special occasions such as New Year's Day, weddings, and certain festivals. 'Western dress is more comfortable,' we were told by the younger women. We lamented the passing of this beautiful attire and, just to please us, one or two of the young women would occasionally wear kimonos to class.

As Pete worked normal office hours, he was faced with commuting during rush hour. At those times, the Japanese people seemingly forget their manners and elbow and shove their way into the already crowded trains with the additional help and support of Sumo-wrestling-type platform staff. While I was fortunate to avoid those particular hours, many was the time Pete would return home looking slightly dishevelled with tie askew, a torn pocket or a button missing! But he didn't care; he was enjoying himself.

Sometimes, when we planned on an evening out, we'd meet at the statue of Hachiko, a well-known landmark beside Shibuya station. The statue was erected in 1934 to honour the dog who accompanied his master there every day. One day, his master died at work and never returned home. The devoted dog continued to wait patiently at the same place for some weeks, fed by passers-by, until he too died, probably of a broken heart.

Frequently, we were invited out by our male students, usually businessmen and, although it was bending the rule of not fraternizing with our students we regarded this as an opportunity too good to be missed. The pattern of events was usually identical each time. First, we would be escorted to one of the many comfortable coffee shops where one could sit for hours over a cup of coffee listening to the music of your choice. Some played only jazz, others classical or popular Western music, and others Japanese. There we would sit for half an hour chatting. Afterwards, we'd be taken to a restaurant for a meal, and finally we would end up in a club or bar.

At all of these establishments, the door would be opened by a bowing, smiling girl and, once inside, the entire staff would chorus, 'Irai shai mase. Irai shai mase' (Welcome. Welcome). Once seated at the bar or a table, a kimono-clad hostess would pour our drinks of beer or sake. No sooner were our glasses half empty than a woman's hand would appear out of nowhere and gently refill it. Despite our protests, our hosts always insisted on footing the bill. After learning nearly all had lavish expense accounts, a common perk, we relaxed more and enjoyed ourselves.

Barely a day or night passed when both Pete and I were not befriended by someone in the street, on a railway platform, or in a restaurant, wanting to practise English. One night in a bar, I fell into conversation with a hostess. She, too, wanted English lessons and from then on, she attended class regularly!

Apart from our students, we both made other Japanese friends who frequently offered to show us around and also invited us to their homes. These visits took on a familiar pattern. After admiring the beautiful, well laid-out gardens complete with stone lanterns and goldfish ponds, off came our shoes as we stooped through the main entrance into the living room. There, we would sit cross-legged or on our knees on the tatami and leaf through the family photograph album while *Oka-san* (Mother) would serve green tea or orange juice and *osenbe* (rice crackers) on a tray. We would then hand over small gifts, neatly wrapped, that we had bought for the occasion and in accordance with Japanese custom. Deep bows by everyone would precede and follow our arrival and departure.

Christmas 1965 arrived, this time a white one, and although it was not celebrated then as in Western countries, there were definite indications that Western-style celebrations would soon arrive. 'Silent Night' and 'Jingle Bells' were being played in some department stores. It would only be a matter of time before Christmas trees with decorations appeared too. On Christmas Eve, while wandering around the neon-lit bar area of Shibuya, friendly drunks came up to us, wished us 'Merry Christmas,' shook our hand and patted us on the back.

Among other things during our stay in this amazingly efficient country, we witnessed many festivals, watched flower arrangement lessons and attended Kabuki theatre and tea ceremonies. Frequently at weekends, we were invited to take a trip into the country with friends. As a result, we saw much of the main island of Honshu, including Kamakura, Hakone, Ise-shima, Nagoya and Kyoto, the fascinating ancient capital. I even visited a Japanese castle with Shigeko, one of my students.

On one such weekend, friends asked me to their parents' farmhouse deep in the rural countryside. During my brief stay, the grandmother died and, as an honoured guest, I was invited to join the family, relatives, neighbours and friends who gathered around the body all night discussing the virtues of the dearly departed! It was another unusual, sombre, and very interesting experience to add to my memories.

Travelling by train was always a pleasure (except for suburban lines during rush hour) as they were extremely comfortable and efficient. While piped music filtered through the compartments, smartly uniformed girls wandered back and forth selling chocolate, fruit and cigarettes from a trolley.

Once or twice, I stayed in *ryokans* (Japanese inns) where, after checking in and removing my shoes (now a habit), I was escorted to my room by a Japanese maid. I would be left alone with a pot of green tea for a few minutes to slip out of my clothes and put on a freshly laundered happi coat or kimono. A gentle tap on the wooden door would follow as I was beckoned to take my bath. On my return half an hour later, the futon, my bed was ready on the floor.

The Japanese bathing ritual is a daily event and usually happens just before going to bed. *Ofurus* stood on just about every neighbourhood corner (many people did not have their own full bathrooms). Frequenting the baths was another of the many local customs Pete and I greatly enjoyed. Late at night, we donned our *yukatas* (lightweight kimonos) in which we now slept, and *geta* (wooden clogs), and shuffled and shivered in the cold wind to our nearby *ofuru*. We must have made an amusing sight, westerners clip-clopping along the narrow streets with our little plastic bowls containing soap and *tenugri* (a dual-purpose cloth used as a flannel and towel) in our hands.

At the *ofuru*, we paid 50 yen (about 20 cents) to the female attendant who sat in a booth at the entrance surveying the swinging doors separating the men's and women's sections. After undressing and self-consciously covering ourselves with our tiny towels fig-leaf fashion, we entered the steamy bath area, which was lined with shiny, tiled walls and mirrors. There we sat naked among the Japanese men on short three-legged stools before rows of taps, where we vigorously washed and lathered every part of our bodies. Then, filling our small bowls with warm tap water, we rinsed thoroughly.

To dirty or somehow cloud the actual bath was a cardinal sin as the water had to be kept clean for everyone. It was, of course, sterilised by its own heat, which was sometimes as hot as 144°F. Indeed, it was so hot that initially we found it impossible to ease ourselves in and we sat carefully on the edge, dangling only our legs in the water. But, through sheer persistence, we managed to immerse our complete bodies, usually close to the cooling tap in one corner of the bath.

After 15 to 20 minutes of soaking ourselves, which was all we could bear, we returned to the apartment, with any cares or worries melted away and our bodies tingling, totally oblivious to the freezing night air. At home, sapped of all energy and feeling pleasantly lethargic, we fell immediately to sleep to the melancholy sound of the flute of the noodle or hot chestnut man making his rounds.

'Bloody hell!' exclaimed Pete late one night as the whole building shook and trembled for about ten seconds. We knew immediately that it was one of the many earthquakes which occur frequently in Japan, something that initially scared us, but after a while we learned to live with them. When the tremors were only slight, they didn't bother us, but if reasonably severe, we hurriedly donned our coats and stepped outside, where it was safer.

It was at Pete's school that he met a young woman who was to have a profound impact on his life. Her name was Hoshiko, and she was a student in one of his classes and a secretary at the Australian embassy. In her mid-twenties, she was attractive and softly spoken. In fact, she personified all the virtues and attributes of Japanese womanhood. Their relationship started casually enough when the whole class invited Pete to join them at weekends for visits to temples and parks. After a while, though, they had eyes only for each other and went out as a couple.

Whenever we could find some spare time, Pete and I liked nothing better than to stroll around the Ginza and its fashionable department stores and shopping arcades. There, we amused ourselves toying with the numerous gadgets, electronic equipment and new devices. We played 'Pachinko' in the numerous one-armed bandit parlours which were always packed with off-duty office staff, businessmen and housewives. At Fugetsudo, a popular coffee shop and hangout for travellers in Shinjuku, another of Tokyo's busy entertainment districts, we would sit over coffee for an hour or two, talking with travellers and fellow teachers.

One day, Masayo, one of my students, invited me skiing with her and some friends. I wasn't much of a skier, but jumped at the chance, little knowing what I was getting into. We left Tokyo for the Japanese Alps in a bus loaded with university students. Several hours later, we checked into an inn high in the mountains.

The next morning, after a Japanese breakfast of miso soup, rice, and pickles, we walked some distance to the ski slopes. I was tired from this initial effort and hardly felt like more exercise. But I tried to keep up with my enthusiastic Japanese friends as we skied for a couple of hours. Back at the inn, they asked me to give a talk about my far-flung travels. It would have been embarrassing to admit I was too tired, so I complied.

That night, totally exhausted, I dropped into a drug-like sleep. The following day, it rained, but that didn't deter the others. Reluctantly I joined them. We boarded the ski lift in a blinding blizzard and skied for hours, halting only for a brief lunch. The day concluded above a very steep slope where I stopped. Masayo then challenged me with the words, 'But you are a man!' Not wishing to lose face, I went down the hill half on my skis and half on my backside! We skied all the next day. How I managed to maintain this pace, I'll never know.

About that time, there were a series of tragic airline accidents, the worst in Japan's history. In the space of one month, three aeroplanes crashed, with a significant loss of life. A domestic All Nippon Airways Boeing 727 plunged into Tokyo Bay. Shortly after, a Canadian Pacific Airlines FC-8 smashed into a retaining wall in fog at Tokyo International Airport and, two days later, a BOAC Boeing 707 crashed near Mount Fuji. Although Japan suffers frequently from earthquakes, fires and floods, the airline tragedies worried and concerned everyone deeply. For weeks, many people were afraid to fly. As it neared our departure date, all our friends urged Pete and me to travel only by boat! Little did they know about my Achilles heel!

I'd now been in Japan nearly six months, and Pete had been there seven. We'd saved enough money for the next leg of our marathon journey. We were both now 27 and, much as we'd liked to have stayed longer in this amazing country, it was time to resume our world travels. I booked a passage on the MM's *SS Vietnam* to Hong Kong, but Pete couldn't make up his mind and was reluctant to make any decision until the last moment. Two days before the ship was due to sail, he suddenly decided to join me after exchanging solemn promises with a tearful Hoshiko to stay in touch.

Like Pete, I disliked the idea of leaving Japan but consoled myself by looking forward to fresh challenges and new adventures. Before leaving, I introduced my private students to colleagues at school to ensure continuity of their lessons. After being showered with presents from just about everyone we knew, we bade *Sayonara* to all our friends.

Over the last few weeks, I'd become quite friendly with Kumiko, our very pleasant young receptionist, speaking to her between classes almost every day. On my last day, when I went to say goodbye, she burst into tears, which took me by surprise. I never did learn the reason, but Dan, an American colleague, suggested she might have had a crush on me!

The first of the season's magnificent white and pink cherry blossoms were coming into bloom when we left Yokohama on March 11, 1966. As we stood on deck and the big ship slipped slowly out of the harbour, a young Japanese man leaning over the ship's rails casually asked Pete, 'What is your country?' Without a second thought, he replied, 'Japan.' And so it was at the time.

CHAPTER 11:

A Romantic Interlude

*'Travel isn't always pretty. It isn't always comfortable.
Sometimes it hurts, it even breaks your heart.
But that's okay. The journey changes you; it should change you.'*
— Anthony Bourdain

After a couple of calm days sailing through the South China Sea on the steamship *SS Vietnam* during which, thankfully, I wasn't seasick, we arrived in Hong Kong. I disembarked for my next adventure while Pete stayed on board. He had decided to remain on the ship until Bangkok from where he would head to Cambodia to see the famous jungle ruins of Angkor Wat. He would then travel to Singapore via Malaysia, where he caught a ship almost immediately for Australia. As usual by now, when we parted, neither of us had any idea when and where we would next meet. In this case, it would be Sydney, Australia, in a couple of months.

Hong Kong was impressive in every respect, being equally or more spectacular than many other major port cities. I wished Captain Elliot of the Royal Navy's *HMS Sulphur* could see it today. He was the one who described it as 'a barren rock' when he claimed it on behalf of Great Britain in July 1841. There had certainly been some changes since Queen Victoria's time. Having been in Hong Kong once before, albeit ever so briefly, it felt quite familiar; however, in contrast to Japan, where there is no tipping, in Hong Kong everyone expected to be tipped!

After checking into a small, inexpensive hotel in the Wanchai district, an area full of bars — some seedy, some cosy, and some so small there was no room to swing a cat, I started exploring the British Crown Colony by bus, ferry and on foot. During my ten-day stay, I went to almost every part of Hong Kong Island and Kowloon. With so much to see and do, it was difficult to know where to begin. After getting very sore feet from so much walking, I decided on a leisurely ferry ride to one of the small nearby islands, Cheung Chen. There, I spent a pleasant afternoon wandering around, visiting one or two temples, peering into the tiny houses of the residents, and observing the local Hakka people tilling the earth.

I'd arranged with a Japanese student of mine, a businessman named Junji, to meet up in Hong Kong, and I greeted him at Kai Tak Airport when he arrived. A kind Cathay Pacific Airline agent, Mr Lee, had booked Junji into a hotel in busy Nathan Road, and then suggested that we both join him and his wife for dinner. That evening, they drove us to the Carlton Hotel, about which Ian Fleming wrote in his book *Thrilling Cities*, 'Dinner at the Carlton unfolds one of the world's most memorable panoramas: the jewelled lights of Kowloon, the harbour and the island.' There, we were treated to a sumptuous Chinese meal of 10 or 12 dishes! All washed down with Chinese tea.

Over the next few days, Junji and I explored together. We rode the old trams that clanged up and down the bustling streets and visited Tiger Balm Gardens and the fishing village of Aberdeen. Another time, we swam at Repulse Bay and lazed on the fine, white sandy beach. On a bus ride out to the New Territories, we strolled around one of the many duck farms and gazed from no man's land at communist China looming in the distance. It was 1966, the beginning of Mao's Cultural Revolution. Tragically, over a ten-year period, more than 20 million people would lose their lives during the upheaval.

Window shopping was fun, for there was nothing that the free port of Hong Kong did not display or sell. Nights, too, like all Oriental cities were fascinating. People, young and old, strolled around or ate their evening meal at one of the hundreds of outdoor restaurants. Every time we passed one of the countless bars, we were hissed and waved at by girls in *cheong-sams* with long slits down each side, revealing beautiful legs. They quietly, but not always politely, asked if we wanted 'sexy girl' or to see 'naughty movie'. I couldn't help but think of Richard Mason's *The World of Susie Wong*, for much of what I was witnessing was exactly as he'd described in his best-selling book. Yet another evening, after strolling around Temple Street Market, we sat down outside on the ubiquitous three-legged stools to a delicious meal of curried fish, fried noodles, bok choy and mango pudding in Sham Shui Po, a densely populated working-class district known for its food stalls and flea markets.

After Junji left, I made the most of the rest of my stay. I rode the funicular up Victoria Peak and went again to the New Territories, this time by train to see more of the countryside. On different nights, careful to watch out for pickpockets, I attended a Chinese opera; went out of curiosity to see one of the highly popular 'kung-fu' movies; enjoyed a delicious eight-course seafood meal at one of Aberdeen's famous floating restaurants; and bartered for knick-knacks at one of several 'poor man's nightclubs' — the open-air night markets. When I'd just about seen everything and covered all the main highlights, I took ferries to some of the many outlying islands.

As I wandered around this city, it was easy to see why it was called 'The Concrete Jungle' for wherever you looked, there were skyscrapers in every direction. By then, I'd seen my share of slums and poverty-stricken areas, but I was aghast at the awful 'cage homes' — tiny metal-framed living quarters with room for only a bed and a few clothes — clinging precariously to the outsides of buildings, exposed to the elements. With only one communal kitchen and bathroom for ten men or women to share, I desperately pitied those miserable residents.

Everywhere I went there was a buzz of activity, both on land and in the harbour, the unwritten slogan 'work or die' being the order of the day. The tremendous squalor and large numbers of beggars did not escape my inquisitive gaze. I often saw people spitting, belching and snorting in the streets where nonchalantly, sometimes in bare feet, they would rub their spittle into the ground. They also liked to push and shove. Appalling habits, I thought, but I'd seen worse in India where everything and anything goes. Besides, I agreed with an irate reader of the *Hong Kong Standard* who had written an angry letter to the editor in response to many complaints, 'Such habits do not make or break a nation.' A few years later, after Singapore introduced strict cleanliness measures, Hong Kong followed suit.

Almost every day, I travelled on the Star Ferry from Victoria to Kowloon and back. At ten cents a ride, it had to be the cheapest ferry in the world. It seemed to me that one could travel across the harbour a thousand times and see something different each time — junks and sampans; vehicle and passenger-only ferries; small and large cargo ships; as well as massive ocean liners. It was a fascinating scene. One day, I looked over the newly opened ocean terminal, which had created a lot of interest among the locals. At that time, the five-year-old *SS Canberra* of the P&O line was moored alongside, making it the biggest ship ever to dock in Hong Kong. At 820 feet long, it carried roughly 2200 passengers and several hundred crew members. The sixties was the heyday for such ocean-going vessels, but within a few years, most would be taken out of service and replaced by even bigger cruise ships.

Also in my jam-packed itinerary was a visit to Macao, the tiny Portuguese colony on the tip of Red China. The 40-mile, 75-minute trip by hydrofoil was comfortable enough, but the sea was distinctly choppy, and although I wasn't seasick, I felt decidedly queasy. In Macao, I stayed in a fifth-floor hotel room overlooking the Pearl River and the People's Republic of China.

Macao was fascinating, with its tiny plazas, cobbled streets, attractive gardens and scruffy Portuguese soldiers who strolled around with their hands deep in their pockets. In contrast to the hustle and bustle of Hong Kong, the pace of Macao was relaxed and leisurely. But the harbour was a different story. Alive with activity and assorted junks and sampans, Red Chinese and Portuguese gunboats took it in turns to ply back and forth in the middle of the Pearl River. Thanks to Sam, a pedicab (bicycle rickshaw) man, I saw just about everything of interest, but I balked at the idea of eating the snakes and frogs, which he pointed out and solemnly assured me 'taste good'.

Macao was where half the gambling-mad population of Hong Kong regularly went to spend their hard-earned wages. Every casino was packed with people of all ages playing cards, working the slot machines, one-armed bandits or other casino games. Mahjong, as in Hong Kong, was played everywhere — outside wherever they could set up a table, inside shops and offices, and even on street corners. The Chinese never missed a chance to play. They also loved Tai Chi, and strangely enough, Western ballroom dancing, which I noticed was practised in every park or spare piece of land.

Returning to Hong Kong, I travelled on the *SS Fat Shan*, one of the large ferries that operated between Hong Kong and Macao. In 1971, she capsized and sank with 80 passengers on board during Typhoon Rose. But in March that year, the weather was perfect, the trip taking three very pleasant hours.

By the time I left Hong Kong for the last time, I'd seen this remarkable place in all weathers and from all angles — from the top of Victoria Peak, aboard the Star Ferry, the quiet rural setting of the New Territories, and the always busy streets of Mong Kok and Tsim Sha Sui.

I had hoped to visit China but, when I heard about the trouble brewing, I decided instead to visit Taiwan, then better known as Formosa. When I went to book my passage with one of the few shipping lines that serve this small island nation, an official informed me that Europeans were not permitted to travel deck class, which seemed strange. I was flabbergasted, but no amount of persuasion would make him change his mind. Company policy, I guessed. So, with no choice, I settled for second class and a tiny, two-berth cabin. Fortunately, I had it to myself.

The *SS Anking* was an ocean-going vessel which mostly carried cargo and about 100 passengers. The two-day voyage across the Taiwan Strait, part of the South China Sea, was very rough, and I spent most of the time below in my cabin feeling distinctly unwell. Happily, the trip turned out to be well worth it.

Formosa means 'Island beautiful,' as named by early Portuguese explorers. It was a Japanese colony for 50 years from 1895 until the end of World War II. As I would soon see, it has some of the kindest and most hospitable people and impressive scenery in the Orient.

Stepping ashore at the main port of Keelung, I was surrounded by a swarm of reporters and photographers. International travellers were few and far between, and it was me with my rucksack whom they chose to interview! The next day, a story about my travels along with my photo appeared in a couple of the Chinese language newspapers in the capital city, Taipei.

I had been interviewed by the press a few times previously when friends recognised my rather unique and unusual experiences and got in touch with their local newspapers. Later, in Sydney, I would be interviewed by a staff reporter for *The Australian*, one of Australia's major newspapers, and in the article, I was referred to as a modern-day Phileas Fogg. While I didn't detect any resemblance between Jules Verne's famous fictional character and myself, the reporter did. I, too, was British and making a unique journey around the world using a variety of different transport. The main difference, however, was that my journey was not 80 days long; unbeknownst to me at the time, it would come close to 80 months!

In Taipei, I checked into the New York Hotel, a small central place which cost only 40NT a night. Like many places in Taiwan, it was named after an American city. I spent a couple of interesting days in the bustling capital, including a visit to the acclaimed National Museum to see artefacts more than 2,000 years old. On a trip to Wulai in the countryside, I enjoyed an exhilarating ride in a handcart pushed along a narrow-gauge railway track by a Chinese peasant. I then set off to travel around the island.

As I sat on one of the city's commuter trains, a young man and woman travelling together recognised me from photographs in the papers, and we struck up a conversation. Chou was a primary school teacher and his companion, Chen, was a designer and assistant lecturer in architecture at Tunghai University in the city of Taichung. As we chatted, Chou invited me to take a look around his high school; so, at the next stop, we got off. After a brief tour of his school, during which Chen barely uttered a word and walked demurely behind us, we had lunch. Chou invited me to come back anytime. He also suggested that I look up Chen when I reached Taichung. 'She will be very pleased to show you around,' were his parting words as Chen nodded in silent agreement.

The following day, I travelled by bus again, this time to Hualien. The journey took four hours to cover the 40 miles, as we stopped a few times for refreshing glasses of sugar cane juice, and the road was one-way only and subject to avalanches. Winding along the Pacific Coast, the road and view had to be seen to be believed — sheer cliffs with a 500-foot drop into the sea. The bus was packed with passengers, mostly soldiers standing in the aisle, and like so many times when travelling on dangerous mountain highways, I was scared stiff.

It was late afternoon by the time we reached our destination. No sooner had I checked into a local hotel than I met Mr Wong, manager of the local taxi service, who immediately suggested I take a cab ride to see the local aboriginals. However, when I explained that was not my usual method of transportation and that I was not well off but just an ordinary fellow, he offered to show me around for nothing on his motorcycle. Being well known in the area, he had many friends, including the director of the famous Ami aboriginal dancing troupe and a cinema manager, Mr Lien.

After watching a film which was constantly interrupted by hoots and shouts from the audience — the Chinese being great moviegoers and among the rowdiest people on earth — we joined Mr Lien for dinner. The meal consisted of mushroom soup, rice, bok choy greens, water buffalo steak, fried lobster and sweet and sour pork. Of all the different dishes and native foods that I sampled around the world, Asian food remained one of my favourites.

The next day, we went to see a show put on by members of the Ami tribe, original Taiwanese people who tattooed their faces and bodies. As I stood entranced by their delicate movements to the music, one girl detached herself from the others, padded quietly up to me and gently led me by the hand, her way of inviting me to join in. While I happily cooperated, I never enjoyed such occasions, being very self-conscious and always afraid of making a fool of myself.

A day or two later, while looking around a temple and pagoda, the chief priest invited me to stay, which I did for two days. An elderly, shrivelled-up little Chinese man, he was both interesting and interested in everything. Although nearing 80, he read Tolstoy and Shakespeare 'determined to master English'. He was also an inveterate and able walker. On my last day, he invited me to join him and some members of his kitchen staff ('my disciples', he called them) for a six-mile walk across rope bridges and along mountain tracks. When everyone else seemed tired, he looked as fresh as when we had started.

After only a few days on the island, it seemed to me that the place was swarming with American GIs. The Vietnam War was well underway, and many soldiers were based there. All seemed to be six feet, six inches tall and sporting crew cuts, which made them instantly recognisable. I noticed wherever they were based or went, the prices of everything jumped. There were also a number of entertainment establishments where they were forbidden to eat or drink and which carried signs like 'Off limits to US Personnel'. I guessed those places had seen too many drunken brawls.

Crossing the famous and spectacular east-west highway was rather unnerving. The road was extremely narrow and tortuous, with sheer drops of 2,000 to 3,000 feet. Despite that, I marvelled at the breathtaking scenery. That night, I stayed in a youth hostel high in the mountains. There, I met a group of 25 Chinese youth corps workers and joined in some of their games.

After rising early with the others, we all set off by bus in a blanket of mist, newly fallen snow, and bitter cold. We had gone only three miles when we were forced to stop to deal with yet another landslide. The others turned back (they had limited time) while I waited for the road to be cleared. Six hours later, it was. At about the same time, an army convoy drove up with a general in command, and I approached him politely for a lift. After examining my passport, he instructed me to share an American Jeep with a young lieutenant. Like just about everyone else I met, he was cheerful and friendly. The convoy stopped several times while bulldozers cleared different sections of the road which had just experienced a snowstorm, especially ironic when the coast was sweltering in a heat wave.

Often, while sitting in small restaurants, I was befriended by local people who sat and talked to me for a few minutes, then left. And much to my surprise, when I went to pay my bill, I found it had already been paid. These gestures always meant so much to me where the people usually have so little in the way of material things and yet are always willing and happy to share whatever they have. In every such circumstance, I felt very humble and always stopped to think how much we in the Western world could learn from them. But it was their patience and humility that impressed me most. We in the Western world are so often intent on teaching others what we arrogantly think is the 'best way', or 'the right way', seldom acknowledging that others have no need or desire for what we have to offer. The more I travelled, the more I realised this to be true.

Earlier, I'd telephoned Chen, and we met at the gates of Tunghai University. This time, she was more talkative. She showed me almost every part of the large sprawling campus. From the rooftop of one of the buildings, we sat and chatted and watched the sun go down. Her full name, she told me, was Ching Chou Chen. *How very Chinese*, I thought.

Aged 22, she was quite pretty with large brown almond eyes, shoulder-length dark hair and a slight, slender figure (she would say 'skinny'), and

A Romantic Interlude

she spoke fluent English with a soft Chinese accent. She had a captivating smile, revealing even white teeth. Much to my delight, she also liked sex, as I would soon find out. She had been born on the Chinese mainland 20 years previously and had fled the country with her parents as a young girl when they noticed social unrest taking its toll. It had been a good move as both her mother and father were teachers and considered intellectuals. Had they remained, they may well have been murdered in the great uprising by Mao's fanatical henchmen, the Red Guards.

After a couple of days spent mostly on the campus of the university, it was time for me to continue my trip around the island. Over these few days, Chen and I had developed quite a close relationship and discussed many things.

It was about eight o'clock in the evening, and we had just taken a walk. As we sat on the steps outside the students' cafeteria, Chen suggested that tomorrow she could come with me. She explained that as she knew the language and the country, she should join me as my guide. Besides, she wanted very much to come. At first, I said no, but she insisted, then pleaded with me. So I agreed. We then separated for the night, she to sleep in her usual quarters and me in one of the male dormitories.

Early the next morning, we boarded a bus from the city centre to the next place on my itinerary, Sun Moon Lake, two hours from Taichung. Once there we hired a small boat. As we rowed around the lake, we talked and talked. Before long, though, the fine sunny weather and romantic atmosphere got the better of us and we slipped down into the bottom of the little craft where we kissed passionately as we drifted around. Half an hour elapsed as we lay in this position. Then we got up and I took both oars and rowed steadily to the nearest land. There we beached the boat on the muddy bank, linked hands, and set off to explore. After walking around for about 30 minutes we came across a clearing in the shade of some trees. There, overcome by our emotions and seeing nobody else around, we made love.

Back on shore, we took another bus ride to the town of Chainee where Chen realised that she had forgotten her ID card, unforgivable in what was then more or less a police state where everyone's movements were closely watched. The authorities wanted no infiltration of spies from the mainland. We then had to visit the police station to get authority for Chen to continue the journey and to travel with me. Fortunately, permission was granted after

I answered a few questions. Her name and address were taken to be checked and cleared with records. They weren't taking any chances, I realised, but it was galling that we were initially viewed in the same light as the GIs, bar girls and pimps. Fortunately, all the fuss and bother, which I thought ludicrous, would prove to be worth it, for we had some ecstatically happy times together over the next few days.

Our next destination was Lishan, a mountain resort and nature reserve. After reaching the tiny township by way of the Forest Railway, a funny little cable train that rocked back and forth, we booked into a small Japanese-style hotel with tatami mats, sliding doors and mattresses on the floor. During our short time together, we stayed at clean and comfortable Japanese-style inns, where every night we made love and fell asleep in each other's arms.

We remained in the picture-postcard area for two days, strolling among the 1,000-year-old cedar trees and refreshing ourselves by gulping in the cool mountain air. It was biting cold in the mountains at night but sunny and warm by day, unlike the coast which seemed to suffer from fiercely high humidity in April. Chen proved to be an ideal companion with a fine sense of humour. She was, in fact, an excellent guide, describing many topics of interest to me.

Descending on the train, we sat in a compartment with a large group of student teachers who, after quizzing Chen about me, requested that I make a short speech. I then stood up and introduced myself as the little train rocked and swayed on its way, making it one of the most unusual places I've given a speech.

En route to the town of Tainan by bus with the same student teachers, I was asked to describe my peripatetic life with the aid of the bus microphone. On reaching Tainan, Chen and I were invited to spend the night at the home of Mr Chow, who had been one of the friendly passengers. There, we enjoyed a delicious four-course meal, followed by huge slices of watermelon and pineapple.

After making arrangements to meet again, Chen and I separated; she went back to university while I returned to Taipei. There, my attempt to send a parcel home both irritated and amused me. The contents had to be seen and wrapped up under the close surveillance of a customs official. After that,

I went to a barbershop where it was my pleasure to have my second shave by a female barber and, although the razor was not particularly sharp, the experience was enjoyable. Gentle and precise.

Back in Hualien, I took Mr Wong to dinner, after which we went to see the Ami dancers again to take more photos. I left the next morning in the cab of a truck filled with water buffalo and once more travelled along the treacherous Pacific highway, a perilous journey at the best of times. Several times we made stops due to the uncertain conditions of the road and once or twice for meals. We were also forced to stop when one of the animals attempted to break loose.

From Taipei, I returned to Taichung by train in the company of a new-found friend, a Taiwanese army captain. Chen was waiting for me at the station, this time with her ID card, which she waved to attract me. Together we travelled by bus to Nantou Hsien high in the mountains. There we swam in hot springs and spoke to peasants working in the fields. We walked back to our little hotel at the end of the day wearing little more than our underclothes as it was so hot. Two wonderful days later, we returned to Taichung where we parted company at the bus station. It was a tearful occasion, and I felt disconsolate. I tried to comfort both Chen and myself with the words, 'The world is very small. And not to worry, for we'd meet again.' Back in Taipei at my hotel, I tried to eat but couldn't, too upset at parting with Chen.

I was just about falling asleep when there was a knock on the door. I opened it and there stood the desk clerk. 'Girlie downstairs want to see you,' he said, and I wondered who it was. It was drizzling when I got to the main door and there stood Chen, soaking wet and smiling. 'I had to come,' she said. 'I felt so bad when you left.' My immediate reaction was one of delight, then annoyance. It had been an exciting, fun-filled and romantic interlude (for both of us, I assumed) and I was grateful to have had such delightful company. But it had also been an emotional farewell with tears of sadness at the thought of never seeing each other again. Now we would have to go through the same painful process again.

Chen stayed with me in Taipei during my last couple of days in Taiwan. She had taught me much about the Chinese way of life, including some of their many customs and proverbs. I liked one by Confucius that appropriately suited the travelling fraternity, 'To travel 100,000 miles is the same as reading 100,000 books.'

I left her at the very busy Taipei bus station and, with tears streaming down our cheeks, we embraced for the last time. Just before parting, Chen presented me with a Chinese seal inscribed with my Chinese name 'to sign your letters with'. Feeling quite lonely and sad, I made my way back to Keelung to board the *SS Anking* again.

Chen and I corresponded often over the next few months, and her beautifully handwritten letters always showed the same concerns. 'Do you have a girlfriend?' (She professed to want me to have one, but I didn't.) 'Are you eating properly?' (I wasn't for a while.) 'Have you had your shoes repaired to keep out the rain?' (It took me a while to get around to it.) And with regard to my broken watch, 'Does it work now?' Finally, 'I'll pray for sunny days.'

For my 27th birthday, she sent me a lovely card which she herself had made. Inside were several hand-drawn pictures with her hopes for me! A luxurious hotel; a 36-22-36 girl; a money tree; thousands of your favourite dishes; a magic pill that gives you a happy long life; a jet to Formosa. It was obvious that Chen had been very fond of me and I was deeply touched.

A year or so later, she told me in a letter that she had married an American teacher, one of her colleagues at Tunghai University, and had moved to the United States. We exchanged a few more letters. Then I never heard from her again. Like so many other delightful people, she had come into my life, and now was only a memory. We had enjoyed a very close and romantic relationship during our all-too-brief time together, which almost certainly neither of us would ever forget.

I travelled back to Hong Kong on the *SS Anking* then boarded the *SS Laos* bound for Singapore. I was allowed to go aboard early when the chief steward recognised me. As before, there was the usual array of travellers. By coincidence, there were several I had seen or met in Japan, including Dorothy, our always cheerful next-door 'aparto' neighbour.

We docked in Bangkok for a day, but because it was so stinking hot, I didn't feel up to any more sightseeing, but I did manage to grab a quick last

A Romantic Interlude

massage! Eventually, we reached Singapore where I checked into the YMCA once again.

Now almost broke and anxious to escape the constant humidity of Singapore, I purchased a ticket on the *SS Malaysia* which carried cargo and passengers but was not leaving for several days. The heat and humidity seemed even more intense than before, and I sought constant relief by drinking fresh fruit drinks, sometimes as many as six or seven a day during my perambulations around the city.

There were several other travellers staying at the Y, most of them coming or going to Europe, Australia or New Zealand. Those who had travelled the well-worn route from London to Singapore were keen to reach 'the land of Oz'. Some were just about destitute and others quite well off. One German youth had only £10 and was existing on bread and jam, but at least he possessed an onward ticket, while a Frenchman had only £5 and no idea where he was going, nor how he was going to get there.

The first few days at sea were very calm, and I swam and played deck quoits with other passengers as we passed some of the many Indonesian islands. We also saw lots of flying fish leaping in and out of the water and a few porpoises. At Port Moresby, the capital of New Guinea, everyone swarmed ashore to explore the town and visit the marketplace.

In all, there were 60 passengers on board the *Malaysia*. One of the passengers was a middle-aged Jewish cartoonist named David Fuhn who once worked for Disney Studios. He drew wonderful Mickey Mouse and Donald Duck pictures and extremely good portraits of people — for a fee. My cabin mate on the ship was an Englishman about my age named Geoff Lowe. As fellow travellers, we had much in common and got on well. He'd been teaching English in Kuwait for two years and couldn't say enough good things about it. He simply loved his time in that hot and barren desert country made rich by its oil, and sang its praises to anyone who would listen.

Somehow word got around that I carried colour slides, and I was asked to give a lecture about where I'd been. The chief steward produced a slide projector, and that night found me talking to a crowded second-class lounge about my travels. They were an enthusiastic audience, and everyone seemed to enjoy my impromptu presentation. Afterwards, I was warmly applauded. It was my very first public travel lecture using my slides, and I realised that my story and pictures were of interest to others. During my talk, I spoke about some of the amazing places I'd been, the interesting people I'd met and the adventures I'd had. And based on my own hard-earned experience, I happily gave out tips and advice.

Little did I know after giving that well-received talk on board the *SS Malaysia* I would shortly be repeating the exercise at the youth hostels where I stayed. These were ideal venues because most of the hostellers were kindred spirits and we shared similar interests and values — primarily a love of the outdoors and curiosity about the world.

Based on the always positive reaction of the audiences, I knew that I had something unique — my personal experiences — to share. I'd already become a better photographer and, with more practise, I would hone my skills as a speaker. In all my presentations, I emphasized the importance of seizing opportunities and overcoming adversity. All subjects about which I now knew quite a lot.

These illustrated talks would be of major importance to me in the months and years ahead. I'd stumbled across a possible way of funding my future travels. In one fell swoop I'd become a motivational speaker, albeit something of an accidental one!

What seemed to impress my audiences the most was that Pete and I were doing the trip completely on our own. Many places and people we'd visited, often 'by shanks pony' (on foot), had also been visited by well-equipped expeditions (including film and TV camera crews) who were sponsored by large companies such as National Geographic and the BBC. They travelled in relative comfort with good food and accommodation included and all expenses paid, while Pete and I invariably travelled with no onward tickets or contacts and often alone. Pete and I knew that we had become rather unique in the travelling world being away for so long without returning home to save or rest up. We had travelled long and sometimes difficult routes alone or with only each other without resorting to begging, borrowing, or stealing.

Although Pete was happy to share ideas and advice, and apart from teaching in Japan, he wasn't interested in giving talks of any kind. No matter how challenging it might be, he'd rather find work and get busy in a regular job to save enough money to continue travelling. There was always a demand for his line of work, so he found jobs quickly. Perhaps the biggest difference in our thinking was that I was more interested in new experiences while Pete was more interested in making money.

A day or two after leaving Port Moresby, we were caught in the wake of a hurricane and the wind howled and raged, whipped up massive waves and threw spray against the porthole. It didn't take long before I lay moaning and groaning in my berth.

Bad weather prevailed, and the 12-day nightmare voyage turned into 14 days. And, as usual when at sea, I couldn't wait to reach dry land. The lousy weather continued until we reached Sydney Heads. Once there, I joined the others in somewhat shaky condition on deck, thankful and only too pleased to have come this far.

We glided under the great Sydney Harbour Bridge to the tune of the Dambusters March over the ship's loudspeaker and then docked in Australia's most beautiful harbour.

Pete came to the docks to meet me, having been notified that the ship would be late due to the hurricane. According to the shipping office, he said there was one brief time when it was thought we might have capsized due to the loss of radio contact. It was just as well that the passengers hadn't been told; otherwise, many people might have worried.

Fortunately, there were no specific requirements for people with British passports to enter Australia and nobody was concerned whether I was penniless or not, which, as usual, I more or less was. As the customs forms which we had to complete stipulated no firearms or other weapons, I had no alternative but to declare my swordstick. It was confiscated and impounded, but the customs officer assured me it would be returned upon my departure, whenever that would be.

By now, Pete and I had each had enough close calls and narrow shaves and didn't need any more for a while. Once again, we were together, but not for long. Within days, we would go our separate ways Down Under to pursue different employment in the largest country in Oceania.

CHAPTER 12:

A Pommie in the Outback

> *'We must go beyond textbooks,*
> *go out into the ... wilderness and travel and explore*
> *and tell the world the glories of our journey.'*
> — John Hope Franklin

For five of my first six days in the 'land of sunshine,' it rained incessantly. In many ways, Australia was much as I had imagined, though I was to be rudely awakened about some things!

My intentions at this stage were to work on a sheep or cattle station and visit a Royal Flying Doctor base. A couple of days of job hunting in the drizzling rain took me into the Sydney Stock Exchange building where, striking up a conversation with a stout, elderly man, I casually mentioned that I was a newcomer to Australia and seeking work. He suggested that I call on a friend of his, the general manager of Northern Estates Ltd., a large property and cattle company whose offices were in the same building. 'Mention my name,' was his parting remark as we went our separate ways.

I did so, and in the short space of half an hour, I was interviewed and offered a job as a station hand at Victoria River Downs (or VRD as it was commonly called), one of the company's cattle stations in the Northern Territory. This greatly appealed to me and I accepted immediately.

No company planes were leaving in the immediate future, so arrangements were made for me to travel by truck from Toowoomba in the Tablelands district of Queensland. A train journey to Brisbane, followed by a bus ride, would get me to Toowoomba in a day and a half and my expenses were covered for the journey.

Half an hour out of Sydney, the sun broke through the clouds, and I soon saw my first wild kangaroos, dozens of them, leaping and jumping along the railway track. The route was lined with eucalyptus trees, and their strange, sickly sweet smell permeated the air. I knew then I was in the Australia I'd heard and read about.

On arrival in Brisbane, I took a bus several miles to the Lone Pine Koala Sanctuary, where I saw dozens of the cuddly little creatures. I learned they could be ferocious little beasts when cornered. I also got my first look at another of Australia's unique fauna, a duck-billed platypus. Shy and elusive, this is surely one of the most peculiar-looking animals in the world. The next day, after staying overnight in a guest house in Brisbane and, with several hours to spare, I hitch-hiked to Surfers Paradise on the Gold Coast, little dreaming that one year hence I'd be working there.

At Toowoomba, I met Jerry, the truck driver with whom I was to travel to 'the terra-tree' (the Northern Territory or 'The Territory' as most people called it). Aged about 35 and an immigrant of Dutch origin, he'd settled in Australia with his family as a boy. Short, muscular and fair-haired, he had become a 'dinky-di' (fine), Australian in speech and mannerisms. A typical Australian 'truckie', he was opinionated and talked nonstop, obviously enjoying the sound of his own voice. It wasn't long before I became tired and hoarse trying to hold a conversation with him over the thunderous noise of the giant Mack engine!

Accompanying us was a young fellow named Dick who acted as Jerry's 'off-sider' and who kept us rather amused by his peculiarly positive attitude. Whenever he spoke, he made only statements. At 19, he knew it all.

Because of Jerry, it was a very slow journey to Katherine, one of the Territory's principal towns. It took nine days instead of the usual four or five. Throughout the trip, he frequently stopped merely to make a point while talking, and he took what he called 'pep pills' (probably amphetamines) at regular intervals to keep himself awake. He once went without sleep for two days and nights! *He was so high, he shouldn't have been driving* — were my thoughts at the time — but he kept going. He finally succumbed, drugged with tiredness. He was also an inveterate drinker, rarely passing a pub without stopping. One day, he spent 15 Australian dollars on grog, an extravagant amount in those days.

Driving that enormous vehicle was hard work at the best of times, and the hot, dusty conditions made it worse. Several times we stopped when Jerry recognised mates of his on the road. On top of everything, we experienced occasional punctures. The truck itself was greatly overloaded, piled high with steel beds and rolls of barbed wire. Each time we approached weigh stations, he skirted them by leaving the dirt road and driving through the bush.

During the day, all three of us sat in the cab, but at night as the truck rumbled on, Dick and I took it in turns to sleep on a mattress squeezed in the back. Once, I showed impatience to reach our destination, which annoyed Jerry immensely. 'What the hell did a silly Pom like me want to go to the Terra-tree for, anyway? He was in no effing hurry,' were some of his surly comments. I told him I wanted to start work and earn some much-needed money, which made him irritable and sullen. *It was a strange mentality no doubt due to the loneliness and isolation*, I told myself, but one I would experience repeatedly in the Australian outback.

At the time, all English people in Australia were called 'Poms' — the more respectful terms 'Brit' and 'expat' had not yet entered the everyday Aussie vocabulary. So, because of my British background, just about everyone referred to me as a Pom or Pommie which may have originated from 'Prisoner of Her Majesty,' a reflection of their own early ancestry, or in reference to the fruit pomegranate because of the red complexion most new settlers had due to exposure to the strong sun. As far as I was concerned, the way these expressions were used was usually quite the opposite of flattering. The Aussies conveniently forgot that they were related to the original Poms. I didn't! Not that it made any difference. I took it all in stride, like everything else.

Gradually as we headed north, the weather became hotter, and the heat of the engine and our sweaty, cramped conditions made things worse. Every time we stopped, we were immediately surrounded by Australia's worst enemies — flies. I was discovering that no matter where you go in that vast country, you can never escape them.

But it wasn't all bad. Some good things happened. I saw much of the vast country in our 2,000-mile drive, most of which resembled a mixture of

American prairie and African bush. Indeed, some of my best photographs were taken in the red sand of the Australian outback. I couldn't help but notice that although it was harsh, cruel and unforgiving, it was in its own strange, wild way hauntingly beautiful. But definitely not a place to be lost or stranded.

I couldn't help thinking, as I had in Africa and elsewhere, of the brave and hardy early pioneers who had overcome the many hazards, including poisonous reptiles and extreme weather, and settled this harsh, untamed land.

Often, all that lay ahead of us for dusty mile after mile was the red corrugated dirt road, clumps of spinifex and thousands of empty beer cans, thrown carelessly from car and truck windows. The outback towns were mostly identical, consisting of a post office cum general store, a garage, a small hospital and at least several pubs, all of which were fitted with swinging doors like in the Wild West. Wooden sidewalks, hitching posts and people in Stetsons, jeans and high-heeled boots completed the effect.

Once, we diverted from the main road to visit the spot where Banjo Patterson had composed his immortal 'Once a Jolly Swagman'. There, in the middle of nowhere, next to a small billabong, stood the famous coolibah (eucalyptus or gum) tree. The setting was tranquil. As we headed further north, the price of food leapt. 'Freightage,' we were told, but it was ridiculous paying $1.75 Australian for steak and eggs in cattle country when it was half the price everywhere else!

As time wore on, Jerry's attitude gradually became friendlier, and he even invited me to try my hand at driving the big Mack. With 16 forward and six reverse gears, it meant changing with both hands. After crunching the gears a few times, I acquired the knack and, seeing the somewhat surprised expression on Jerry's face, I realised that I was no longer considered 'such a useless Pommie bastard' after all!

After eight days and nights, we broke down. There was nothing we could do except wait for help, so we lay under the truck to escape the blazing heat. When a driver in a Ute — an open-backed Holden station wagon and Australia's most popular vehicle — finally pulled up, Jerry half sympathetically and half sarcastically suggested that, as I was 'in such a bleedin' hurry,' I should go in it. Then, with thumbs up and the reassuring words of 'You'll be right' (the nicest words he'd said to me in over a week), he waved me goodbye. That evening found me gratefully completing the last stage of that particular section of the journey.

Another day and a half was spent in Katherine, a small town noted for its abattoirs. During this time, I sent a telegram to the station, informing them of my arrival and then waited for a lift. That same night, I slept on a bench in the transport depot where I met a grizzled old prospector who searched for uranium with a water divining rod. He apparently had struck it lucky a few times, but only went to town every three months to restock with supplies.

It took a further two days to cover the 170 miles to the boundary of Victoria River Downs, my ultimate destination. This time, I travelled in an empty road train. Some 200 feet long, it consisted of three 'dogs' attached train fashion to the main vehicle, another Mack. Fortunately, Sam, the driver, was less moody and more cooperative than Jerry, which made the journey only slightly more pleasant than before. After opening and closing umpteen wire gates — to prevent cattle from roaming — and slowly and laboriously rattling over rocks and ploughing through deep bull dust, a very fine, choking red dust common in the outback, we finally reached the station boundary fence.

Leaning against it waiting for me stood Paddy, a middle-aged Irish Canadian, a long-time resident of Australia, and a boundary rider by profession whose lonely life was spent endlessly checking and repairing the exceptionally long fences. 'Just sling yer swag in the back, mate, an' we'll get going,' he said as we shook hands. Then we were off to Pigeon Hole, one of the three outstations on VRD property. Paddy talked for most of the two-hour ride, mainly about the virtues of outback life, as we ploughed through more thick bull dust, while I was content to merely sit back and listen.

It was late afternoon by the time we reached Pigeon Hole, a collection of metal huts and wooden paddocks. Here, I was introduced to Abe, the head stockman, who told me to join the camp at the foot of the adjoining hill. Camp consisted of one tent for the cook and several camp beds strewn around the area, together with the men's personal effects. It was deserted except for the cook, who was busy baking bread.

Falling into casual conversation with him, I soon formed an opinion which was to be repeatedly confirmed. Camp cooks in the outback enjoy a certain reputation. Drifters, like many station hands, they are usually articulate, moody, sensitive and invariably short-tempered. To a man, they have a fund of yarns to relate, which they do at the drop of a hat.

Just then, some stockmen appeared on horseback, three whites and four Aboriginals. Covered in red dust and wearing Stetsons and jeans, they reminded me of numerous cowboy films — until they opened their mouths and cursed in the broadest Australian twang. Half an hour later, I sat on the wooden railings of the yard, watching as they loaded up the recently arrived empty road train. They separated the wild bullocks amid loud shouts and swirling dust, forcing them up the ramp using battery prodders and stout sticks. It was nearly dark by the time the job was completed. After returning to camp, we helped ourselves to massive corned beef sandwiches, exchanged desultory conversation, and then fell asleep under a clear, star-filled sky.

The following day, after helping Paddy repair a windmill, we arrived at the main station in time to hear the clanging of metal against metal. 'It's tucker time mate,' said Paddy. 'Eat first an' we'll get yer settled in later.' Seated in the cookhouse a few minutes later, I chatted to a couple of hands, and over great mouthfuls of steak and gulps of tea, they explained some facts about VRD — my home for the next few months.

VRD lies roughly 160 miles south of Darwin, the 'Top End's' principal city. Reputedly, it was then the second-largest cattle station in the world and the largest in Australia, covering 5,200 square miles with a main station and three outstations named Pigeon Hole, Moolooloo and Mount Sandford.

The main station was essentially a small town with a post office, general store, garage, blacksmith shop, a small clinic cum hospital (staffed by a young, single registered nurse), and a recreation hall used for a weekly film show or, in days long since passed, dances and other social events. Age and lack of maintenance had made the latter little more than a broken-down shed. Other buildings were the cookhouse-cum-canteen, butcher's shop, and various accommodations consisting of single quarters of corrugated aluminium, and metal and timber bungalows for the married staff. In the centre stood 'the big house' with a couple of neglected tennis courts alongside. Behind the courts flowed the murky, silent Wickham River inhabited, I was told, by crocodiles or 'salties' as they are known.

About a quarter of a mile away was 'the compound,' a collection of tin shacks housing some 150 Aboriginals — men, women and children. At about the same distance in another direction was the overgrown station cemetery. There, on a couple of the fading, unkempt tombstones, the following epitaphs could vaguely be read: 'In memory of John Anderson, one of the Northern Territory's finest horsemen who died after a severe fall from a horse' and 'Jim Smith ... who died of blackwater fever.' Throughout the entire station, according to rough estimates (for nobody knew for sure), 30,000 wild cattle grazed.

Neighbouring stations, all at least 60 or 70 miles distant, possessed unusual and colourful names like Wave Hill, Monteginni, Hubert River, Pine Creek and Mataranka, to mention a few.

In that part of the country, there are two seasons — 'the dry' and 'the wet'. My stay coincided with the dry when hardly a cloud appeared and not a drop of rain fell for months, making the ground parched and brick hard, and the grass almost white in appearance. But at least work could be accomplished, whereas during the wet, the rainfall is so abundant that virtually everything comes to a standstill. Air drops become necessary, and the boundary riders are forced to exchange their vehicles for horses.

On my first day at work, I was asked to clean out the garage storeroom. Full of rusty nuts and bolts and various tools and spares, I swear the place hadn't been cleaned for 20 years. After two days, everything was clean and neat, except for me — I was filthy!

After that, for what seemed an eternity but was actually about three weeks, I worked on the saw bench, cutting up timber to be used as firewood. Together with two 'hands' and a couple of Aboriginals — referred to as 'blackfellas' — I worked eight hours a day, handing down timber from a gigantic pile and sawing it into small logs. After such a boring, repetitive task, we were all glad to see the last log disappear when we distributed the firewood to the various families.

It didn't take long to adapt to the station routine. Rising early at 6:30 a.m., I joined the others for a breakfast of steak and eggs. At 7:30, we assembled around the garage to receive our instructions and duties for the day. As the nights and early mornings were usually very cold like in most desertlike terrain, we stamped our feet and burnt rubbish in old petrol drums to keep warm as we hung around.

The days were broken with meal times and 'smoko' breaks, and at first I enjoyed steak for every meal but, as time passed, its attraction faded, especially as it also appeared in 'smoko' sandwiches every day!

Jeans and Stetson became my regular dress, not to emulate the others, but for practicality. Besides, there was a limited selection of things to buy at the station store. Jeans were tough and serviceable, and a Stetson provided excellent protection to the neck and shoulders in the very hot, penetrating sun.

It wasn't long before 'the Pommie Jackeroo' became my nickname. Upon seeing me, some of the hands would burst into the then popular song of the same name. But 'Pommie bastard' was more frequent — a term of endearment I was frequently told! Other daily greetings were, 'Tally-ho old chap,' or 'G'day mate. 'Ow yer goin'?'

Not surprisingly, in that environment, I unconsciously adopted a similar way of speech. Words like 'crook' (sick), 'Sheila' (girl), and 'beaut' (fine) became part of my vocabulary. Other common expressions were 'strewth', 'fair dinkum', 'my oath', and 'no way in the world.' With the strong nasal accent of the locals, it was like a whole new language to me, and it was sometimes necessary to ask someone to repeat what they had said!

It became very evident that most Australians living in the bush or outback miles from anywhere — some say they suffer from xenophobia — resent and dislike strangers, especially new immigrants, or 'wogs' as they were called. Initially, I had a hard time trying to prove that I wasn't such a silly Pom fresh from 'the old country'! Fortunately, with a friendly attitude and after proving my worth, I was as near as possible accepted.

To illustrate this point, one day, the assistant manager Ian asked me to take the three-tonner to Mount Sandford. Blackjack, an Aboriginal, was to be my off-sider. Graham, a garage hand — well known as the chief 'bludger' (idler) — overheard and promptly exclaimed, 'You, a Pom, can't drive that!' Not only did we reach our destination safely, but we got back more quickly than everyone anticipated. From then on, my duties became more varied and ranged from driving the various trucks and tractors to meeting the mail plane from Darwin.

Other days were spent loading and unloading trucks, whitewashing the houses, and depositing the refuse. Once, an absorbing afternoon was spent with Ian operating on a heifer. Some were menial, dirty jobs and others very interesting but, regardless, it was always a relief to get under the shower at the end of the day, clean off the dust and sweat, and dress in clean clothes. The usual workday was over by 4 o'clock.

By 4:30, I was often saddling up one of the spare mares to explore the surrounding area. It was on those occasions that I saw kangaroos, wallabies, brumbies (wild horses) and, once in a while, a dingo and even the odd emu or wild camel. Most of the wildlife which gave me such pleasure were considered vermin and were shot on sight by the stockmen, a sad but essential necessity because the cattle needed the little grass available at that crucial time of the year. Often, when exploring the local area, I stopped to admire the barren, arid wasteland with its awe-inspiring scenery and colourful, dramatic sunsets. Like much of the Australian wilderness, it was a desolate and hostile environment, extraordinarily beautiful in its own way.

Much to my delight, the minute I had stepped onto Australian soil, I saw a variety of new and exotic birds. As in Africa, the avian life was phenomenal and abundant. On my long road trip to VRD, I'd already seen many colourful birds, and there were even more around the station, at the river, and in the surrounding landscape — laughing kookaburras, black and white cockatoos, lyrebirds, kingfishers, bee-eaters, lorikeets, finches, herons, egrets, storks, and pigeons, to mention only a few. But I never saw anything to match the great flocks of green and blue budgerigars. Sometimes as many as 10,000 at a time darkened the sky.

The account of my outback days would not be complete if I omitted some of the characters and personalities I lived and worked with, most of whom were much as one reads about. To me, they were far from fictional! James McCauley, one of Australia's eminent 20th century poets, described his fellow countrymen thus, 'The men are brave, contentious, and ignorant.'

The manager of the property was a bachelor called George Lewis. He and his assistant, Ian Michael, shared the big house with Ian's wife and small daughter. Both were considered gentlemen as far as the hands were concerned. Under their combined supervision were approximately 60 hands whose jobs varied, often from one day to another, and half of whom spent most of their time mustering and branding cattle and sleeping under the stars. The blackfellas were employed as stockmen, while a few did odd jobs around the main station.

George Lewis, then 66, was a small, wiry, dapper man who considered himself undressed if he didn't wear a hat and who abstained from just about everything except work. Having lived in the outback all his life, he possessed a reputation throughout the Territory for his wide knowledge of the cattle industry. He was a great advocate of demonstrating a point by example and, like many celebrated people, he was quiet, reserved and unassuming. He looked much younger than his years and was actually fitter and stronger than most men half his age, as I witnessed one day.

I'd been at the station some weeks when George politely invited me to join him on a boundary ride. A couple of hours later, his strength and stamina became obvious as he casually felled trees and chopped stakes to repair a fence. While I sweated from the heat and exertion, the work didn't seem to be affecting him in any way. He barely uttered a word until it was time for a 'bully beef' lunch, then he reminisced freely over the 'billy' (campfire cooking pot). By the end of that particular day, I was exhausted and hoped against hope not to be asked to join him again!

Slippery was George's best mate. Like George, he was a *bona fide* bushman who took great pride in the fact that he had been born in the outback. Originally from northern Queensland, he had been a sheep and cattle drover all his life. A slender, gnarled fellow aged about 68, who smoked like a chimney, he had a lined, wrinkled face and a deep, gravelly voice. And an awful cough.

Knowing more true tales of the outback than any other man I met, Slippery also recited bush poetry like Olivier quoted Shakespeare. Some evenings, he sat in my room, entertaining me with talk about his beloved younger days. From him, I learned the difference between cows, steers, and bulls; and the techniques of trapping, drafting, and branding. Like several of his mates, he was also very bow-legged, the result of spending years in the saddle.

Geoff, the station accountant, was third in seniority after George and Ian. Sandy-haired and in his mid-thirties, he was married to Gillian, a pretty, quiet woman of about the same age. They had their own house and a blonde, blue-eyed toddler whose name I forget, but whom everyone adored. Geoff and I got on well, and a few times I was invited to dinner at their place.

Art, the cook, was a middle-aged, fattish, grumpy man who boasted a regimental voice which was occasionally given to loud, infectious laughter. In the cookhouse, he was the undisputed boss. When he gave orders, everyone did as he said. During my second day at the station, as he stood at the head of the table, I made the mistake of approaching him for a second helping. Seeing my plate empty as I walked up to him, he bellowed, 'What d'ya want?'

Feeling rather like Oliver Twist, I meekly answered, 'I'd like some more, please.' With that, he almost tore the roof down! 'Where the fuck d'ya think you are? On yer daddy's yacht?' he shouted, 'You've got to think of yer mates, ya know.'

My lesson learned, I never returned for seconds, but periodically he would rebuke me for taking one too many potatoes or an extra slice of pineapple or piece of apple pie.

Heaven help the man who arrived at the door before the gong sounded — Art shouted and cursed at them, 'You got nothing better to do than 'ang around 'ere.' Most meals, therefore, were usually eaten in silence, which he would break with moans and groans about the things he had done for most of his life, 'drink and work, and too much of the latter'. We all agreed he'd missed 'talk' in his diatribe. Still, he was an excellent cook, providing different variations of menus and taking great pains over his desserts. Often, there were as many as a dozen choices, from jellies and jam tarts to fresh fruit and various puddings. He wasn't just good at his job, he was superb.

Cynical Elmore represented a character in one of Dickens's books. Thickset with enormous bushy eyebrows and a chin that was seldom shaved, he walked with a limp — the result of a heavy fall from a horse. He had been a fine stockman for many years until the accident crippled him for good. Now a driver, it was he who, when I asked a question out of curiosity and interest, casually rebuffed me with the words, 'What are yer, a flamin' policeman?' This and similar comments made me realise that you don't ask questions in the bush, you only make statements. And my enquiring nature was subdued temporarily.

Chris, the barrel-chested 'chippie' (carpenter), kept us in stitches with his quips, most of which were about his ex-wife. 'My wife had more faces than a four-sided clock' and 'She ran off with me best mate. Cor, I miss 'im,' were some of his amusing references.

Nick, the Italian blacksmith, was one of three married men. His accent was often jokingly and sometimes rudely imitated — even more so than mine. He had been some 12 years at VRD. He and his Italian wife maintained a beautiful garden filled with vegetables and fruit, which he sold to the blackfellas at exorbitant prices.

Doug, a middle-aged, mixed-race boundary rider, was very quiet and reserved but, like everyone else, enjoyed his grog when he could get it. Born at the station 'when the blackfellas were still wild', he reputedly knew every creek, hill, and gully in the district.

Graham, nicknamed 'Cocaine' (slow dope), was the greatest talker and bludger of all who spent most of his working day 'shooting a line' or driving to and from the airstrip in an attempt to keep out of Ian's or George's way!

Hennie, an immigrant from Denmark and chief groom, was a quiet man in his early thirties who looked after the horses and kept very much to himself.

Wayne, at 16, was the youngest hand. He came from the big city of Sydney and was sent north by his parents to toughen him up. He had no specific duties but helped out wherever he was needed. He spent his off-hours playing with stick insects and shooting birds, mostly pigeons, for the pot.

Then there was 'Sister', the station-nurse. In her mid-twenties and red-haired, she was relatively attractive but somewhat masculine in her ways. Being the only single woman — we seldom saw the few married ones — she always had a long queue of suitors and patients who complained of ailments as often as they could. Indeed, as time wore on, she became better-looking by the day.

The only other single girl I saw during my stay at VRD was an attractive air hostess when I went to meet the Darwin-bound plane. I blushed self-consciously and felt both embarrassed and pleased at seeing such a pretty face after what seemed like several months but was in fact only about eight weeks.

Most of the other blokes were quite young but still characters in their own right. With the exception of George and Ian, who tried to set an example, all swore, smoked, and drank as only 'fair dinkum' Aussies can. Only some could read and write, and their conversation was invariably limited, consisting of the crudest language which punctuated every sentence. All used identical expressions which, although highly colourful, grew dreadfully boring.

Most were moody, laughing one minute and complaining the next, but from them, I learned about their way of life in the bush, including rolling my own 'smoke-ups' and relaxing on my haunches in the cattleman's crouch.

Of the Aboriginals, most of whom were considered 'urbanised', King Brumby was the head. An elderly fellow, his duties consisted of settling occasional labour disputes and chopping firewood. Grey-haired and thin with bloodshot eyes, he was polite and friendly to everyone and spoke amusing Pidgin English. Unlike many Aboriginals, he didn't care much for 'walkabout', preferring to remain at the station and let his friends come to him. He also took great pride in wearing his badge of office around his neck, a piece of old metal with his name inscribed which dangled on a chain. At his invitation, I attended a 'corroberee' (an Aboriginal party) once during my stay. It began in the early evening and continued throughout the night. Aboriginal singing and dancing took place to the accompaniment of the strange music of buffalo hide drums and didgeridoos. It was an interesting experience, but a couple of hours were enough for me.

None of the station employees received very high wages, but remuneration was adequate, especially as all food and accommodation were provided and there was little or nothing to spend money on. Regardless, most men managed to blow their wages all at once on booze when the occasional opportunities arose!

At the end of the day after dinner, most evenings were spent in conversation, limited though it was, or playing cards, shove halfpenny or illegal 2-up sessions.

There were many mosquitoes around so, come nightfall, I was forced to flee under my net as they buzzed around. In bed, I curled up with a book, reading anything I could lay my hands on, including the bestseller *They're a Weird Mob* by John O'Grady; *Beyond the Black Stump*, stories of Ned Kelly, the infamous outback bandit; and *We of the Never, Never*. The latter was an account about life at Elsey station near Mataranka, also in the Northern Territory, written at the turn of the century by the beloved Australian author, Jeannie Gunn. My sleep was sometimes interrupted by the howl of dingoes or loud snoring which penetrated the flimsy walls of the building.

Occasionally, a driver or lone stockman would arrive at the station, stay the night, and then move on. Being so isolated, there were few visitors to that remote corner of the world. Our only outside contact was limited to the scheduled plane and the two-way radio, which was routinely tuned in twice a day and in cases of emergency.

One Saturday, Art and two mates drove to Top Springs, the nearest township, a distance of 68 miles. The town boasted one garage and a pub. When they returned to the station quite drunk, they proceeded to distribute grog to the blackfellas. All night long, the noise from the compound was deafening as one and all became intoxicated. The next day, Art was noticeably absent from meals, and we had to fend for ourselves. The commotion did not go unnoticed, and George summoned him for a talk. Art had broken a cardinal station rule — no drink to the blackfellas — and he was fired. He stayed on for a while until a replacement could be found.

Alcohol was sold in limited quantities at the station store once a week, but it was not always sufficient to quench some monumental thirsts. I never saw anything to match the amount of liquor imbibed on a regular basis at VRD.

It was early one evening when five desperate-looking individuals rode in on horseback. Dirty, dusty and unshaven, it seemed that at any moment they would reach for guns or resort to iron fists. These were some of the 'centre camp boys', paying one of their rare social calls — most of their time being spent mustering cattle. An hour later, freshly showered and shaved, they took on their real appearance of 18- to 23-year-olds who delighted in dressing up in brightly coloured shirts and new Stetsons, who constantly worried about their images and whose sole conversation consisted of horses and cattle! To celebrate their unbroken five or six weeks in the bush, they drove to Top Springs to collect their long-overdue grog. They returned merry and drunk about 4 a.m. Over the next two days and nights, they casually downed 26 x 24 cans of beer and numerous bottles of gin and rum. Even when asleep, their hands clutching a bottle, they would take further swigs of plonk.

While climbing over a wall of his room, where the centre camp boys had locked him in for fun, Stu, the butcher and a chronic asthma case with a wheezy cough, cracked a rib and was ordered to hospital. Before he left, he asked me to take over his duties because, as he put it, he didn't trust 'those other cranky bastards'. All of a sudden, I was faced with the job of station butcher, and I'd never butchered anything in my life!

In taking possession of his battered old Land Rover, I also took on his

Aboriginal assistants, named Sunshine and Killer Jack, and sometimes Big Mike, all of whom were familiar with butchering. Thereafter, each morning for the next fortnight found me distributing milk to the families and delivering meat from the cookhouse to the compound. My new duties also included feeding the ducks and chickens and collecting the eggs. At that time, a bantam had turned broody, so I shut her in a separate cage on top of six eggs and looked forward to being a proud father in 21 days!

Kills were made every three or four days. Together with Killer and Sunshine, I'd set off by Land Rover or horseback to muster a few 'killers' (cattle destined for butchering).

Apart from being excellent horsemen, Killer Jack and Sunshine, like many Aboriginals, were both brilliant trackers. After a wave from Sunshine — signalling stop — Killer Jack would then dismount, stare intently at the ground, then point. Moving on in the direction indicated, the wild cattle would suddenly loom into sight. A rough, hard chase over rocks and hillocks ensued in an attempt to drive two or three into a pen where they would be quickly dispatched with a .22 bullet through the brain. My first shots went wide and only injured the poor animals, but Sunshine put them out of their misery with one well-placed bullet, at the same time teaching me exactly where to aim.

After cutting their throats with a vicious-looking carving knife, Killer Jack and Sunshine began the process of skinning and carving using razor-sharp knives and a huge axe. The meat was then carted to the back of the cookhouse, where it was hosed down and then hung in the freezer for a few days. Nothing was wasted, including the entrails, which were fed to the always-hungry mongrel dogs.

The person mainly responsible for bush kills was Gilbert, the 24-year-old centre camp head stockman. Bush kills occurred when killers were too far from the station to be quickly mustered. A superb horseman, Gilbert's greatest excitement was riding among the long-horned cattle and attempting to get close enough to the stampeding animals to bring one down with a .38 revolver. The butchered animal fed all the staff and the Aboriginal families living in the compound.

Gilbert also invited me to participate in the occasional horse-breaking event. Like the cattle, they were also quite wild, having been let loose to graze freely for several months.

As several of us sat on the railings surrounding the pen, our own private rodeo began. To whoops, shouts and facetious comments from the sidelines, the horses bucked and reared in a crazy attempt to unseat their riders. Time and again, each man was thrown, except Gilbert. After wiping off the dust and spitting and muttering profanities, they remounted. This continued until each horse was subdued and could be led quietly away. When my turn came, I was trembling from apprehension, but I was determined to try my luck and prove myself. Incredibly, after clinging desperately to my mount for all I was worth for about a minute — which seemed like ten — I wasn't thrown. But I was stiff and sore for days!

Six months passed surprisingly quickly, and, although I could have stayed at the station indefinitely, I knew it was time to move on. The day I left, I helped bury an Aboriginal toddler who had died of pneumonia. A regular coffin was too big, so Chris knocked up a pint-sized wooden box. It was moving to see some of those rough, tough individuals remove their hats and bow their heads in the station cemetery at the short service around the tiny grave.

As I left, I shook hands with my mates and dumped my belongings in Graham's Toyota. We pulled away to shouts of 'Good on yer' and 'See yer, matey.' Within minutes, we were skidding our way through bull dust in the direction of Darwin. Once again, I was leaving a place that, despite a multitude of things I didn't always like, I'd learned to love.

Meanwhile after some months, Pete had left Sydney and hitch-hiked the long, arduous route to Cape York Peninsula in the far north, where he got a job at one of the open bauxite mines as a mechanic on heavy machinery — bulldozers and earth movers. In his latest letter, he commented that the money was good, but he didn't care much for the 'hot as hell' oven-like temperatures.

In Darwin, the Northern Territory capital and its largest town, my first stop was the dental clinic to have a troublesome tooth extracted. The tooth had quite a history — drilled and filled in Ethiopia, filled again in Ceylon, and yet again in Japan! I wasn't too impressed with Darwin, so after just two days decided to move on.

On board a small mail plane, it first touched down at a couple of isolated stations, then deposited me in Wyndham, my next stop. Situated in north-west Australia in the Kimberleys, another wild and remote region, its two-page tourist brochure stated, 'the highest temperature, the biggest thirst, and the roughest golf course in Australia'. A policeman I chatted with suggested that I try the wharf for 'a well-paid job'; so that night I slept under the jetty to be one of the first to apply for work. The tide woke me the next morning when seawater lapped over my feet. I got a job manhandling bags of cotton-seed into nets which were transferred by crane into waiting freighters.

The first few days were awfully difficult as the bags — at 70 to 80 pounds — were more than half my weight. Being slightly built, I wasn't cut out for such strenuous manual work. It was physically challenging, and the 112°F heat was unbearable. Several times, I thought I would faint. The hours seemed exceptionally long, from 7:30 a.m. to about 9 p.m. with a couple of short breaks for tea and an hour for lunch. Surprisingly, I stuck it out for nearly a month and was paid handsomely. In fact, it was the most I'd ever earned in my life. Nevertheless, at the time, I hated it, and each evening I contemplated quitting. But the next morning, I was back at the sheds, refreshed and ready to make more hard-earned cash.

The 'wharfies' were from all parts of Australia, some being New Australians, including Poles, Czechs, Italians, Dutch, Scottish and Irish. Most were big, tough and fairly friendly and like all 'true blue' Aussies believed that Australia was 'God's own country'. As in any large group of men — the wharf employed about 200 — there were a few difficult types, one of whom it was my misfortune to encounter. When I innocently went to pick up a spare net my first day, a huge, brawny individual shouted, 'Leave that fucking thing alone,' followed by, 'Git yer own bloody thing yerself.' My apology failed to placate him, and he continued to shout abuse and profanities at me through-out my stay. I learned to ignore it for tempers frayed easily in that brutal heat, and curses and foul language were to be expected as part of the day's work!

Because of the arduous nature of the job, nobody spoke while exerting themselves, saving all their energy for lifting and handling the bags. One or two men were hospitalised with sunstroke, and all of us constantly perspired until we were soaking wet. Every time we had 'a spell' (break), we raced to the nearby fire hydrant to douse ourselves with water for relief. On advice from a mate, I took salt and vitamin C tablets to avoid dehydration. When I was assigned to lifting bags of urea weighing 110 lbs each, they were just too heavy. I ricked my back and had to take two days off.

Throughout this period, I was the guest of Everitt Bardwell, director of the local Royal Flying Doctor base in Wyndham. We met through a mutual wharfie friend and, recognising my interest in the unique service, Everitt invited me to stay at the base.

Everitt was one of those people everyone likes. Middle-aged, genial and warm-hearted, he had established a reputation in his field, having been in the Royal Flying Doctor service for more than 30 years. During his career, he had played an important part in the organisation and opened up two new bases. A couple of times, I joined him hunched over the microphone and, in the course of several interesting conversations, he told me about the service.

Inaugurated in 1925 by Reverend J. Flynn of the Presbyterian Church, known as 'Flynn of the Inland', the service was established to assist the isolated people of the outback in times of sickness or distress. In the beginning, ancient pedal transmitting sets and early aircraft were used. Appropriately known as 'the mantle of safety', the service then controlled more than a dozen radio bases scattered throughout the country, which included the essential 'School of the Air'. Every remote station or homestead had an airstrip and radio transmitter.

At Tennant Creek, a tiny, one-street town in the Northern Territory, which I had passed earlier en route to Katherine and VRD, there is a memorial commemorating the Very Reverend John Flynn, OBE who died on May 5, 1951. On the memorial is a bronze plaque which reads:

> *His vision encompassed the continent; he established the Australian Inland Mission and founded the Flying Doctor Service. He brought to lonely places a spiritual ministry and spread a mantle of safety over them by medicine, aviation and radio.*

It was now early November and still pretty hot when another mail plane ride took me to Broome, a delightful, tropical Penzance on the Arafura Sea where I spent two pleasant days beachcombing, wandering around

Chinatown and watching the pearl fishing boats coming and going. After deciding to hitch-hike, since the bus service only operated once or twice per week, I reported to the police to inform them of my intentions as my route ahead to the southwest consisted of 200 miles of desolate bush road.

It wasn't long before I was picked up by a friendly truckie who, like most of his counterparts, talked constantly for mile after mile. When he sighted a large turkey in the middle of the dirt road, he drove straight at it, killing it instantly and smashing his radiator at the same time! *'Serves him right,'* I muttered to myself. After losing a few hours patching up the radiator as best we could, we continued on our way. As for the poor turkey, it was shredded to pieces!

In Port Hedland, a rapidly expanding port town, I enjoyed an ice-cold beer while sitting in a deck chair at an outdoor cinema. At a construction site I was offered work. Many times the offer of employment came my way before I even opened my mouth, especially in pubs, which often acted as a labour exchange. 'Looking fer work, mate?' or 'Want a job?' strangers would say at a bar, in a shop and even hitchhiking. I also encountered numerous fellow travellers on the road. It often seemed that everyone in Australia was coming or going. Some were drifters unable to stay in one place, but most were young people lured by good money, which jobs in the bush invariably provided.

En route to Perth, I teamed up with a tall, young, bearded guy named Phil from Queensland. Unlike many Australians, he was taciturn and didn't have much to say, but we got on well for the couple of days we were together.

As we sat by the road, a huge truck with a Holden car on top, halted and the driver called out, 'Can you drive a semi?'

Without hesitation, I replied in the affirmative.

'Well she's got six forward and she'll cruise at 40. Oh, and she's got no brakes.'

I gulped and asked how the road was.

'OK. She's straight as a die,' the truckie replied casually. He then climbed up into the Holden to sleep. Taking it in turns at the wheel, Phil and I covered another 300 miles. Fortunately, only once did a slight problem arise. As we neared a narrow bridge, a car approached from the opposite direction. With no brakes, we couldn't stop, so I waved frantically from the cab window at the oncoming driver who conveniently stopped just short of the bridge.

Several times, we skirted great bushfires, common throughout Australia due to the extreme heat and bone-dry terrain. We also made a point of stopping, by easing off the accelerator and rolling to a halt, at the famous '80-mile beach', which was completely deserted in both directions.

On another ride in a Holden Ute, the driver stopped at a remote store for a cool drink. While we were inside the dust-covered wooden building, a six-foot snake came from nowhere and slithered onto the hood of our vehicle. 'Crikey!' exclaimed our driver as his eyes focussed on the interloper. After getting over our initial alarm, I had Phil take a picture of me posing close to it. Our driver then informed us it was a black-headed python and non-venomous. Thank God!

Phil and I went our separate ways when we reached the town of Geraldtown. I continued to Perth, Western Australia's lovely ocean capital, where I stayed at the YMCA for two days. Elsewhere, when on the road in Australia, I stayed in small, cheap guest houses or private homes that hung boards outside advertising 'Rooms to Rent'.

A mining engineer gave me a ride to Kalgoorlie, the old gold-mining town. He was proud to point out that apart from no fewer than 26 pubs — there were 60 in the gold rush days — the town still had the only legal brothel in the country.

For the next two days, I sat on the road just outside the small town of Norseman, sleeping in the bush at night, while waiting for a lift across the formidable Nullarbor Plain — a vast, flat, inhospitable, treeless tract of land that takes two days to cross by truck, car or train. One young fellow I spoke to had been waiting four days for a ride. Regarding it as an exercise in futility, I gave up when my patience waned, deciding in the long run it would be quicker by train.

The Transcontinental Express proved to be very comfortable and delivered me in a night and two days to Adelaide, the beautiful garden capital of South Australia. I liked Adelaide immediately; it impressed me as much as Sydney.

My mother had urged me in a recent letter to look up her brother Frank, if and when I made it to Adelaide. One of five brothers and four sisters in the

Williams family, he'd left home for Australia in 1937 at the age of 16 to seek his fortune Down Under, when my mother was expecting my sister and two years before I was born. Soon after the Second World War broke out, he'd enlisted in the Australian Army, served in the Far East, was captured by the Japanese, and spent three years in a POW camp.

Taking a long-distance bus from Adelaide to the small, quiet, coastal town of Millicent, I went to visit him. In his modest home, I met his wife, Faye, and their four children, ranging in age from ten to 16. Until then, I was the first in the family to see this 'missing' relative who, from most accounts, hated letter writing.

Now, thanks to Faye, who had initiated correspondence with my mother, we were having a something of an emotional meeting as I brought them both up to date with who's who in the family.

When we met, 21 years after the end of the war, Frank was a well-built, slightly balding, quietly spoken man who worked as a foreman on a nearby farm. I got the impression during our conversations that he might have gone a lot further in life if the war had not cruelly intervened. That said, we never spoke about his wartime experience. Like many Allied servicemen, he'd almost certainly suffered terrible indignities at the hands of his captors. He never mentioned it and I didn't ask. He seemed content enough with his lot.

I spent about a week with my Australian relatives during which time Robert, the eldest son, took me gliding, not once but three times. Having a bird's-eye view of the farms and fields of South Australia was absolutely exhilarating, and I enjoyed every minute of it. There I was in the tiny cockpit, my 16-year-old cousin piloting from behind, with no sound but the whistling of the wind and nothing between us and the sky but a plexiglass canopy. It was magical. After the first flight, I was hooked and made a note to take gliding lessons in the future, but, like so many good intentions, I never got around to it.

All too soon, it was time to be on my way, and after saying goodbye to my Ozzie family, I boarded a bus for Melbourne which, as I would discover, vies with Sydney as the country's most beautiful and liveable city. While staying at the YMCA, I spent a couple of interesting days exploring the town and visiting the beautiful Botanical Gardens and Captain Cook's house, originally built in England in 1755 and transplanted brick-by-brick to Australia in 1934.

A leisurely hitchhiking trip followed with several pleasant and accommodating Aussies who gave me rides through the rolling green hills of Victoria and New South Wales. Before I knew it, I was back in Sydney roughly nine months after leaving the city, having travelled right around Australia.

I had stayed in touch with Robin Preece, a good friend whom I'd met in training as a sales rep for the Nestlé Company. He'd immigrated to Australia and was living in Sydney, so I looked him up. I found his little flat in the trendy suburb of Paddington after being directed by locals who quoted pubs as landmarks. We celebrated our reunion with dinner in Kings Cross, Sydney's SoHo. Robin escorted me around the city for the next few days, during which we crossed the beautiful harbour via the Manley ferry and ogled the many gorgeous, bikini-clad girls at Bondi Beach.

When I made enquiries to book a flight to New Zealand — the next stop on my itinerary — the airport staff inconsiderately went on strike closing down all the airports. So, once again, I was faced with travelling by ship. I'd written to Pete, informing him of my whereabouts. His prompt reply indicated he was hot on my heels. It was November 1966. Since leaving the UK, we'd been travelling on and off for slightly more than four years.

Victoria River Downs, Northern Territory, Australia

Lone stockman driving horses into pen

Loading cattle road train

Gilbert breaking in a horse

Riding with Tom (stockman) and Hennie (groom)

King Brumby chopping wood

Aboriginal women and children

New Zealand

Partaking of Christmas Day lunch

Fishing boats and Mitre Peak, Milford Sound

Errol, skipper of the Da Vinci, *Milford Sound*

On the road, South Island

Pete and fellow hikers in Southern Alps

Taking a dip in Southern Alps

Pete (in car) with friends near Motueka

Apple picking, Motueka, New Zealand

Carol and Joy

Joy

Marie loading apples

CHAPTER 13:

Sheep, Mountains and Apples

'To move, to breathe, to fly, to float,
To gain all while you give.
To roam the roads of lands remote,
To travel is to live.'
— Hans Christian Anderson

The Japanese-owned and crewed 30,000-ton liner *SS Oriental Queen* was packed, mostly with young Australians bound for a working holiday. It was almost summer 'Down Under'. New Zealand was preparing for its annual invasion of visitors. Most of the passengers, like me, were bent on acquiring casual summer jobs and travelling around the country. Although I'd heard both good and bad reports about every other country we'd visited, of New Zealand I'd heard only good things from other travellers, and neither Pete nor I would be disappointed. Fortunately, as a member of the British Commonwealth, entering the country back then was relatively easy. No visas were required. As long as you held a British passport and had no criminal record, you could get in and obtain work without difficulty.

The crossing of the Tasman Sea from Sydney to Auckland took four rather rough days during which, as usual, I was as sick as a dog and had only three meals. As I lay like a corpse in my berth, I told myself it would all be worthwhile once I reached 'Aotearoa', the land of the long white clouds — the original Maori name for New Zealand. Boasting landscapes similar to the Alps of Switzerland, the fiords of Norway, the beaches of Australia, and the jungles of Malaysia, New Zealand is undoubtedly one of the loveliest countries on earth.

On arrival in Auckland I made various enquiries about jobs, only to be told by everyone, 'Nothing doing 'til after Christmas.' It was then I learned that absolutely everything closes down for a two- to three-week period during the end-of-year festivities, so I made the decision to see as much as possible while waiting for businesses to re-open.

I'd been in New Zealand just two days and was travelling by bus from Auckland to Rotorua, the town famous for its geysers, mud holes and hot springs, when we came across an overturned car by the side of the road. Upside down inside the vehicle and calling for help were a middle-aged Maori couple. Both were rather drunk but not badly injured. After I helped the bus driver extricate them from the damaged car, which was more or less a total wreck, the couple giggled and laughed, dusted themselves down and acknowledged our assistance with a 'Thanks, mate.' We left them happily sitting against their car, taking turns to swig from a bottle of wine.

In the pouring rain, I made my way to the youth hostel in Rotorua, only to be turned away. It was full, so I stayed in a nearby guest house down the road. New Zealand has many fine youth hostels in picturesque places, and I would take advantage of their excellent amenities. With good roads and courteous people, the country was also ideal for hitchhiking.

The majority of New Zealanders I met loved their country; however, they preferred to keep a good thing quietly to themselves.

I had no problem hitchhiking to Wellington, the capital, a hilly city situated around one of the world's loveliest harbours. There, I made further enquiries about work and was again told, 'In the New Year.'

It was warm and sunny when I took the car ferry from Wellington to Picton on the South Island. Fortunately for me, the three-hour voyage across the normally turbulent Cook Strait was nice and calm. On board the 9000-ton *Wahine* — which 18 months later would capsize in a violent winter gale just outside Wellington harbour with the loss of 51 lives — I stayed on deck, leaning over the ship's rails, enjoying the ocean breeze, and taking in lungfuls of fresh, salty air. And as always, wondering what lay ahead.

At the Labour Department in Nelson, I made more enquiries about work and was given the name and address of a sheep and tobacco farmer in

Motueka, a small, rather sleepy town known for its fruit, hops and tobacco farms. It was also the gateway to the Abel Tasman National Park. The fellow behind the desk suggested that I contact the owner directly, which was exactly what I did.

On the way there, I got a ride with a charming and friendly woman. Dorothy Gaskell lived on the outskirts of Motueka and, curious to learn more about me and my adventures, she insisted on inviting me to her home where I met her husband, Derek, over tea and strawberries and great dollops of cream. As I relaxed in their comfortable home, little did I know then that this pleasant meeting would lead to a lengthy pen pal relationship with Dorothy and her family. They would follow my travels eagerly for many years.

After enjoying the Gaskells' typical Kiwi hospitality, Dorothy drove me to the Moss farm. Mr Moss, the 'cockie' (farmer) was not at home, so I introduced myself to his buxom, fiftyish wife, and told her I was looking for work. She told me rather curtly to occupy one of the huts 50 yards from the farm house and to report for work at 8.30 in the morning.

Most New Zealand farms have quarters for their workers, which vary in comfort and size and are referred to, like their weekend cottages and chalets, as a 'bach', meaning bachelor quarters. Most consist of one room furnished with a table and chairs, one or two beds, cooking utensils, and an oven. This one was filthy and neglected, full of dust and cobwebs, with a broken windowpane, an old bed with sagging springs, and a soiled mattress. I wasn't deterred, however, for it seemed that I now had a job, so I set about cleaning up the place.

Bright and early the next morning, I met farmer Moss, a fat, red-faced, middle-aged man, who very quickly proved himself an exception to the 'friendly New Zealander'. 'Where yer from?' he asked brusquely. When I told him, 'the UK,' he replied cynically, 'That's no recommendation.' He then put me to work lifting heavy bales of hay onto a truck in a field and stacking them into the barn. Working with me was a young Maori named Rusty, who reassuringly told me, 'Take no notice of 'im,' when I mentioned my somewhat cool reception.

Like most Maoris I met, Rusty was a strong, happy-go-lucky fellow who joked and sang as he worked. He occupied the bach next to mine, and every evening I fell asleep to the sounds of his guitar and lusty singing. At the end of my first day as a 'roustabout' (jack of all trades), I was dead tired. After cooking myself a meal, I fell into my lopsided bed, oblivious to my rather depressing surroundings.

The next day, we worked together in the huge shed, sorting wool into various grades. In the afternoon, we were told by Mr Moss to assist the shearers who had come to do the crutching (removal of wool from the tail and rear legs for hygienic purposes). The actual shearing would be done in the spring.

Back then, the sheep shearers of New Zealand worked exceptionally hard but also enjoyed very high wages. Normally, they worked in gangs of five or six and were constantly on the move from farm to farm the length and breadth of the country. Paid on a contract basis, the more sheep they handled, the more they earned. Big beefy fellows who dressed in shorts and singlets, they sweated continually from the exertion as they were totally immersed in the demanding, physical job at hand.

As they toiled tirelessly for hour after hour, clenching the struggling sheep firmly between their legs, Rusty and I were expected to keep them busy. This meant stepping outside the shed and into the pen, selecting any sheep at random, and chasing it into a corner where a tussle would follow as we attempted to turn it on its back to drag it up the ramp and over to the shearers. As each sheep weighed close to 150 pounds, and hopped, skipped and jumped from one corner of the pen to the other, I tired rapidly merely from chasing them. After a couple of hours and twice losing my footing and being thrown on my back, I was totally exhausted. For the shearers, this was all in a day's work, but for me, it was as tough as any job I'd had so far. Meanwhile, Mr Moss casually watched and made caustic comments.

At least half a dozen times, a frisky animal escaped from my grasp just as I was about to hand it over to the shearers. Mr Moss bellowed at me, 'Get 'im!' When I stopped to take a well-earned breather, he sarcastically remarked to Rusty, 'What's His Nibs doing?' When he complained that I wasn't working fast enough, I quietly suggested that he make allowances as I was new to the business. That shut him up.

After a couple of days, the crutching was completed, and we were assigned to hoe tobacco in the fields. It, too, was a backbreaking job, monotonous and repetitive, and the hours dragged by. Meanwhile, Dorothy Gaskell had taken it upon herself to keep a motherly eye on me, and late afternoon when work was finished, she visited me every second day bringing magazines and homemade scones and cakes. It went without saying I very much appreciated those visits.

Just about every day, I approached Mr Moss and explained that I was far from satisfied with my bach. The roof was now leaking, and the electric kettle had broken. After more than a week had elapsed and nothing was done, I told him I was quitting. While I was quite prepared to work hard, I considered it only fair to have certain minimal comforts. Only then did Mr Moss apologise and promise to make amends. But it was too late. I'd had enough!

Mrs Moss had overheard the conversation and remarked with great spitefulness as she wiped her hands with her apron, 'I don't think this kind of work suits you anyway. I said only yesterday in the Labour Department that you should be in the house, not in the fields.' That stung. But she didn't care. Nor did she realise there was much more to me than met the eye. I may have been skinny, but I knew I was every bit as tough as others of my age and size. Very coolly, I replied, 'Looks are very deceptive,' and walked away. She was by nature a rude, insulting bitch, a mean-spirited person who liked to criticise. As I packed my things, Rusty told me that my predecessor was so fed up, he 'snatched it' (ran away) not even bothering to get paid.

One consolation about working on that farm was hearing the vibrant birdsong, not only early in the morning but all day long. New Zealand may not have the large variety of birds like its neighbour, Australia, but it has many unusual endemic species, including the tui, bellbird, takahe, and wood pigeon as well as several types of penguin and the nocturnal weka and kiwi.

After learning that the fruit-picking season wouldn't start for a while, and as I now had enough money, I decided to see more of the country and return to the area for the beginning of the harvest. A letter from Pete confirmed that he, too, was now in New Zealand and was presently employed on a sheep farm near Hamilton on the North Island. I wrote to him suggesting that, circumstances permitting, we rendezvous in Motueka in a couple of weeks.

Travelling south down the west coast to the town of Greymouth took exactly nine lifts in the course of one day. Once, as I sat beside the road, an ambulance with siren sounding screeched to a stop. A nurse rolled down the window and told me to jump in. The siren continued as we raced over the next several miles until I was told to jump out. Another ride was with a young newlywed couple and, although their 1957 Hillman Minx was already full, they insisted that I squeeze in. Other rides were in the old, lovingly preserved, and highly polished vintage cars seen all over the country (because new cars were outrageously expensive).

Riding along the West Coast between Greymouth and Westport with a young New Zealand doctor, we stopped at Punakaiki Rocks to admire and take pictures of the pancake rock formations and blowholes, which in years to come would become a very popular tourist attraction.

At the picturesque village of Te Anau nestled along the southern arm of the 33-mile-long lake of the same name, I stayed at the local youth hostel. There, I met a petite, chubby woman named Pauline who invited me to join her in her little white Fiat 600 on a ride to world-famous Milford Sound, one of New Zealand's popular tourist attractions. The car, of course, immediately reminded me of Anna-Maria and my happy safari days.

Along the way, we stopped here and there to admire the scenery, including near the entrance of the Homer tunnel, a three-quarter-mile-long marvellous feat of early engineering, all done with shovels and pickaxes. There, for the first time, I became acquainted with one of New Zealand's naughtiest and cheekiest birds, the kea. A species of large parrot with olive-green plumage and orange on the underside of its wings, it lives in forested alpine areas. Because of its antics and interaction with people, it is referred to as 'the clown of the mountains'. Quite tame, they pick through rucksacks, sneak up to people and snatch food from their hands. Mischievous as they were, I couldn't help but be amused by them.

Situated in the heart of Fiordland, an unspoilt 3,000,000-acre national park of forests, fiords, lakes, waterfalls and rugged, majestic mountains, Milford Sound is a major highlight for those visiting New Zealand. After a spectacular ride through exceptionally beautiful scenery, we pulled up outside the large, impressive Milford Hotel at the end of the sound and gazed at Mitre Peak, the highest point overlooking the sound. Moored to the small jetty close to the shore sat half a dozen brightly coloured fishing boats, creating a picture-perfect postcard setting, and I couldn't resist taking a few photographs. As I took in this idyllic panorama, it occurred to me to try my hand working on one of the boats. It would be a marvellous way of seeing more of this incredibly beautiful and inspiring area, and might be the answer to conquer my seasickness. '*It's worth a try,*' I told myself as I made my way to the jetty.

The boats were empty when I approached them except for one fisherman busy repairing nets. 'Someone might need a mate,' he told me when I asked about a job. 'Try the pub.' There, I chatted to several skippers, one of whom was awaiting a new mate. If he didn't show by 6 p.m., I was told, the job was mine. The next two hours passed interminably slowly as I hoped against hope the new man wouldn't turn up. At precisely two minutes after six, I returned to the pub to be greeted with, 'You're on,' by Errol, the young skipper of the *Da Vinci*, one of several crayfishing boats moored nearby.

In his late twenties, Errol was tall, sturdy and ruggedly handsome with a mop of curly dark hair. We hit it off immediately. I was quite thrilled at the prospect of working on a fishing boat, although apprehensive about my sea legs. This job would be the perfect test. I admitted to Errol that I was always seasick to which he laughingly told me, 'No worries, you'll be OK.'

The 28-foot *Da Vinci* fishing boat ran on diesel and could easily be handled by a crew of two. It basically consisted of a wheelhouse, a tiny galley, and a small cabin with two bunks and a closet. A heavy winch, booms, nets and pots were neatly positioned in various places on deck.

After a good night's rest on board, we set off up the sound very early the next morning. Everything was fine as I made the tea and fried some eggs during the ten-mile journey to the mouth of the fiord where we bobbed up and down in the open sea.

No sooner had I washed the dishes than that all too familiar feeling enveloped me. Seconds later, up came my breakfast. While I vomited time and again, Errol casually busied himself with lowering and raising the pots. Between spasms of sickness, I attached bait to them, but soon I just lay inertly on deck waiting for death to engulf me!

As Errol casually went about his job, he told me in a matter-of-fact tone, 'No worries, mate. You'll come good.' However, two hours later, my condition was in no way improved, so Errol turned about and returned to Anita Bay at the tip of the sound to sit in calm water for a while. But that made no difference; I was just as sick. I asked to be put ashore, but Errol refused, convinced that sooner or later I'd 'come good'. The rest of the day, I continued to be violently sick to the point that my whole body ached. After dropping anchor, the night brought no relief, and just about every half hour I had to leave my bunk to throw up. When morning dawned, I felt slightly better, and we headed out to sea again. That day was worse than the previous one, and I implored Errol to put me ashore. I was a liability, not an asset, I told him pathetically.

Reluctantly, Errol once again turned around and headed back to Milford. If anything, he seemed more disappointed than me. He admitted that he'd never seen anyone so sick in his life!

We pulled in close to the shore of secluded Anita Bay, and I jumped ashore to while away the rest of the day while Errol continued cray fishing. He told me he'd pick me up later in the afternoon and take me back to Milford. The small bay was eerily quiet and deserted except for an empty deerstalker's hut at one end of the sandy beach. All alone like Robinson Crusoe on his deserted island, I amused myself by searching for green stone (New Zealand jade) in the small streams that trickled into the sound. Soon, however, I was viciously attacked by swarms of sandflies, blood-sucking, biting midges, and I had to wade knee high into the water to escape them. That evening, Errol returned me to Milford, where I spent another night on board. I was fast asleep when another young skipper came searching the cabin with a torch; his wife, he said had gone 'with some other joker'. And I thought I had problems!

It was with enormous regret that I had to give up such a promising and interesting job and feeling none the worse for wear but greatly disillusioned that I still hadn't overcome my peculiar affliction, I continued on my way in glorious sunny weather.

I was headed for Dunedin, a predominantly Scottish town. In the youth hostel there, I met a young Australian named Malcolm. He seemed terribly homesick, having never been away from home before. We teamed up for a couple of days and were still together on Christmas Day as it drizzled with rain. As we shared a cold Christmas pudding and some chocolate and bananas by the side of the road, nobody seemed interested in picking up two wet, young hitchhikers. Trudging forlornly through one little town, we glanced with envy into living rooms decorated with Christmas trees and tinsel, where happy families tucked merrily into roast turkey dinners. It was one of the few occasions when I, too, felt homesick for family and friends.

In Alexandria (Otago Province), known for its orchids and the 1860s gold rush, Malcolm and I spent a day picking strawberries at a local farm, then went our separate ways. As I headed out of town, a friendly housewife leaned out of her window and asked me if I'd like a cup of tea. I'd no sooner downed it and continued on my way when another woman on the outskirts of town sent one of her small daughters running after me with a plate of food. Always hungry, I gratefully wolfed it down while waiting for a lift.

My next destination was the popular resort town of Queenstown located on the northeastern shore of lovely Lake Wakatipu, surrounded by the snow-clad Remarkables mountain range. It would soon be known as New Zealand's 'Adventure Capital'. There, I checked into the youth hostel. Moments later, into the lobby walked Pete! His job on a sheep farm had only lasted two weeks while he replaced a sick labourer. Now he, too, was seeing as much of the country as possible before heading for Motueka.

For the next three days, we were as happy as schoolboys as we explored the Mount Cook region deep in the Southern Alps in perfect weather. This was the place where Sir Edmund Hillary, conqueror of Mount Everest, and George Lowe, another great climber whom I would meet in South America, had both cut their mountaineering teeth, often hiking and climbing together.

Our base was the youth hostel in Mount Cook National Park and, although it was full, the warden allowed us to sleep on the floor in the common room. After packing a sandwich lunch, we set off each morning to hike through the mountains. During this ecstatically happy period, we hiked across glaciers, scrambled up steep rocky slopes and swam in icy-cold mountain pools.

After climbing to a high ridge, we would sit on a huge boulder and munch our sandwiches. We didn't talk much, allowing the splendour and silence of the mountains to envelop us. The air was clear and refreshing. As always in such an idyllic natural environment, and inspired by our surroundings, our senses always felt sharper, and we felt totally relaxed and at home.

We were lucky. Some visitors never even catch a glimpse of Mount Cook due to unpredictable weather. As in all high mountain terrain, dense mists and cloud sometimes descended on the area for weeks at a time. Throughout our stay in New Zealand during the summer of 1966–67, we experienced nothing but sunshine and clear blue skies.

During the afternoon of our third day in the Southern Alps, we took a ski-plane flight, a sensational and unforgettable experience well worth the cost as we sat, noses pressed to the windows, thrilled to bits. As our 6-seater Cessna brushed walls of granite and ice and swooped between high, jagged peaks, we completely forgot our fear of flying and became increasingly excited. Coming in to land on the Tasman Glacier, we banked sharply, then levelled out and bumped to a halt on the packed snow. After a five-minute snowball fight between us, our three fellow passengers and our young pilot, the return flight was just as exciting.

When we reluctantly left that splendid alpine wonderland behind, we kept looking back, watching the mountains recede until they eventually dropped out of sight.

Our next stop was Wanaka, the tiny picturesque town nestled at the end of a glistening jewel-like lake of the same name and dominated by Mount Aspiring. There, we did a ten-mile hike through the mountains, occasionally skirting isolated, high-country sheep farms and pausing to admire the wild, rugged scenery and to listen to the whisper of the wind. Our keen young eyes scanned the landscape. Apart from a solitary shepherd who was scrambling up a steep hillside in the distance accompanied by his faithful sheep dog, and a herd of red deer silhouetted against the skyline, we were alone. The mountains of the magnificent Southern Alps belonged to us.

That evening in the lakeside youth hostel, we joined several other young people in discussion. A young English overlander with a mass of badges on his rucksack started boasting about his travels and how he had sold blood in Kuwait (then a well-established way of supplementing funds when travelling through certain parts of the Middle East). He'd been to 15 countries, he loudly announced to his audience, and by the way he spoke knew all the answers.

After he'd been boasting quite long enough, I quietly asked him how long he'd been on the road.

'About nine months,' he declared, his chest puffed out with pride. 'How about you?' he asked.

'About four years,' I replied quietly.

With that, he got up and disappeared into the male dormitory. We didn't see him again for the remainder of the evening!

Gradually we made our way north to Havelock, a small village where an old schoolhouse had been converted into a youth hostel. There, we took a boat trip through Pelorus Sound and in the evening climbed a nearby hill to see a stunning display of glow-worms.

As the fruit season was about to start, we made our way to Motueka. The Labour Department gave us the address of an apple orchard at Mariri, some five miles from Motueka, owned by a middle-aged couple named David and Jill Templeton. Over a cup of tea in their spacious, tin-roofed, whitewashed bungalow, David and Jill happily took us on before showing us to our bach.

Considerably better than my previous one, this bach consisted of a bedroom, living room, and kitchenette with an outhouse toilet and shower. Sparsely furnished, it was nevertheless quite luxurious compared to the other place, I told Pete. When describing the work, the Templetons told us that we would be working with eight Australian girls, which perhaps more than anything else helped us make a decision. But as events turned out, we would be somewhat disillusioned.

As we were the first workers to arrive, we spent a couple of days thinning out trees that were deformed or too heavy with apples. It was a tedious task, but one we initially tackled with zeal and enthusiasm. Then Joy, a voluptuous 22-year-old Australian brunette, together with 21-year-old blonde twins Carol and Marie arrived on the scene, followed shortly after by five other Australian girls, also in their early twenties which made things more interesting. 'This is going to be fun,' commented Pete in the privacy of our bach. I agreed, totally unaware of the giant strides 'women's lib' had taken over the past couple of years!

At 8:30 the next morning, we all paraded outside the fruit packing shed, the girls dressed in shorts and blouses and us in our shorts and t- shirts. David Templeton, a soft-spoken, somewhat moody and dour Scot, briefed us on our duties while Jill told us about the different varieties of apples and how best to pick and handle them to avoid bruising. Wearing canvas aprons to hold the fruit, we would climb five-foot steel ladders and began to strip the trees bare. David's role was mainly supervisory, fixing things here and there and ensuring that everything worked smoothly, while Jill spent most of her time with the staff of girls, packing apples in the shed.

While Pete and three of the girls assisted Jill in the shed, the rest of us started picking apples in earnest. My main job was to drive the tractor and trailer to the work area, help with the picking until the trailer was filled, and then drive to the shed. There, the apples were gently dumped onto the conveyor belt where they were sorted and packed by Jill and the girls. Pete was responsible for keeping them supplied with boxes, a task easier said than done, for as the season progressed the girls' skill and speed increased. When the boxes were filled with each apple neatly wrapped in tissue, Pete nailed on the lids, wired and marked them before helping to load them onto a daily truck, which then transported them to a refrigerated ship at the nearest port, destined for export abroad.

At first, working in the orchard was fun as everyone chatted and joked and enjoyed the warm sunshine. The Templetons had invited us to eat as many apples as we liked but, as they knew only too well, it wasn't long before we didn't like picking them let alone eating them. A couple of weeks went by, and we had picked all the Jonathans, so we started on the Delicious followed by the Golden Delicious, then the Sturmers. By the time we had picked the Coxes, Statesmen and Granny Smiths and also had them stewed for dessert, I for one felt I never wanted to see another apple again!

As time passed, the job became increasingly monotonous. Compared to some of our other jobs, it became sheer drudgery and was certainly a time when my personal quest for 'experience and adventure' didn't pay off! Equally bored and frustrated were the girls. They started arguing with each other, and then started picking on me. Their breaks gradually became longer and more frequent. This put the pressure on me. I was expected to do a full day's work, I diplomatically pointed out, so why not them? They merely made excuses. When I told David, he shrugged his shoulders as if to say 'Too bad' and shook his head.

It was with some slight relief that Pete and I swapped places now and again but, if the work in the orchard was exceptionally mundane and boring, it was extremely hard in the shed. Lifting the 40-pound boxes, stacking and loading them all day long made us perspire continually.

Once, after a particularly difficult day in the orchard with the girls, Pete returned to the bach quite depressed. None of the girls felt like working at all, which perplexed and annoyed him. 'They don't want equality,' he grumbled. 'They want superiority.'

During this discussion, it dawned on us who the ringleader was — Polly. Judging by her actions and behaviour, she sorely regretted not being a man.

A real tomboy. It was she who was always the most argumentative. When I'd invited the girls to take turns driving the tractor, it was Polly who wanted to drive it all the time. So we decided to be firm and stop her using the tractor completely. Although she made a great fuss, it solved the problem. When Polly realised she couldn't have her own way, she became rather morose and subdued. That decision made both Pete and I rather unpopular for a time but, by remaining firm, the girls cooperated much more.

We continued to dislike the work intensely and although the weather remained mostly warm and balmy, the days dragged. For six days a week except Sunday, we worked from 8:15 in the morning until 5:30 in the afternoon, stopping only for lunch and midmorning and afternoon tea breaks. The wages were extremely basic, and several times we told ourselves it simply wasn't worth it. But we had told David and Jill we would stay for the entire season, and we felt obliged to keep our word.

Apart from only a few rainy days when work in the orchard was called off, and we worked in the shed, the days got hotter and everyone became extremely tanned. If nothing else, it was a healthy life with plenty of fresh air.

Greatly relieved when the working day was over, we returned to our bach, showered, changed and prepared dinner. Most meals consisted of mutton or lamb, which was plentiful and inexpensive in a country of 30 million sheep. Our appetites were tremendous after we stopped eating apples, but for a while our meals didn't seem complete without a sprinkling of apple sauce!

Fortunately, there were compensations after work. As the entire area was inundated with seasonal workers — Motueka swells from a normal population of 3,000 to 8,000 during the fruit and tobacco harvest — there were nightly dances and parties held at different villages and farms. On those occasions, we borrowed the Templetons' second-hand Morris and filled it with as many of the girls who could get in. A twice-weekly film was shown in the open on a lawn of a nearby church, which we occasionally watched reclining on the grass with blankets and pillows. We also had barbecues on a nearby beach at night when we sang and danced until the early morning, then refreshed ourselves with a moonlight swim. Suddenly, six weeks had passed, and we were halfway through the season. It was only thanks to the hectic social life that the next six weeks were endurable!

Other times, we took long walks or visited friends on neighbouring farms, and late Friday afternoons we drove to Motueka to do our weekly shopping. There, after sharing a jug of draught, we stood outside the pubs in time to watch the 'six o'clock swill'. As daily licensing hours only went to 6 p.m., it seemed that every man for miles around was downing as much grog as possible before closing time. At precisely 6 p.m., the pubs emptied and drunken men staggered and stumbled into the streets.

Mail, an eagerly awaited event, arrived every day except Sunday, and everyone took it in turns to run down the drive and collect it from the box. Every second day, Pete heard from Hoshiko, and he replied to her almost every other night. We also heard from Mike, telling us he was now in Australia having sailed by liner from Durban six months before. He had a job as a timekeeper at a construction site at Mount Tom Price, a huge iron ore mine in Western Australia and, for once, he found himself without a girlfriend. Some time later, we heard he went to Japan and then back to England where he married, had a family and settled down.

One evening we attended the annual 'vice-versa' dance held in a nearby village hall when all the men dressed up as women and the girls as men. I swapped clothes with Joy, who was roughly my size. While I dressed in her skirt, blouse and high heels, she wore my trousers and shirt. She applied makeup to my face, and I painted her upper lip with a moustache. After breaking two of her bras filling them with foam rubber, I was ready. Meanwhile, Pete had changed clothes with Marie. That night after being pinched numerous times in obvious places by Maori girls, who repeatedly joked, 'Now you know what it's like to be a woman,' Joy and I walked off with first prize, while Pete and Marie came second.

Towards the end of our stay we heard over the radio that two young men in the area had been run over by a car. No names were given out until the next of kin had been notified. As Pete and I had become quite well known in the district and the descriptions and ages of the two men resembled ours, many people thought it was us, especially as we both took frequent walks at night, including where the fatal accident had occurred. Several people phoned the Templetons offering their condolences until they learned we were still very much alive! However, we knew only too well it could have been us.

Finally, the fruit-picking season came to an end, and only then did we feel a little sad and nostalgic. Weeks before, everyone had been counting the days, wishing for the end to come as soon as possible. When it came, it brought mixed feelings of relief, sadness and excitement. Autumn was creeping up on us. The leaves were beginning to fall, and the days were becoming shorter and cooler. It had been a bumper season for David and Jill, in fact, their 'best year ever', but there were no bonuses for the staff.

After fond farewells, everyone left in pairs and headed in different directions all over the country. Having already seen most of the more interesting parts of the South Island, Pete and I decided to return north. Back on the road, every driver who picked us up insisted on stopping at a pub for a 'snort'. And to buy us one. Then when they dropped us off somewhere along the road, they invariably shouted, 'Cheery,' or 'See you, sport.'

After two weeks, we were back in Wellington. Pete decided to return to Australia to 'make some good money'. He then flew to Melbourne where he quickly found a tool and die making job in a local factory.

After making exhaustive enquiries about ships sailing for South America to no avail, I contemplated which route to take next. The icy-cold winds of winter sweeping through Wellington made me bend over double and long for sunshine again, so I too decided to return to Australia. A day later I boarded an Air New Zealand jet for Brisbane in Queensland.

I had been in New Zealand just about six months, long enough for it to become my favourite country so far and for me to regard it as my 'second home'. I loved everything about this extraordinarily stunning place, and I never missed any opportunity to tell people all about it. In later years, I would return several times, bringing tour groups to discover this glorious land of natural wonders for themselves.

I would gladly have settled in New Zealand and did in fact apply to immigrate some years later. But I didn't have the required qualifications and my application was turned down! I was tremendously disappointed at the time.

CHAPTER 14:
A South Sea Sojourn

'Enthusiasm is the greatest asset in the world.
It beats money, power and influence.
Enthusiasm is no more or less than faith in action.'
— *Henry Chester*

I acquired two jobs in Brisbane; one was to work in the steaming heat of the New Guinea jungle and the other in the hinterland of Queensland. However, as neither started for some weeks, I went instead to Gladstone, a port town on the Queensland Coast, close to the Great Barrier Reef, after being told jobs there were easy to get. I took the train and, on arrival, went into a 'greasy spoon' restaurant where I began chatting with two Englishman, a middle-aged man named Bill and a younger one called John. Both were 'brickies' (bricklayers) by trade, John being a sub-contractor and Bill, his sidekick. They were working on a motel in the vicinity and invited me to join them as a labourer in their gang. Despite disliking the idea of heavy construction work, I decided to give it a go. Accommodation was supplied by a local couple who ran a boarding house. Gladstone itself was an uninteresting aluminium refinery town but was expanding like many others and boasted many bars. It soon reminded me of *Peyton Place*, where everyone seemed to know everyone else's business.

For the next 10 weeks, I learned about trowels, scrapers, pointers, levels and lines. I also learned how to mix and make cement and about the weight of bricks in wet and dry weather. Of the eventual total of 80,000 bricks used in the construction of the building, I felt sure that I'd handled 75,000 of them!

The actual job was an excellent way to become even more fit and, like my days on the Wyndham wharf, it was tough. We started work at 7 a.m. and usually finished around 6 p.m., stopping only for tea and smoko breaks. The money was excellent, which helped to make up for the unpleasantness of the work. And there were other compensations. As in so many of the places I visited or lived, I saw much of the surrounding area.

The construction gang consisted of three Poms, four Australians, an Italian, a Dutchman, a Yugoslav as well as a few other foreigners employed as casual workers on the site. Just like the 'wharfies' and the outback station workers, all shouted, swore, gambled and drank. And, to a man, every payday, they immediately headed to the nearest pub to down as many beers as they could. One of my new mates was an Australian named Brian, usually called 'Bri', with whom I spent much of my free time. He was a Sydneysider and proud of it.

Keeping a gang of bricklayers supplied all day with bricks in the sweltering heat was no easy task. God forbid if you got behind. There was hell to pay keeping them waiting for more bricks or cement. I was again called 'that Pommie bastard' on a regular basis, but I was used to it by now and accepted it like everything else. It was the Italians, Greeks and Yugoslavs that I felt sorry for. They were called every rude and obnoxious name one could think of! But they shrugged it off like everything else they faced. It was all in a day's work.

I soon discovered that wheeling a barrow-load of 35 bricks up a narrow, sloping wooden plank and then along a platform ten feet above the ground without stumbling or falling while sweating through every pore of your body for eight or nine hours a day was very challenging. It required a certain amount of dexterity and expertise one could only learn on the job. Handling and sorting the bricks was such hard work that, even wearing thick workman's gloves, my fingers bled so profusely that they often had to be bandaged with heavy-duty duct tape.

Thankfully, on Sundays, our one day off, Brian and I could escape in his small motorboat. We'd go to Heron Island, a small, pristine coral cay. There we amused ourselves for hours at a time, exploring the island, picking wild lemons, and digging out oysters embedded in the rocks. Like hundreds of islands in the Great Barrier Reef — another of the great natural wonders of the world — it was a real South Pacific island with tall palm trees and rock pools where one could observe tropical fish. And not a soul to be seen. It was 1967, and the island had yet to be developed, so there were no hotels or resorts. We had it all to ourselves. We wandered at will and swam and beachcombed to our heart's content. We also fished a few times for barramundi and perch, and on those boat trips, we observed manta rays, reef sharks, dolphins, and sea turtles close up in the crystal-clear waters.

Another of my co-workers was Cliff from Melbourne. In his late thirties, he had a very red face which suggested to me that he drank a lot. And a nasty piece of work he was at the best of times. One day, we clashed over some trivial matter, and Cliff got angry. I told him to 'sod off,' which made the situation worse, and he threatened to 'fucking flatten' me. I didn't sleep well that night, afraid of what the morning might bring. At breakfast, Cliff told me he was going to 'knock the shit' out of me. Luckily, John got wind of the upcoming altercation and told Cliff firmly he would have 'none of that' and to cool off. That was the end of that.

I disliked confrontations and avoided potentially ugly situations as much as possible, believing instead in diplomacy and discretion. No matter what, though, I would not allow myself to be bullied or beaten up. I'd had a few judo lessons, boxed in the army a few times (albeit reluctantly), and had faced a lot of rough and tumble at boarding school. So I knew I could and would, if necessary, stand up for myself.

Every day on the building site, my muscles ached and my body burned from the hot sun. As I'd discovered much earlier, I wasn't cut out for such hard labour. After a while, I considered giving up, but John, the boss, persuaded me to hang on. 'Stick with it, lad,' he'd say. Those words of encouragement kept me going despite my better judgement and would much later ring in my ears. But there were some advantages to working on the Queensland Coast. I slept well, ate well and saved a fair amount of money.

I spent my 28th birthday on that construction site and, like so many others, it passed like any other day. But it reminded me to keep moving, for time was fast slipping away. Previously, it seemed Pete and I had all the time in the world, but it was about then that we realised we couldn't travel indefinitely. We had no intention of letting the trip last ten years, which it so easily could. More than once we'd run into older travellers, fellows in their thirties, who seemed to be wandering aimlessly, with no roots, no families and no home. We didn't want to be like those drifters with no set plans.

With the eventual completion of the motel, we transferred to Surfers' Paradise, Australia's premier holiday and beach resort, to begin work on a new shopping centre. A great rainstorm had recently caused severe flooding, leaving hundreds homeless. Coastal erosion had set in, spoiling the glorious beach for some time to come. It was, I read, the worst flood disaster the state of Queensland had experienced in 74 years.

John invited me to stay at his spacious apartment overlooking the ocean for the month I worked on the new site. Frequently on days off, I took walks with him, his wife, and their two children along the dunes and fabulous white, sandy beach. With him and his family, I attended a rodeo; several drive-in restaurants for cool drinks, ice creams and hamburgers; a bird sanctuary; and various stretches of the Gold Coast. Having now spent another three months in Australia, I reluctantly left for Sydney to catch either a boat or a plane to South America via some of the South Pacific islands.

Back in Sydney, I stayed with some of Robin's friends at Cronulla, a suburb of Sydney and close to a beach which had also been damaged by adverse weather. I then made the decision to fly to Fiji, hoping to find a ship there to Panama. After updating my typhoid vaccination I flew in a Qantas jet to Nadi, Fiji's main airport.

Stepping off the plane in Fiji, I discovered a quintessential South Pacific paradise where the sun shines every day and flowers grow all year round. And immigration asked me to produce my onward ticket. Not knowing my precise plans, I didn't have one, but before officially entering the country, I had to purchase one. This entry requirement always posed problems for travellers like me who were not sure of their exact plans or next destination. Despite carefully explaining my situation, I had to buy a ticket back to Sydney to satisfy the authorities. I promptly cashed it in two days later, losing a few hard-earned shillings in the process.

On Viti Levu, Fiji's principal island, I'd booked into an Indian-owned guest house. The next day I travelled to Suva, Fiji's capital. The bus took seven hours with frequent stops to go the 140 miles along a dusty, winding red-earth road past massive banana and sugar cane plantations. En route, we picked up people with their chickens and fruit and once two young Kiwi beachcombers. One of them told me he had been working as a 'powder monkey' on a construction site but had given it up when a co-worker doing the same job was literally blown to pieces.

The New Zealanders suggested that I stay in the same hotel as them. The South Seas Hotel was on a hill overlooking the beautiful harbour. The proprietors of the hotel had coincidentally lived in East Africa for many years. So we had a lot to talk about.

'Be interested to know how many times you get accosted,' said Lawrence, one of the Kiwis. 'The place is full of whores.' Sure enough, typical of most ocean ports, Suva had its fair share of 'ladies of leisure' who tried to entice me to buy them a drink or take them back to my room.

I was very struck by a great resemblance between Fiji and East Africa — the people (Fijians, Indians and Europeans), the undulating landscape, the red-earth roads, and the amazing variety of tropical flora.

Among my observations, I noticed most of the islanders shared similar attitudes and lifestyles. Blessed with a year-round temperate climate, they were extremely happy-go-lucky people who spent much of their time laughing, singing and dancing in the idyllic settings of blue skies, tall palm trees and long, sandy beaches. They lived a simple, contented life.

As I discovered, throughout the islands, everyone spoke good English, and their obvious happiness was quite infectious. Everywhere I wandered, through villages and towns, I was greeted with broad smiles, 'Hello, what is your name?' Upon telling them, they would reply, 'Follow me. I take you to my village.'

I was invited into many mud and thatch houses, where I was always invited to enjoy a drink of *kava* (made from the roots of coconut trees). Fiji's national drink is drunk from a coconut cup, and one is supposed to shout, 'Bula' (cheers). It has an anaesthetising effect on the mouth and tongue, making it a little difficult to speak. Once, with a couple of Fijian friends, we were all chatting and drinking merrily when I suddenly noticed the conversation had stopped — no words were spoken as we just sat grinning at each other for the longest time! The kava was having an effect!

Being extremely able and lusty singers, most of the islanders sing constantly about anything and everything including sports, politics, women, and if there is no known song for a certain occasion, they make one up there and then. Fiji is often referred to as 'the Friendly Isles', and I found it difficult to imagine that Fijians were cannibals not so very long ago.

I met many people in Fiji, including a pretty, young New Zealander named Jane Burrows, a secretary from Christchurch. She was about my height with shoulder-length blonde hair and beautiful green eyes, and was spending her annual holiday there. Since we shared similar interests, we decided to explore the island together. Hungry for pleasant, platonic female company, I happily joined her to see as much as we could. It was a totally innocent and brief friendship — we didn't even kiss. Over a couple of days, we went swimming and took one or two bus trips to different parts of the main island. Once, on our way back to the hotel from the beach, we decided to walk in only our swimsuits. This was just as well, for a violent tropical thunderstorm erupted. Amid loud thunder and lightning, we ran to the hotel, arriving laughing and soaking wet. Another day, we went by local bus to Korolevu — a beautiful beach of golden sand fringed with coconut palms — where we wandered hand in hand and swam in the warm crystal-clear Pacific Ocean.

At that time, the weeklong Miss Hibiscus festival was under way, which meant that everybody was determined to have a good time, including us. At Albert Park, we rode on the big wheel, watched the girls' rugby match and the car racing, and strolled around numerous stalls.

My last day in Fiji was spent racing around Suva purchasing new things. I always went on something of a spending spree in a free port. I bought some new clothes to replace others that had shrunk in the wash or worn out.

A few days previously, I had purchased a ticket to Panama on a freighter which was to be my home for the next 25 days. Travelling by ship was still much cheaper than flying and allowed me to see more of the South Seas. And I thought I might finally find my sea legs this time. What wishful thinking on my part!

Having been told that the *MV Sprucebank* would leave at 6 p.m., I hugged Jane farewell and made my way to the Suva docks. The ship actually sailed at midnight. How frustrating!

Two days previously, I'd seen the liner *SS Oriana*, the large, luxurious passenger liner which plied back and forth between Australia and the UK. The *Sprucebank* was less than half her size and an insignificant tub by comparison.

With my fingers crossed, the first few days at sea went very well, and I lost no time exploring every part of the ship. A rusting old freighter, the *Sprucebank* was manned by British officers and an international crew and was picking up copra from some of the islands. Copra is the dried meat or kernel of the coconut and used for a variety of products.

As one of four passengers on board, I occupied the cabin of the third apprentice who was on leave. It was situated amidships on the port side, quite spacious and comfortable with a wardrobe, bed, desk, wash basin, and porthole — probably equivalent to a first-class cabin on a liner — but without a bathroom. I also had a cabin steward, a cheerful Filipino named Alex, who made my bed and brought me a cup of tea every morning.

Captain Allan was from Northern Ireland. Upon learning that I'd spent close to 18 months in his part of the world, we got on well together. The chief engineer was a Scot; the others (engineers, apprentices and radio officer) were a mixture from all over the British Isles.

Two of the passengers were a middle-aged couple who, I discovered, had worked at VRD a few years earlier. An extraordinary coincidence — for how many people had even heard of Victoria River Downs, let alone worked there? The husband was now working in Western Samoa, where they were headed. The fourth passenger was a white-haired, elderly civil servant who had been a vice-consul somewhere or other and had been awarded the OBE for his services. He, too, was bound for Western Samoa and spent much of the time on board drinking whisky and soda or gin and tonic, an easily acquired habit in the tropics.

Arriving in Nuku'alofa, capital of Tonga, our ship was docked for two days while copra was loaded. During this time, almost everyone went ashore. I went to see the late Queen Salote's tiny palace and also had a ride around her estate in her old London taxicab. I also befriended two young male Tongan schoolteachers who showed me around their village and grass hut homes.

As the officers and crew (being typical seamen) all enjoyed drinking and womanising, it was inevitable that every time we docked, various women came, or were brought, on board. On our second morning in Nuku'alofa, Jack, the youngest apprentice, aged about 17, told the officers what had happened to him the previous night. To quote his very words, he was 'rudely awakened, raped and robbed' then he went on to say his trousers and one or two other items had been stolen. I did not go unscathed. Someone had rifled my clothes and emptied the pockets of my pants; luckily, he or she didn't touch my rucksack, presumably afraid of waking me. Fortunately, none of my valuables were taken.

After docking at Vava'u, I went ashore with some of the crew, where we joined some Tongan policeman in the one and only social club. Shortly after, a few of us set off in a police vehicle to see more of the island and stopped at a beautiful cove with a sandy beach. A hundred yards or so down the beach sat eight dusky island beauties, including one named Elizabeth, clad in brightly coloured *lava lava* (sarong) and blouses. They were off-duty nurses picnicking on the beach and listening to the latest western hits on their record player. They promptly invited us to join them for lunch. Very soon, we were all happily eating cooked yams, pork and chicken followed by fresh fruit. Suddenly, sandflies attacked us in droves, and we had to beat a hasty retreat into the water to escape them.

I returned to shore the next day, hoping to see Elizabeth, but no luck. As I was heading back to the ship on the launch, I finally spotted her and a girlfriend and invited them on board the *Sprucebank* and showed them around under the close scrutiny of an enormous Tongan police officer stationed at the gangplank.

When we parted, he quietly said to me, 'Nice girl, smart girl, nurse. She love you now. You come back?'

I shrugged. 'Perhaps,' I said. 'I sincerely hope so.'

Although my stay in the South Pacific islands was relatively brief, it was there that I felt I had found my utopia for the first time. *But the role of a beachcomber, strolling up and down the beach and living on seafood and fruit, would only be appealing for so long*, were my thoughts at the time. A Robinson Crusoe existence would undoubtedly become very boring after a while.

As it was customary at sea for guests to be invited to the captain's cabin for drinks on Sunday morning, I duly went and sat down with a Coke. Almost immediately, I had to excuse myself when that all too familiar feeling of nausea swept over me. Fortunately, I was feeling slightly better when we docked in Apia, the capital and largest town in Western Samoa, with a population of only 25,000 and like all the South Pacific islands, not a tall building in sight. Somewhat confusingly, after crossing the International Date Line, we had arrived at our destination two hours before we left our last port!

Whenever we waited for the pilot before entering a harbour, islanders from near and far came on board or sat alongside in their flimsy canoes, displaying their merchandise. Tapa cloth, shell baskets and reed mats were among their usual handmade products. I purchased a *lava lava* in Apia, which would come in useful when relaxing at night.

I was struck by the many churches in Apia and their large congregations with the parishioners all neatly dressed in white. I took a refreshing swim in the pool at Aggie Grey's Hotel made famous as an R&R stop for hundreds of American sailors and soldiers during World War II.

Interested in seeing the grave of Robert Louis Stevenson, I set off by taxi with Clive and Bill (ship's radio officer and 1st apprentice, respectively) to drive the three miles to the foot of Mount Vaea, atop which lies Stevenson's grave. Leaving the taxi, we were joined by half a dozen young, sarong-clad Samoan boys who offered to act as our guides. We told them, 'No thank you,' but that didn't deter them. Before starting to climb the 600-foot hill, we passed the old colonial-style home of the famous author, then the home of the British High Commissioner. As it was so hot, we stopped and drank some of the soft, sweet water of a nearby stream from which Stevenson himself may have drunk.

The rough path leading up to the tomb was twisted and almost completely overgrown, so at times we left it and scrambled our way straight up through dense green jungle, hauled and pushed by the eager boys. After roughly half an hour, we reached the summit. It was blazing hot in the open sun. I was giddy, my body ached, and my breath was coming in short gasps. It had been a strenuous climb in the middle of the day. I sat down to rest on a fallen tree and looked across at the nearby grave. It seemed such a forlorn, ill-kept tomb, an ugly, concrete coffin resting on blocks on the ground. At one end was a bronze plaque commemorating 'Tusitala', teller of tales to the Samoans, as RLS was affectionately known. On the plaque beneath his name and dates (1850–1894) was this excerpt from his most well-known poem:

> *Under the wide and starry sky*
> *Dig the grave and let me die*
> *Gladly did I live and gladly die*
> *And I lay me down with a will.*

*This be the verse you gave for me
Here he lies where he longed to be
Home is the sailor, home from the sea,
And the hunger home from the hills.*

A second plaque next to his commemorated his wife, Fanny.

After recovering from the exertion, my companions and I set off down the winding, hand-hewn footpath, pausing at the bottom for another sip from the stream.

Later, as we walked along the coast road, we were befriended by some young women who invited us to one of their houses, a large, open-thatched hut accommodating a family of four. There, they entertained us with songs of the South Pacific. Wearing leis and grass skirts, they demonstrated their dancing skills to the accompaniment of a single ukulele. We sat enchanted for one-and-a-half hours, oblivious to the sudden torrential rain outside.

The *Sprucebank* stayed in Apia for three days and, as ships were fairly infrequent, we were visited by many of the townspeople wanting to look around. Meanwhile, the labourers worked all day carrying sacks of copra from truck or shed up the ramp onto the ship. For their strenuous efforts, so we were told, they received a can of meat and a loaf of bread at lunch time and a small wage.

One day as I walked alone, I met a young Samoan girl who waved to me from across the street. Mary was her name and she beckoned in the usual island way by extending the arm with the palm of the hand downwards and at the same time emitting a 'psst' through the teeth. In her large matted-grass dwelling, I met no fewer than 15 members of her family who invited me to lunch and made a great fuss of my fair hair and blue eyes. Lunch was delicious and consisted of large portions of chicken and pork along with the catch of the day, which was seabream — but could have been red snapper, sailfish, black merlin or tuna — together with sweet potatoes, yams, taro, cassava and plantains followed by coconut, bananas, papayas and mangoes, now my favourite fruit. It was a tasty and typical meal enjoyed throughout the South Pacific islands.

After spending three enchanting weeks among the islands, I was sorry to leave. When we pulled away from Apia, our final port of call, I realised I had left yet another piece of my heart behind.

The other three passengers had left the ship, leaving me the only guest on board. I settled down to what would be 18 days at sea during which, fortunately, the weather was mostly quite mild and the sea only a little choppy. I busied myself writing of my recent experiences in my journal or reading. We crossed the equator without fanfare. Over that period, I learned that life on a ship can be extremely monotonous.

Once a week during our long voyage we had fire boat drill. Twice we developed engine breakdowns, which left us drifting helplessly for a few hours. One evening, the lights suddenly went out, and the engines stopped. The engineers struggled for half the night to repair the problem so we could get under way. At the same time, the air conditioning broke. Bugs living in the copra started to create a nuisance — infiltrating, stinging, and biting everything at hand. Fortunately, they quickly died off when the air-con was repaired and the ship reached cooler waters. Also, a peculiar squeak developed in the engine room which nearly drove everyone mad. It continued throughout the trip, despite intensive investigation by the engineers.

One day, I inserted a short message into a bottle requesting that the finder contact me, stating where and when it was found. I signed it, 'itinerant wanderer currently travelling the world', and threw it overboard. But I never heard from anyone.

For exercise, I took walks on deck usually just as the sun was setting, watching shoals of flying fish and schools of porpoise from the bow. Those occasions were quite exciting, not only for me, but also the others on board, who considered anything interesting if it broke the monotony. Often, too, I chatted and joked with the ship's officers who seemed very pleased to have a new face at the dining table. One evening I was introduced to the world of hi-fi by one of the ship's crew, an enthusiast who possessed all the equipment and kept himself happily occupied this way.

By this time, as proof of my cosmopolitan identity, I was in possession of my third British passport (having filled two), a New Zealand driving licence, a South African identity card, a Chinese lucky token, as well as my Kashmiri swordstick and clothes from different parts of the world.

Being a dreadfully poor linguist with little aptitude for picking up new languages, I had neglected to study any Spanish despite the fact that I was soon to visit Latin America. However, before leaving the ship, I was presented with a small English-Spanish phrasebook by one of the apprentices in which he'd written, 'Good luck on your travels.' Encouragement that I was pleased to get.

A few days out of Panama, I was in a buoyant and excited mood, eager to see new places and step once again on *terra firma*, and for the umpteenth time, I spent restless nights wondering what lay ahead. Two days from Bilbao, we sighted our first other ships, and on nearing the coast, we joined several awaiting canal clearance. A small launch brought the pilot to the ship, and that evening after dinner we passed through the famous Panama Canal.

Another leg of 'the trip' had come to an end, and another notebook had been filled with observations and impressions about my latest peregrinations.

CHAPTER 15:
Thieves and Serenatas

> *'I have wandered all my life,*
> *and I have also travelled;*
> *the difference between the two being this,*
> *that we wander for distraction,*
> *but we travel for fulfilment.'*
> — Hilaire Belloc

I disembarked in Cristóbal the next morning knowing little Spanish except *ola, adios, por favor* and *gracias*, adding *buenos dias* to my vocabulary as the immigration officer stamped my passport welcoming me to Latin America. Eventually I would learn more Spanish and I even started to gesture, speaking with my hands and arms, like a true Latin.

Before boarding the train for the short ride to Panama City, I took a walk around the port town. I'd walked only a few blocks when I was confronted by a policeman dressed in jungle-green uniform angrily brandishing a revolver and shouting a torrent of incomprehensible Spanish. Obviously, in a no-go district, to proceed further was out of the question and, as I turned around, I realised that yet again I was in a strange new world.

Later, as I plodded through the bustling heart of the city, perspiring from the humidity, I looked for a small budget hotel. Two policemen just scratched their heads and shrugged when I asked if they knew of such a place. Noticing my predicament, an elderly Spaniard dressed in a white tropical suit intervened. After introducing himself as Charles, he offered to accompany me to Hotel Centrale, a comfortable and inexpensive hotel in the old part of the city.

Charles waited for me on a bench in an adjacent square as I checked in, and then took it upon himself to show me some of the sights and recommend some reasonably priced restaurants. Between describing places of interest, he complained bitterly about the squalor, poverty and corruption in his adopted country, where 'the reech get reecher'. He also warned me to take great care of my valuables. Good advice, for the place was seething with dodgy-looking individuals. Thanks to years of travelling, I'd become fairly 'street smart' with a sixth sense to stay out of trouble. But, despite my hard-earned survival skills, every so often I still came close to being 'taken'.

In a small restaurant, Charles suggested I try *bistek* (beef steak) which arrived with rice, salad and spaghetti, a dish that became more or less my staple diet early on while travelling throughout Latin America.

The people of Panama intrigued me, for nowhere else had I seen such a hodgepodge of humanity. Just about every race, creed and colour was represented including *gringos* (a Latin term for all foreigners, but literally meaning Yankee). So nobody took a second look at me.

Inside the lobby of my hotel, I chatted to a couple from the United States who were on a leisurely world tour. Both seasoned travellers and obviously quite well off, they had purposely incorporated Panama into their itinerary to purchase some of the famous, succulent oranges found in that part of the world.

Mister 'Johnson's the name' was in his late sixties, a retired American engineer and, like me, had travelled since his early days. He proceeded to show me exactly what he carried on his travels. 'The lighter the better,' he declared. From a zipped jacket pocket, he produced his pocketbook (a purse cum wallet) which contained the following items: one miniature toothbrush, a pair of nail clippers, a needle and thread, a small pocket knife, a twelve-foot nylon clothesline neatly folded up, a couple of safety pins, and two laxative tablets. Back home, when hotels asked for his luggage, he produced only his pocketbook, which must have raised a few eyebrows. All his clothes were drip-dry for laundering overnight. As this trip was longer than usual, with changes of climate, he also carried an airline shoulder bag. All his clothes were nylon, neatly folded and tightly wrapped in rubber bands.

He then continued to proudly show me how he carried his passport, driver's licence and traveller's cheques, all of which he could produce instantaneously. Everything was concealed in zip pockets, sewn by his wife and unnoticeable from the outside. One pocket contained crisp, one-dollar bills while another contained a wad of tens. Another, twenties. Finally, he produced a small pen-gun loaded with one tear-gas bullet, which he gladly admitted he had never had to use. His wife was equally well organised and carried an identical airline bag and a small attaché case. Full of tips gleaned from years of travelling, they were the most indefatigable and well-equipped travellers I would ever come across.

I met several other travellers in Panama, either in the street or at my hotel, mostly Americans. Nearly all had depleted their funds, and for most of them, that was the end of the line because there was no overland route south through the impenetrable Darien Gap. To proceed south, one had to fly. Or take a boat.

After purchasing a plane ticket to Bogotá on Avianca, Colombia's national airline, I arrived at the airport and sat down in the lounge to await my flight. A tall, harassed-looking young Englishman flopped down next to me and asked if the Bogotá plane had left. When I replied it hadn't, he breathed a sigh of relief and said, 'That makes a change. Nothing runs on schedule in this sordid part of the world.' Jonathan was a chartered accountant, he told me. Although based in London, he was seldom there for he enjoyed working and accepting positions abroad. He had travelled widely with all expenses paid and with prearranged jobs.

The flight left 15 minutes late because the embarking passengers were all shouting, pushing and elbowing each other for a seat. Normally, I would have been flabbergasted at this display of rudeness and attitude of every man for himself, but here in Latin America, it seemed to be part of everyday life. There was no seat allocation — everyone found their place with what seemed to me the maximum amount of noise and fuss. I'd never seen such an excitable crowd.

As Jonathan was, for once, on a fairly tight budget, we decided to share a hotel room in Bogotá, the capital city in the eastern mountains of the Andes, well known for gang violence, pickpockets, drug dealing and as one of the most dangerous cities in the world. Fortunately, I didn't know that at the time.

Together we checked into a small pension in Avenida Siete, the main street. Our room possessed no windows, smelled musty, was stifling hot, and only after using the toilet did I discover it didn't flush. Later, when I went to take a hot shower out trickled a few drops of cold water. It was by then too late to look for an alternative place, and it was dirt cheap, but already I was beginning to learn that Latin America could be very frustrating at times.

Spanish-style architecture greeted us the next morning when we took a look around town while fast, impatient drivers and stray donkeys hogged the streets. Perhaps, not surprisingly, everyone made the sign of the cross each time they stepped off the pavement. The many churches were doing a roaring trade, packed to capacity, while poverty-stricken peasants and beggars sat or stood outside with hands outstretched. For the first time in my life, I saw businessmen commuting to work on roller skates!

Jonathan typically stayed no more than two days in a place and was soon on his way, so I moved into another small hotel further down the street. My new room also had no windows, but it did have a door to a balcony. It was slightly cheaper than the other, but the plumbing was of much the same standard!

Slowly my Spanish improved, which made it easier to ask for directions and order meals. Fortunately, accommodation and food were very inexpensive in Latin America. Regardless, I once again needed to find work. So, between spates of sightseeing, I made various enquiries. Everywhere I went, I was politely cautioned to guard my valuables. Once, in a rundown part of town, an old peasant woman came running after me, merely to advise it would be dangerous to walk further. Even a shoeshine boy laughingly told me, 'The thieves here are so smart they can remove your socks without removing your shoes.' A gross exaggeration, I thought, but after many warnings, one I took to heart.

Outside the posh Tequendama Hotel where I paused to read my map, a young student named Fernando Peralta introduced himself. Within a few minutes, he suggested that I stay at his home, 'to help my English.' After talking it over with his parents who thought it a great idea, I moved in with the Peralta family who very quickly regarded me as one of them.

Fernando's father was an accountant; his mother, Maria, was involved in charity work. They lived in a large, rambling house in a fashionable suburb of the city. Of their three children, Fernando at 18 was the eldest. Next came

Elizabet, who was 15 and like her mother very pretty, and last but not least, there was Fabio, aged seven. Typical of many well-to-do families, all spoke fluent French in addition to Spanish. Like most Latin children from good homes, all three were extremely obedient and well-behaved and addressed their parents as 'Señor' and 'Señora.' Their house stood in the well-to-do area of South Chapero and was staffed by two maids named Graciella and Anna Lucia. I became 'Señor Da-veed' to everyone. In return for their gracious hospitality, I would give the entire family English lessons. It was an ideal arrangement that suited me fine. At least for a short while. I got my room and board. And the family got to practise their English on me.

Nearly a week had passed, and I still needed to make some money. Fortunately, like in Japan, a number of English conversation schools had recently sprung up, and I struck it lucky. One needed a part-time teacher immediately.

My classes at school were larger than in Japan, with 20 to 30 adult students. At first, it was hard to pronounce some of their names: Adolpho Clavijo, Maritza Escamella, Clara Casenada, Gabriel Diaz, and Maria Christina Barbosa. Most of them were excitable, fidgety, and talkative — more characteristics of the Latin people! They tried my nerves more than once until I hit on an idea to keep order in class. It was simple but effective. Every time a student spoke out of turn or interrupted the lesson, I fined them the equivalent of 5 cents. As none of them were well off, and a tiny amount could buy a bar of chocolate, the system very quickly had the desired effect. The money that I collected I distributed to some of the many street beggars. When a nun joined my class, I felt sure her presence would have a subduing effect. Alas, she was more excitable, talkative, and giggled louder than anyone else.

The situation reminded me of Lawrence Durrell, the famous English writer who recounted in his book, *Bitter Lemons*, about teaching on the island of Cyprus where his students were nice but very unruly. He tried several things to discipline them, but nothing worked. Apparently, the boys ignored him, and most of the girls fell in love with him.

All of my students, however, were extremely friendly and several of them invited me to parties or offered to show me around in their spare time. One named Sylvia, a vivacious, curly-haired woman in her mid-twenties, invited me to a party where I first tasted *aguardiente* (an extremely potent anise-flavoured alcoholic beverage made from sugar cane). I'd never been any good at dancing, feeling particularly awkward and self-conscious on the dance floor, but with Sylvia's expert instruction and feeling heady from the effects of too much drink, I managed quite well and danced most of the night.

Numerous invitations followed, not to celebrate any special occasion but simply because Latins love all kinds of merrymaking. And use any excuse for a fiesta. Fortified by *aguardiente,* I would dance the night away and then stagger back to my apartment quite dizzy and exhausted. Picnics, too, were popular, and several times I drove with new-found friends into the surrounding countryside for a barbecue — Colombian style. Huge steaks and jacket potatoes followed by strawberries and ice cream were washed down with fine Argentinian or Chilean wine to the accompaniment of music from a solitary guitar. Invariably, halfway through such occasions, I developed a fierce headache and had to take a break. I was never sure whether to attribute it to the altitude — the Bogotá environs being several thousand feet above sea level — the alcohol or the hectic Latin lifestyle.

Much as I enjoyed staying with the Peraltas, I preferred my independence, so Fernando helped me find a suitable *apartamente* not very far from his home. During my brief stay with his family, Graciella and Anna Lucia had spoilt and fussed over me, and when Fernando explained that I was moving out, they showed their displeasure with tears! Having learned that many Latins are sensitive and temperamental, I carefully explained that I wasn't going far and would visit them often. That seemed to cheer them up. Despite the frequency of my subsequent visits, each occasion was celebrated as though I'd been away for years.

I moved into my apartment, part of a large, old house where I could gaze at the surrounding mountains of the lower Andes from my bedroom window. I settled for this place because the price suited my pocket and the proprietress was not only pleasant, young and attractive, but she also spoke a tiny amount of English. Ida was her name (pronounced 'ee-da'), and she, like everyone else, called me 'Da-veed.'

Well past midnight after dinner one evening, I joined four of my male students, all of them carrying guitars or violins. This was to be my introduction to a *seranata*. After clambering into one of their battered old cars, we drove some way before pulling up outside a dark, shuttered house. We followed Alvaro, the leader, and assembled below an upstairs window.

Like an orchestra conductor, Alvaro silently raised his hands, quietly uttered the words, 'uno, dos, tres' (one, two, three) and suddenly the street came alive with beautiful, romantic music. As I stood enchanted, several lights in other houses went on, and neighbours opened their doors or windows to clap, or shout 'Bravo!'

About half a dozen romantic ballads were played or sung, each separated by a short interval, during which there were whispered consultations between the friends on what to play next. Alvaro briefed me about what was going on. The girl living in this particular house was a sweetheart of Pedro's, one of the members of the group, and to show his admiration, he was serenading her. If the feeling is mutual, Alvaro went on, the girl must switch her light on to show her appreciation. 'Eef it remains on,' Alvaro continued, 'she weeshes to be keessed.' We all looked up. A light went on — then off. Pedro simply smiled, shrugged his shoulders and commented, 'Thee night is yong.'

We made several other stops that night to serenade more girlfriends and, I suspect, friends of sweethearts and only once did a light remain on. The lucky Jorge then excused himself and returned beaming several minutes later. My respect for the group's perseverance mounted as we kept at it 'til nearly dawn. I was impressed at their uncanny ability at choosing the right windows. To me, the *seranatas* were a unique and exceptionally romantic courting gesture, and the haunting melody of 'Strangers in the night' sung in Spanish will always remain a vivid memory.

After inserting an ad in *El Tiempo*, Colombia's respected main newspaper, I acquired several private students, three or four of whom came to the apartment for lessons. One businessman preferred his lessons in different coffee shops, and another, a lawyer, requested them at his home. Twice a week, a chauffeur-driven Buick picked me up and took me to the exclusive suburb of Chico, where I gave lessons to the lawyer and his wife in the comfort of their luxurious, modern home. His wife was one of the most beautiful and elegant women I'd ever met. Every time I reported for lessons, she made a grand entrance in a different designer gown.

Friends continued to warn me daily about the high crime rate, which after a few weeks I had good reason to believe. As I walked down a crowded street after school one afternoon, I suddenly felt a hand gently sliding towards my wristwatch. No sooner did I react and turn around than the would-be thief had melted into the crowd.

One evening during a light drizzle, I took a walk with José, another of my students. The streets were busy as usual with crowds milling around when, out of the blue, we were approached by a fairly respectable-looking fellow. Nearing us, he discreetly opened his jacket and pointed to a holstered revolver while flashing an identity card which I couldn't decipher. With a deadpan expression, he jabbered away in rapid Spanish, nodded his head in the direction of a quiet side street, as if to say, 'Let's go,' and then touched the butt of his gun.

After so many dire warnings, I immediately thought it was a stickup and had horrifying visions of being found murdered in the gutter. As my heart pounded from apprehension and fear, I considered which way to make a run for it. Before my imagination ran completely riot, José gripped my arm, 'Daveed, passaporte. Passaporte!' I fumbled for my identification and quickly handed it over. The fellow thumbed through it then handed it back. He then shook my hand and with the same deadpan expression mumbled, 'Gracias,' and disappeared into the night. My alarm was reasonably justified for when I related the incident to a colleague, he remarked, 'You were lucky. That plainclothes police officer was legit. It's not uncommon here for the wrong people to carry guns and fake ID cards.'

It was during my stay in Colombia that a great tragedy took place in a small mountain town not far from Bogotá. Two trucks, one carrying wheat and the other rat poison, were involved in a head-on collision. Nobody seemed to notice or care when some of the poison ran over the wheat, which was duly gathered up and taken to the bakery where it was made into bread. Within 24 hours, 74 people had died from poisoning and 140 were hospitalised, all from the same town. So grief stricken and angry were the victims' relatives that to seek revenge from the town where the rat poison originated, they mixed ground glass with sugar and arranged for it to be sold. Many more people suffered as a result. After hearing that horror story, it was weeks before I stopped scrutinising every bowl of sugar.

In my apartment, I had to contend with several issues. No water when I wanted to wash or shower. No toilet paper in the lavatory. The phone was out of order. A different problem every day!

Invariably, ordering breakfast from the kitchen was an exercise in frustration. Often there were no eggs available so one of the two servant girls, Anna or Carmen, walked to a nearby store to purchase some. While the eggs were being fried or scrambled, I went back to my room to shave. After being called to eat, I sat down only to discover there was no bread, so once again off went Anna or Carmen. By the time they had chatted to friends on the corner and returned, my breakfast was of course cold. I suggested to Ida not to bother in future which upset her a little but why they couldn't plan ahead by purchasing eggs the previous day was completely beyond me.

I asked Ida one day if she knew of a tailor in the vicinity where I could have my jacket repaired after I'd ripped it climbing onto one of the many dilapidated and always overcrowded buses. As she owned a sewing machine, she kindly offered to repair it herself. It would be ready 'mañana,' she said. When tomorrow came, and it hadn't been touched, I asked again for the address of a tailor. 'Don woree,' said Ida. Three days later, it was finally 'feexed'.

Sometime previously I'd given Ida 45 US dollars for safekeeping, and when it was time to pay my monthly account, lo and behold, the money had disappeared. Ida searched frantically everywhere, but to no avail. The police were then called in, and questions asked, and Ida sobbed her heart out. The money was never found. I tried to console Ida as best I could, for she felt this incident would spoil my impressions of her 'contree'. I assured her I had made too many friends to think anything but good of Colombia but it left me with doubts about Ida.

At three o'clock one night, I was awakened by the deep sobs of a man outside. I got up and opened the curtains. Three men stood facing another on his knees in the garden next door. A would-be car thief had been caught in the act and was now receiving Colombian retribution. Interspersed with loud, angry voices, I heard the dull thuds of fists finding their mark. I immediately thought of calling the police, but Ida crept into my bedroom and whispered to me not to interfere. 'The poleece will com, don' worry.' The commotion continued for nearly two hours — sobs and angry conversation for a few minutes, then a rain of heavy blows followed by more cries and groans. Finally, just before dawn, a police car pulled up and bundled the thief, who by then must have been feeling very sorry for himself, into the back.

Two nights later, I was woken at about two a.m. — this time by a group of musicians on their musical mission — another *serenata*. As they stood playing in the street beneath umbrellas in drizzling rain, I couldn't help wondering, '*What kind of country was this, where violence and romance went hand in hand?*'

On top of everything I was coping with during my stay in Colombia, I made appointments with various people, usually my students, only to be let down time and again. Almost to a man, they had no sense of urgency or punctuality, and many were the times I fumed and waited in vain. When I mentioned this to a colleague, he merely commented, 'You have to get used to it or you'll go round the bend.' Despite the many frustrations, I came to adore Latin people, if for nothing else but their zest for life and tremendous charm.

Pete duly arrived in Bogotá and immediately moved in with me. From Australia, he'd flown to Tahiti where he spent a couple of delightful days and then caught a ship to Panama. Together, we made plans to travel down the Amazon. While he went sightseeing, I set about saying, 'Hasta luego' (until next time) to my friends. The three months since my arrival had flown.

Prior to leaving, I was invited to a bullfight by a student named Sebastián. After queueing up with hundreds of other spectators who pushed and shoved from all sides, we passed through the stadium turnstiles where we were frisked by tough-looking riot police. Once inside, I forgot the long wait and got carried away by the atmosphere like everyone else. To a fanfare of trumpets, El Cordobes, the world champion matador at the time, flanked by lesser-known toreadors, marched into the ring. A massive roar from the crowd greeted the hero before everyone settled back to watch.

While this was going on, Pete was outside desperately trying to purchase a ticket, but he gave up after twice being hit with truncheons by mounted police who had their job cut out controlling the boisterous crowd.

It hadn't taken me long to notice that Latin women, besides being well-groomed, were beautiful and apparently passionate. Noticing the excited expressions of many in the audience, I couldn't resist commenting to my companion 'I suppose the ladies lose their hearts to the matadors,' to which Sebastián casually replied, 'they lose everything.'

After bidding 'Muchas gracias' and 'Adios' to my many new friends, I left a parcel of warm clothes with the Peralta family to forward to an address I would send later. Had I known how long it was going to take to arrive at its destination, I wouldn't have bothered.

Despite Colombia's alarmingly high crime rate, which I did my best to ignore, the country was well known for its two magnificent coastlines (the Pacific and Caribbean), coffee, emeralds, spectacular scenery, and extraordinary flora and fauna. I hadn't done much reading about it before arriving and, as much as I enjoyed my stay there, I regretted not doing full justice to it. In other words, I should have stayed longer and ventured further afield.

CHAPTER 16:

Pete Saves a Life

> *'Do not go where the path may lead,*
> *Go instead where there is no path*
> *and leave a trail.'*
> — Ralph Waldo Emerson

Pete and I were, as usual, in extremely high spirits to be back on the road as the feeling of complete freedom overcame us. Our destination: the Amazon. We travelled by dilapidated long-distance bus to Cali, where we booked into another ridiculously cheap hotel. Two dollars per night was all it cost, and we had to admit it was certainly great value.

An armed guide travelled on the bus with us in the event of being attacked or held up by bandits. We stopped several times at roadblocks where everyone had to leave the bus while it was searched and we were frisked for concealed weapons. Amusingly, long knives, machetes, and swordsticks were considered alright, but not arms or ammunition.

As we climbed and snaked our way through the Andes, the road became extremely steep and winding. Time and again, Pete and I were to glance nervously at one another, gulp and hope for the best. Before then, we had both travelled over what were undoubtedly some of the most treacherous roads and passes in the world, but those in the Andes were equal to or even more hazardous than most we had travelled. We seldom knew how high we went, never bothering to ask, intent only on reaching our destination — hopefully in one piece.

On what seemed like every bend, there stood one or two white wooden crosses. These we were told did not necessarily represent one or two people or cars, but invariably whole busloads or truckloads of people! And as everyone made the sign of the cross and held their breath, we made our way slowly and carefully along the most hair-raising stretches. Frequently we stopped at roadside shrines where everyone prayed; a collection was taken and offered to the Virgin Mary.

On another perilous mountain bus ride, we had fabulous views of snow-capped Chimborazo. At just over 20,000 feet, it is Ecuador's tallest mountain and slightly higher than Africa's Kilimanjaro. I knew what Pete was thinking as we looked at each other knowingly. He was fully recovered from his Mount Fuji ordeal, and we were both longing to stop and climb it.

In the town of Pasto, Colombia, on the border of Colombia and Ecuador, we reported to the Immigration Department which, throughout Latin America, was then controlled by the police. An armed sentry took us to 'El Capitán'. Smartly dressed in riding boots and khakis, Captain Santa Cruz clicked his heels together, saluted and bid us, 'Buenos dias.' He then repeated it in English, adopted a Napoleonic pose, obviously very pleased and proud of himself. He continued in broken English, saying how much he respected the British 'poleece'. We spent a few pleasant hours in the company of the captain and some of his fellow officers, who kindly insisted that we stay for lunch. During the meal, they excused themselves one by one until only the captain was left. 'Security duties,' he explained as he recounted some of his experiences. Because the police in Colombia also perform military duties, much of their time was spent on standby, always ready to be called. He himself spent much of his time on mountain and village patrols hunting terrorists and bandits and had seen several of his colleagues die in ambushes.

With our British passports, we had little difficulty in crossing most borders in South America, and we crossed into Ecuador easily enough. However, some Latin countries at that time had strict customs, regulations, and road and rail checks at their borders and well into no man's land. The first we knew of this was when most of our impassive Indian peasant fellow passengers — on yet another creaking, rickety bus — suddenly came very much to life. Most had been shopping across the Colombian border and, as we neared a roadblock, one very fat woman with a pigtail and gold teeth thrust two parcels containing linen into my lap while another dropped a new transistor radio onto Pete's. As they sat back grinning, we cottoned on, winked at each other, and settled back to watch the fun. Another woman then placed several pairs of new cheap shoes around our feet, and yet another handed out enormous sombreros to all the men. I didn't mind wearing one hat, but when another was plonked on top, I felt ridiculous!

When we stopped, a uniformed customs officer entered the door, and a hush fell over the bus. After walking slowly up and down the aisle, prodding various sacks and bags with a stick, he seemed satisfied.

No sooner had the engine of the bus spluttered into life, than an old Indian man in front of us turned around and in his faulty Spanish and sign language indicated, 'One down. Two to go.' By now, everyone had become very excited, and we were doubled up with laughter. The excitement reached its peak as we approached the second customs post with little or nothing confiscated and nobody fined.

As two or three passengers obviously lived between the second and third posts, they started throwing bundles out of the windows, intent on walking back when they left the bus to collect them. This highly amusing farce ended when we pulled away from the last customs post, and everyone remaining returned the articles to their original owners! As if to cap everything, all this took place under the eyes of an off-duty customs officer. Even funnier when his colleagues had finished searching the bus, they were, in turn, searched by fellow customs officers! By then, Pete and I were beside ourselves. It was hilarious.

We witnessed these antics several times in South America and always cooperated good-naturedly with our bus companions. We had little or nothing to lose, and the passengers had something to gain. They relied on us gringos knowing that rarely would we be searched, and this strategy always paid off, providing us with endless fun at the same time.

Another Christmas on our marathon journey was fast approaching as we neared Quito, Ecuador's mountain capital. It seemed strange hearing 'Hark the Herald Angels Sing' and 'O Come All Ye Faithful' sung in Spanish from the packed churches. An even more unfamiliar experience was seeing old peasants, both men and women, carrying Christmas trees home on their heads.

As it was now our intention to travel down South America's most famous river, a fellow traveller had suggested the best way to get there was by the once-weekly military plane, so we applied to the Ministry of Defence for information. There, after consulting one army officer after another, a colonel suggested that, since their planes were usually full, we should apply to the Brazilian embassy as they operated a similar service. There, we were informed that there was a possibility, but we would first have to obtain visas. Earlier, we had learned that British passport holders did not require visas for Brazil, but the officials at the embassy insisted that we did. We rushed out to get passport-sized photos. On our return to the embassy, perspiring and annoyed, we were solemnly told the next plane was full! Eventually, upon arriving in Brazil, we learned much to our frustration we didn't require visas after all!

Our hopes of flying dashed, we decided instead to travel to Lima from where we could make our way overland to Pucallpa, a small town situated on one of the Amazon's tributaries. While I watched the rucksacks, Pete went off to look for the bus station. He took so long that I thought he was lost, so I explained my problem to a friendly priest standing nearby, and he offered to take me to the bus station on his Vespa. What a sight we must have been, he in his flowing black cassock and me with my rucksack on my back and Pete's strapped on behind, as we careened through the hilly streets of the capital. Fortunately, we found Pete as expected, making enquiries at the bus station.

Due to the infrequency of the bus services, we actually did some hitchhiking. Thumbing always offered one great advantage — to visit numerous small villages and towns where we would normally never have gone. Not to mention often meeting some very nice people.

In the town of Riobamba, we came across a bearded Scotsman half sitting in the gutter, retching for all he was worth, probably suffering from food poisoning. 'Is there anything we can do for you?' Pete enquired politely to which the Scotsman shook his head.

Having asked us where we were from, he stretched out his arms wide and exclaimed loudly in his Scottish burr, 'Ah, Britain. Decent food.' Then quietly, 'I wish I was there now.'

Throughout our wanderings, Pete and I suffered frequently by sampling local foods. More than once, we both endured Montezuma's revenge. Although it caused us, and me in particular, enormous discomfort at times, we simply regarded it as one of the more unpleasant aspects of travelling, especially in so-called third world countries.

Further south, we were picked up by a bespectacled American butterfly-catcher driving a Volvo. Leaving him when he turned off the main road, we sat by the roadside and shortly two tiny peasant children shyly approached us. Dressed in rags, they stared at us until we produced some sweets. No sooner had we given them some, they raced away across the plain as fast as their little legs could carry them.

In the historic town of Cuenca, we explored the colourful marketplace. It was full of poverty-stricken Indians, most of who had walked down from their homes in the mountains, bent double under sack-loads of produce, taking them to the colourful, lively market. Both Pete and I loved markets because they were invariably busy and interesting and always captured the flavour and atmosphere of everyday life. The weekly event served a dual purpose for the locals — not only a place to display their wares but also a place to exchange gossip. Colourful ponchos, scarves and blankets lay next to green beans, maize, shrivelled, unappetising fruit and a remarkable variety of potatoes.

At one time, we sat on the road for roughly five hours before a truck came along and picked us up. As the back was loaded, the driver insisted we get into the cab, which was already occupied by himself and his assistant. We squeezed ourselves in; then began what was probably the most frightening ride of our lives to date. The sun was just setting over the mountains in a spectacular display of colour as we drove. Higher and higher we went, and although the driver travelled slowly and carefully, both Pete and I felt very nervous because the road was so narrow and exposed.

Gradually darkness fell and before long we were travelling in swirling mist. Unable to see through the windshield, our driver stood up and leaned out of his door with his foot on the accelerator. Pete and I sat virtually blind and immobile while the driver's thickset assistant sat with arms crossed, quite unconcerned. When we passed one particular spot, he pointed down at the sheer drop of *un kilómetro* which did nothing to ease our apprehension. On and on we went, slowly negotiating hairpin bends as we sat petrified for some hours. Eventually, we entered a dimly lit village of wood and mud buildings.

'I've had enough for one day,' I said to Pete, who apparently felt the same way. Feeling limp and drained from our unnerving experience, we asked the driver to drop us off. In the darkness, we rapped on the door of what looked like a small hotel. A woman opened the door with a lamp in her hands and showed us to a dungeon-like room. After one look at the filthy state of the beds, we walked out to the nearby square where we found another place. It wasn't much better, but it was 2 a.m. and intensely cold. So we made do.

The next morning as we walked around, we noticed a statue of José de San Martin, the founder and liberator of the country whose stone busts and pictures were displayed throughout the country in offices, parks and town squares. It was a kind of hero worship similar to other South American countries where, for example, it would be Simón Bolívar in Bolivia and Bernard O'Higgins in Chile.

Invariably, after finding a hotel, showering and changing, we took an evening stroll. In all the towns and villages of Latin America, the plaza is dominated by a massive old church and is the focal point, principal meeting place and main source of entertainment for most of the town's residents. Beggars mingled with courting couples walking around the square. Street photographers with ancient box cameras solicited customers to a constant background of loud music. Practically every shop played records all day long and well into the night, convincing us that Latins would die without music of some kind. And many were the nights we fell asleep to the strains of 'High Noon' and 'Canadian Sunset' from blaring radios nearby.

Generally, we chose our hotels carefully in Latin America and, as always to minimise expenses, we shared a room. Frequently, we had rooms with a balcony and sometimes fine views and, when no *balcón* was available, we were disappointed. For an average price of only a couple of dollars, we seldom complained. Our main daily meal usually cost us the equivalent of 50 cents.

When eating in restaurants in South America, much to our delight, we found that fish dishes always came with sweet potatoes and chunks of cooked white corn. Sea bass marinated in lime and sprinkled with chopped parsley was a Peruvian favourite and especially delicious when downed with a local Cristal or Pilsen beer.

Not far from the Peruvian border, we boarded our most uncomfortable bus to date — by then we thought we'd seen them all. Inside, I guessed the height from floor to ceiling was only four feet, and the aisle ten inches wide. For most of the eight-hour journey, I shared my seat (made for two dwarfs) with two small boys, a large, musty bundle of blankets and one enormous cabbage. Pete sat across the aisle, squashed in a similar position. He'd been complaining recently of severe constipation, and only the previous evening his system had returned to normal. Now, in these cramped and uncomfortable conditions, he was terrified that he might have to go at any time!

We had heard about the 'crazy, drunk drivers' of Latin America, but fortunately we never came across any, not drunk anyway. Most were friendly, courteous and competent; to be fair, most accidents in that part of the world were the result of landslides, road cave-ins, or faulty brakes.

We were able to cross into Peru on a Sunday after paying the immigration officer an American dollar for his 'overtime'. Peru was another of the many countries that greatly excited and intrigued us, with its mountains, vast plains and desert. It also boasts superb sandy beaches and hundreds of square miles of impenetrable jungle — truly a country of contrasts.

At the small border town of La Tina, we had to wait six hours for a ride as there was no bus service. When a truck finally came along, we climbed in the back joining several Indians, a pig, a goat, and two chickens. We travelled four hours this way, stopping five times en route for passport checks at police roadblocks.

After staying in another cheap hotel, we plodded out of Sullana the next morning and passed a shabby roadside café. The elderly proprietor beckoned us over and presented us with rolls and coffee to celebrate 'Navidad' (Christmas). We spent most of the day on the back of a truck travelling through the vast, scorching Atacama Desert. In Piura, home of 1967's Miss World, we celebrated Christmas 1967 with a dinner of chicken and chips, orange juice, and ice cream. Oh, the luxury of it all.

At the small town of Chaclacayo, we chose not to travel by bus for some inexplicable reason, but to continue hitchhiking. Our decision proved to be providential. The bus in which we might well have travelled collided head-on with a truck, killing 12 passengers and injuring 15. We shuddered and were greatly saddened when later we heard this news!

In Lima, Peru's huge, busy capital, we occupied a hotel room which was so small that one had to go outside while the other changed. The walls were made of cardboard, and it took no time for us to realise the same establishment was used as a brothel!

While collecting mail at the British embassy, the receptionist asked, 'Why don't you fly? It's much quicker,' when we casually mentioned we were heading to the Amazon. 'Only when necessary and when the price is right,' we explained. Three days later, high in the mountains, we came across the debris of an airliner which had crashed into the mountains in bad weather, killing everyone on board. Around the same time, an Avianca Douglas C-54 had crashed into the ocean which made us even more apprehensive about flying. We much preferred our feet on *terra firma*.

Outside the town of La Oroya, we saw our first llamas tended by their Indian herdsmen and, as the scenery became more rugged and spectacular, we couldn't resist breaking into song, something we often did on the road, especially when we felt inspired by what we saw. At that time, the song 'Born Free' was popular and, considering it very appropriate, I sang it often and at the top of my lungs. Another of my favourites was 'The Happy Wanderer.' Fa la la!

Another truck ride, this one for eight hours, took us to Huánuco where we had our first view of the snow-capped Andes from a height of over 14,000 feet. On New Year's Eve, after travelling over a particularly dangerous stretch of mountainous road, we counted ourselves lucky to be alive. We shook hands at midnight, wished each other 'Buena suerte' (Good luck) and, exhausted, fell asleep in our sleeping bags under a parked truck.

It took us three and a half days to reach Pucallpa from Lima. Immediately upon arrival in this jungle town on the banks of the Ecayali tributary in the Amazon rainforest, we canvassed numerous launch owners, hoping to find one going to Iquitos, where we hoped to pick up the regular Amazon steamer. After fruitless enquiries, we went to the local police station where we met a somewhat inebriated major — the fifth slightly drunk police officer we'd met in a couple of weeks — and explained our position to him. He suggested we return in the morning. By then, he was completely sober, apologised for being under the influence and, with his cooperation, we found a small launch.

That afternoon after taking a siesta in our hotel room — something we now frequently did because it was a local custom and because the heat was stifling — we visited one of the two air-conditioned cinemas, a rarity for us while travelling, as the life we were living was infinitely more exciting than the big screen.

Our two nights in Pucallpa were unbearable. In contrast to the biting cold of the mountains, it was extremely humid. Also in the dense jungle, there were mosquitoes by the thousands. And, as always, they loved me!

We breakfasted on fried eggs in a little wooden shack on the banks of the river. And as we sat eating, two Swiss travellers came in, stared at us and then at our breakfast, and one said, 'Here ve are in ze middle of ze jungle, and you people order fried eggs for breakfast. You could only be British!'

Hans and Martin were travelling in the same direction as us. All four of us boarded the *MV Cosmos*, one of numerous small boats that plied back and forth along stretches of the river, carrying supplies and passengers. Specially designed for river travel, it had a flat bottom to enable it to slide off hidden sandbanks. We hit a couple but soon slid off. There was no danger involved; we merely came to a shuddering, jarring halt.

Arriving at Ramón Castilla (where the borders of Colombia, Peru and Brazil meet), our boat was engulfed by many Indians pushing and shoving as they sought places to sling their hammocks. We were quickly hemmed in by men, women, children, a small black monkey and a talkative green parrot in a cage.

The boat journey to Iquitos took three days and two nights and passed very quickly as we relaxed in our rented hammocks or on the roof, enjoying the cool river breeze. The boat stopped at several riverside villages to load or discharge cargo. With no roads or vehicles, the river was the lifeline for the local people. Now January, it was the height of summer in the southern hemisphere and being in the heart of the dense rain forest, it was extremely hot and humid.

We eventually arrived at Iquitos, on the banks of the Itaya River, another Amazon tributary. As Peru's prosperous and vibrant northernmost city, it is known for its rustic stilt houses and the only roads leading into the surrounding jungle and a few tiny townships. There, we made enquiries about boats to Belém in the Amazon delta. Fortunately for us, a boat was leaving in a couple of days.

Martin and Hans stayed on board when we disembarked. They were heading for a small town up one of the tributaries. We waved farewell to them as the sun was setting and wished them luck as their little craft chugged out of sight.

In the Iquitos town centre, we met a German airline hostess who was on holiday. While her daily hotel bill came to ten dollars, ours was much less, so she quickly moved to our accommodation. We also told her about a very reasonably priced restaurant. We'd developed an instinct for sniffing them out.

A visit to a fish sanctuary provided some interesting revelations about the so-called deadly piranha fish with the formidable reputation. A carnivorous fish many only five or six inches long with razor-sharp teeth, piranha inhabit the lakes, rivers and floodplains of South America and only attack their victims, usually large mammals, when they are injured and bleeding profusely. Attracted by the scent of blood, the piranhas swarm their victim and shred it to the bone. A slight scratch on the arm or leg would not be sufficient to attract them, we were relieved to know.

We had by this time swum in the muddy, brown waters of the Amazon to escape the tropical heat when the *Cosmos* stopped in what looked like a safe spot. It was perhaps foolhardy and reckless of us, for we were totally ignorant of the hazards of the river.

The real dangers were the electric eels varying in length from one to eight feet. Some small ones were on display in glass aquariums outside a small store. Pete was invited to touch one, and he received a distinct electric shock. As he gasped, we realised what just a slight touch from a much bigger one could do.

We also heard about the black caimans, members of the crocodile family, which grew up to 12 feet long and, although considered dangerous, usually only ate fish, birds, reptiles and small mammals. Seldom human beings!

No matter the tributary, the river current was fast, meaning one could easily be swept away and drown. Giant anacondas inhabit the river and surrounding jungle and are known to squeeze and suffocate their prey to death. A twenty-foot anaconda could, and often did, overturn small boats. It was with that newly acquired knowledge and some apprehension that we bought our tickets for the next leg of our Amazon River trip.

Before boarding the next vessel, we'd learned about the incredible bird and wildlife in the rainforest where we now found ourselves. Indeed, South

America has an extraordinary and dazzling display of flora and fauna. At more than 1,200 species, the birdlife even eclipses that of Africa and Australasia. They include the colourful toucan, a variety of parrots, parakeets, hummingbirds, kingfishers, storks, quail, jays, grebes and flamingos. Seeing just some of these species gave us enormous pleasure.

Even though we hardly glimpsed much of the wildlife, we knew it was there. Caimans, capybaras, tapirs, jaguars, coatimundi, armadillos, anteaters, and nine species of monkeys along with various amphibians and reptiles live and lurk in the river or surrounding jungle. Alpacas, llamas, foxes, and the magnificent condor vultures make their home in the mountains or on the vast plains.

Our next boat, the *SS Lauro Sodre*, was a Brazilian steamer and, once on board, we were surprised at its size. At 5,000 tons, it carried between 300 and 400 passengers, the majority of whom, including us, would sleep in hammocks. While we commandeered some deck space, the ship filled with people including several American and European travellers. Among them were Teresa (Swiss) and Breta (German), both in their late twenties. Our paths would cross again in the not-so-distant future. There was also a young couple from Los Angeles and what seemed like hundreds of native Indians with their cardboard suitcases and huge bunches of bananas. Plus a couple of caged parrots.

At meal times, a bell was rung, and we all had to form a queue outside the galley with our own utensils. The food was dreadful, usually thick tasteless stew followed by black coffee. The coffee cost extra, and we only got it if and when the cook felt disposed to making it. After our first few meals, we were not very happy, for we'd learned this particular trip would take several days.

We had been on board only about 24 hours, and the sun was just setting when we hove to a couple of hundred yards offshore from a small village. A small launch had come out to meet us and was tied alongside when I got out of my hammock and ambled forward to get a better look at the proceedings. Little did I know an accident was about to happen; and I would be an eyewitness.

As I leaned over the rail, I noticed several people transferring their belongings into the small boat. There was a sudden yell. I looked closer. A young Peruvian boy aged about nine missed his footing while attempting to step from our boat into the launch, slipping and plunging into the river. As I watched, he drifted helplessly away — head bobbing, arms flailing, obviously unable to swim. I glanced around. Nobody seemed to be doing anything except stare. I started to run to the stern, thinking of diving in after him — but Pete beat me to it.

He'd seen the entire incident and quick as a flash assessed the situation. With astonishing presence of mind, he first hurled his wallet onto the deck and then, with complete disregard for his own safety, hopped onto the rail, paused for a second, gauged the distance and, as the boy drew level, jumped into the river beside him. It was a leap of faith.

Meanwhile, the two men in the launch were struggling to get the motor going. Seconds later, it roared into life and they chased after the two rapidly receding figures, finally catching up with them some 150 yards downstream.

A rope was thrown to them. Clutching it for all they were worth, Pete and the boy were hauled to safety. Both were soaking wet, somewhat dazed and totally exhausted, and the boy was suffering from shock. When Pete stepped back on board, still catching his breath, I handed him his wallet, and there was a stunned silence followed by a ripple of applause from the European passengers. 'Phew, that was a close thing,' he uttered between gasps. 'I couldn't hold on any longer.'

Later, when he had showered, changed, and recovered from his ordeal, he admitted, 'If that boat hadn't come along when it did, we'd never have made it.' The boy was struggling and Pete was tiring fast, trying to hold both his head above water and that of the terror-stricken boy. He was about to let go of the boy to try to reach the bank on his own. But he held on gallantly for as long as he could and, in doing so, saved the boy's life. It was a heroic, superhuman feat by any stretch of the imagination. In most countries, Pete would have been decorated with a medal for such an act of bravery but, in this remote and isolated part of the world, this daring act passed virtually unnoticed. I greatly admired him for it. And made a point of telling him so a number of times.

What impressed me most was that over the years I'd seen him grow before my eyes into a true man of the world. And I was very proud to be his good friend. The incident reminded me that my trips down two of the greatest rivers in the world had been marred by accidents. Thanks to Pete, this time, a life had been saved. That very same night, listening to the news on my little radio, we learned of the tragic death by drowning of Mr Holt, Australia's Prime Minister — a strange coincidence.

As our river journey continued, we spied grass huts out of which emerged girls in miniskirts clutching their Sony radios. The mighty Amazon had been conquered by the 20th century, but we learned that primitive tribes still lived as they had always done deep in the jungle. And we longed to explore the region for ourselves. But much as we wanted to, we couldn't do everything.

As we steamed downstream, keenly observing the riverbanks, we spotted a lone sloth high in a tree while flocks of scarlet macaws and other colourful parrots flew overhead. We also saw spider monkeys dangling, swinging and hopping Tarzan-style from branch to branch. Well before they came into sight, we'd hear troops of howler monkeys, the loudest primate on earth. We stared in awe at the huge kapok trees, by far the tallest of trees found in the Amazon rainforest, and spotted exotic plants such as the heliconia and pink and purple orchids which grew in abundance just about everywhere.

The bustling metropolis of Manaus was a revelation and is where the great Amazon River officially begins. Like most river towns in that part of the world, with no roads in or out, it is only accessible by air or boat, which means that just about everything has to be shipped in. With stunning architecture, its most impressive building was the massive, opulent Opera House built during the boom of the 1800s when rubber barons had more money than they knew what to do with.

The trip by boat to Belém took eight days and was extremely boring at times because the river was so wide in places with no passing scenery. Sometimes we hove to off an isolated village, and everyone went ashore to get some exercise and have a look around.

On board, we happily passed the time away chatting to our international and Indian travelling companions, playing with the hordes of children, reading and napping in our hammocks. There was one particularly well-developed Indian girl to whom Pete frequently spoke. After her mother had seen them chatting several times, she politely suggested that he marry her daughter. Pete was rather nonplussed at the idea, especially when he learned the girl's age. She could easily have passed for 17 or 18, but was in fact only 12! Of course, he politely and firmly refused the mother's offer.

After a few days, the food on the boat became quite inedible, so we bought bread and jam and fresh fruit from villages every time we stopped. We'd been wearing shorts around the ship's deck for some time when the captain issued instructions forbidding them. We couldn't fathom why — maybe there had been complaints from other passengers — but we complied with the order. Always conscious that we were visitors from afar, we didn't want to offend the locals in any way.

The humidity was dreadful when we finally arrived in Belém, so we made a beeline out of the city. Once more, we resorted to hitchhiking, and it wasn't long before a truck stopped. The driver's name was Josef. Our destination was now Brasilia, a three-day ride south over a red-earth, switchback highway flanked by dense, impenetrable jungle. After several hours, we stopped at a restaurant and ordered 'the meal of the day.' A few minutes later, a feast was served — piping hot roast chicken, pork, rice, macaroni and fried eggs followed by thick, sweet, black Brazilian coffee. Having eaten so little on board the *Lauro Sodre*, we descended on the food with wild abandon and gorged ourselves silly. It was so plentiful that even after three helpings there was still enough to feed two or three more people. And the cost, a mere couple of dollars.

As we continued on our way, Josef stopped to pick up passengers for a small fee, including one small boy and his two pet armadillos. At Anápolis, Josef was going no further, so we took a bus for the next three days, stopping only for meals and to spend the nights in roadside 'dormitorios'. On board our bus was a British engineer named Dennis who related some amusing experiences he'd had in Brazil.

Brazilians, like all Latin Americans, are football fanatics and anywhere, anytime you can watch men and boys practising in the streets, churchyards, waste patches of ground, and jungle clearings. When Brazil was knocked out of the World Cup, the nation virtually went into mourning. It was just after this that Dennis was given a parking ticket, and feeling that he was being victimised, exclaimed, 'Brazil. Bah!' throwing his hands into the air Latin style to show his annoyance. 'Police. Bah.' Then finally, 'You can't even play football.'

'That did it,' he said. The comment was just too much for the police, and he was hauled off to explain himself at the police station.

On arrival in Brasilia, then the most futuristic, modern city in the world, we spent the night under a bridge as all the new hotels were expensive. After a quick tour of the city by bus and on foot, we were back on the road. During the 26-hour journey, the bus skirted the Pantanal, a vast area of wetlands and savannah plains and home to masses of wildlife, eventually arriving in Rio de Janeiro.

'God made the world in six days. The seventh he devoted to Rio,' is a well-known Brazilian saying. Seeing it for ourselves, we soon realised they were not empty words. It is magnificent and undoubtedly one of the most beautiful cities in the world. 'Have you come for Carnaval?' a teller in the bank asked. When we replied, 'No,' he said emphatically, 'But you must stay for it.' Regrettably, it was still three weeks away, and we didn't want to hang around that long.

After each buying an enormous pineapple-flavoured ice cream, we took a walk in the early evening and were amazed at the number of streetwalkers who stood on every street corner or mingled casually with passers-by. 'Ola,' they whispered to prospective customers. There was nothing abusive or unpleasant about their approach to clients, and no one seemed to be shocked, or even stared.

In between sightseeing, swimming, downing cool drinks and lazing on the famous Copacabana Beach, we noticed that every public garden and park was littered with courting couples. Brazilians, perhaps more than other Latins, appeared extremely affectionate and made no attempt to hide their amorous emotions. They kissed anywhere and everywhere at what seemed to us the slightest excuse.

By this time, Pete and I had experienced the best and worst of Latin drivers, but in Rio they all seemed quite mad. Most drove on the assumption that all pedestrians are incredibly agile and aimed their vehicles accordingly, only swerving away at the last minute. They never slowed down, and it seemed that half our time exploring the city was spent dodging traffic.

Next came the giant metropolis of São Paulo after a most enjoyable and comfortable ride in one of Brazil's famous long-distance buses. Modern and equipped with all facilities, travelling on the bus was pure pleasure, especially after some of our previous bone-rattling and frightening road trips.

After quickly seeing some of the main city highlights, we boarded a bus heading south.

Having achieved another of his main goals, which was to travel down the Amazon, Pete now became impatient to get to Canada because it was the country which he most wanted to see. I felt I owed it to myself to see more of South America while I could. So, after firm handclasps and wishing each other 'all the best', we parted company in the town of Porto Alegre destined not to meet again for many more months.

CHAPTER 17:
Latins are Lovely

*'A nomad I will remain for life,
in love with distant and uncharted places.'*
— Isabelle Eberhardt

I fell in love with many cities during my travels, and Montevideo, capital of Uruguay, was one of them. Beautifully laid out, it is a vibrant city with wide tree-lined avenues, countless kerbside cafés, glorious beaches, and a Mediterranean climate. Regarded as a leading South American holiday resort, it caters to everyone's taste and pocketbook. It was in stark contrast to Panama City where one day I'd bumped into a fellow and he'd turned around snarling, wanting to fight. In Montevideo, one only has to inadvertently rub shoulders with a stranger, and he or she bends over backwards to apologise. In shops, offices, and restaurants, if the staff knew some English, they would always reply to my 'Gracias' with 'Eet ees nothing' and a beaming smile.

By the time I got to Montevideo, my feet had started to play up from so much walking, so I spent much of my time sipping cool drinks at a pavement café in Avenida 18 de Julio, the city's main street, watching the girls go by. There one could linger over one cup of coffee in a restaurant all day, and no one asked you to move. I was interested to see the men carrying their *yerba maté* in cups made from gourds and their thermos flasks filled with hot water (for refills), sipping the tea through a metal straw as they walked. They drank it everywhere — in the office, at the beach, waiting for a bus, or watching a soccer game. It was a much-loved universal custom.

A few days later I was in Buenos Aires ('BA' to the locals), the largest city in the southern hemisphere: sophisticated and cosmopolitan with its multitudes of cafés, restaurants, immense office blocks, smart shops and incredibly elegantly dressed men and women. It reminded me of a mixture of Paris, Rome and Madrid. After stepping off the ferry from Uruguay, I met a young Argentinian who was also going into town. He directed me to an area of bars where there were some cheap hotels. After being almost knocked down on the way there by a speeding bus on the main drag, Avenida Corrientos, I found a small hotel where I checked in for my first night.

Not knowing a soul in the city, I decided to visit one or two of the newspapers to tell my story and as I'd done elsewhere, hoping that it would produce some worthwhile results. *Some exposure might open a few doors to opportunities such as a job offer*, I mused.

At the offices of *El Nacion*, I was politely let into a huge, musty library by the receptionist where I was then interviewed by a reporter entirely in Spanish. This amused me no end for he couldn't possibly have understood my very limited Spanish! I then called on *La Prensa*, another well-known and influential newspaper, where Theo, a young staff reporter, invited me to lunch 'to sample your first Argentinian beef.' It was the first of many such meals and consisted of a huge rare steak served with salad, bread roll, and red wine followed by coffee.

Theo wrote a fairly accurate account of my wanderings, including the comment that I had made to him over lunch, 'Your women are beautiful, your beef is excellent, and your wine is the best.' I meant exactly what I'd said.

Both stories were published the next day. The one in *La Prensa* appeared in the centre of all the world news. There, smack in the middle of chaos and tragedy, revolutions and disasters, was a photograph and story about my travels written in an amusing vein. But there were no phone calls.

Earlier, I had decided to try to find work, preferably teaching English in Argentina for a few months, so I began calling on some foreign language schools. Unfortunately, I'd arrived at a bad time as all were closed for the summer holidays. The same day, as I strolled down Calle Florida, a popular pedestrian street, I stopped on a corner to get my bearings.

Just then, a fellow in his early thirties sidled up to me and said, 'Ola, amigo. So you are a traveller?'

'Yes, I am,' I replied quizzically.

Ricardo was his name, and he promptly invited me to join him for a coffee. He had decided that I must be a traveller because of the colourful bag I carried over my shoulder — my shan bag purchased in Thailand in which I carried my passport, maps, camera, etc.

Ricardo then told me something about himself: a native of Buenos Aires and single with a private income which he told me was 'only a leetle, but enough'. Although I was to know him for some weeks, I never found out exactly what he did. He was a man of mystery. As we carried on a pleasant conversation, I was a little suspicious, but he seemed friendly and likeable enough. During our talk he insisted that I meet his friend, who was well off and owned a yacht. With nothing to lose, half an hour later I was saying, 'Mucho gusto' to Alberto Chamé, a 40-year-old businessman in the textile industry who was tall, distinguished-looking and tanned with dark, thinning hair. He was, as I was to find out, very much a man about town. We liked each other immediately. Like many Latins, Alberto was *muy simpatico* (very amiable). When I explained that I was seeking work, he decided there and then to hire me as his private tutor to help improve his English. As a successful businessman who owned his own company, he travelled extensively. Although his English was quite good, he wanted to refine it.

'From tomorrow we start,' he said. 'You come to my house at 7:30. We take breakfast together and then we have a lesson. OK?'

'Of course it's OK,' I replied. Things were looking up. So began what would be another brief but delightful friendship.

For the next three weeks, every day except Sundays, I left my little pension where I now lived in downtown BA at 7 a.m. and commuted by underground to Alberto's home, which he shared with his mother in the fashionable district of Sarmiento. There, we breakfasted together on poached or fried eggs, croissants, and coffee served by their maid, during which we had a conversation in English. This was followed by reading the city's daily English newspapers. As he read aloud, I corrected his pronunciation. Once the lesson was over, Alberto paid me the equivalent of about 20 dollars, good money at the time, and I was very grateful. I also learned that he was a suave bachelor playboy who constantly lamented the fact that he was still single. 'I have not yet discovered a girl I can love enough to marry,' he told me more than once. Usually followed by a big sigh.

Not content with utilising my services each morning, Alberto also requested — insisted was more like it — that I join him occasionally for lunch or dinner. Despite my objections, he was only too pleased to pay for the meal, realising that one of the finest ways to improve his English was having a teacher constantly at hand. Every time I offered to pay, he would dismiss the suggestion, wave his hand, and say, 'No problema'. We visited city cantinas where we heard lively Latin music and almost always ordered juicy inch-thick steaks. On these frequent occasions, I noticed that Argentinians not only enjoyed eating out often and quite late at night, but ate their beef with relish and gusto. With 60 million cattle, outnumbering people two to one, the Argentinians were considered the greatest beef eaters in the world. Most liked their steak rare and oozing with blood. They also had the highest rate of stomach cancer.

Alberto was a gracious host, always concerned about my welfare, ensuring in his own inimitable way that I was enjoying myself. Often, I accompanied both him and Ricardo on drives around BA. They were extremely proud of their fine city, pointing out famous landmarks, including the bullet holes on the building of the Ministry of Agriculture which had been fired upon during the revolution when Juan Perón came to power. Among other interesting sites, we visited La Recoleta cemetery and the tomb of Eva Perón, Argentinian icon and wife of the dictator. Sadly she died much too young.

Twice, upon returning to my hotel after a couple of outings with Alberto, I discovered messages waiting for me. Both were from television studios, requesting me to appear on a couple of shows.

When the date of the first show dawned, I was apprehensive prior to appearing in front of the cameras but, like so many things I discovered in my daily wanderings, there was nothing to be worried about. I forgot my shyness as I talked about travelling, a subject very close to my heart.

My first appearance was on the Channel 7 Network with Antonio Carrizo, host of *Bienvenido, Sábado* (Welcome, Saturday), a weekly variety show during which Mr Carrizo questioned me about where I'd been. After I got into my stride and talked about some of the highlights, the time went quickly. The second interview was hosted by a German baroness whose name I forget, but who had me walking on stage to the sound of African drums to relate some of my safari experiences. Both audiences were receptive and clapped loudly, which gave my confidence a boost. I was paid about $50 each time.

After the newspaper articles and TV exposure, I began to be recognised on the street. Over the next week, I got some phone calls and received several invitations to meet people for a chat and a conversation in English over drinks or a meal. These meetings were pleasant enough, but produced no prospects of any work.

Stepping out at night was always a pleasure in BA. The streets were crowded with throngs of people strolling here and there or just sipping coffee or wine at the many pavement cafés and restaurants. Once, Alberto asked if I had tried Argentinian pizza and when I replied, 'No,' he said, 'I can't believe.' He ushered me into the nearest pizza parlour where we sat down to mushroom pizza followed by pancakes and *dulce de leche*, made from sweetened condensed milk, a favourite dessert enjoyed by children and adults throughout Latin America. And which quickly became one of mine.

For a while, I pounded the pavements of the city in my quest to make some money. It occurred to me that I should try to sell some of my many photographs, so began a few days of calling on magazine companies. At each one, I was told they had their own photography staff or utilised the services of stock photo agencies who quite naturally wanted to keep photos on file for at least six months. I couldn't allow that. By now, my photos had become my most valuable possessions and would continue to be so in the future. Eventually, I found two publications that wanted some. One was a woman's magazine interested in pictures of women and children. The other was a men's adventure magazine. I was paid what I considered good prices and was very pleased indeed.

Having told Alberto that I was available to give illustrated travel lectures, he decided to organise a gathering of his friends for one of my slide shows and for which he insisted they pay. Previously, all my talks had been given free of charge but, as many people had pointed out, taking photographs had cost me a considerable amount of time and money. Alberto assured me that his friends were well off and would be only too happy to pay for the show.

In due course, we all assembled one evening at Alberto's own apartment, a hideaway which he kept purely 'for entertainment' (meaning lady friends who visited for cosy, romantic afternoons or intimate evenings). It was on the 8th floor of a highrise building with a splendid view of the city. With Alberto translating, the event was a great success, and I was rewarded quite handsomely. With this payment and quite by accident, I'd entered the professional lecturing field.

Besides Alberto and Ricardo, I made other good friends in BA, many of whom insisted on inviting me to their homes and showing off their genuine English china and Queen Anne furniture. At the few parties I attended, I saw some absolutely wonderful tango dancing. But without a partner, I never plucked up the courage to try it!

Another new friend was a journalist who had helped me sell some of my photographs to his magazine. Hugo Brown had already covered the Bolivian guerrilla campaign and the Chilean-Argentinian border dispute and been forced to ride 100 miles on horseback once when he'd never before ridden a horse. A big, strapping fellow, he was occasionally sent on assignments to join police border patrols or accompany armed troops in the jungles of Bolivia, which his young wife, Pamela, worried about constantly.

When I stayed with them for a few days, Hugo related some of his adventures. Like Alberto, Hugo punctuated most of his conversation with 'bueno, fenomenal' (good, exceptional) and it wasn't long before I did too. 'Hello' and 'goodbye' was always 'Ciao,' a common Italian expression used widely in Argentina and Chile. General expressions like 'Que tal?' and 'Que pasa?' (How are you? And how's it going?) became part of my daily vocabulary.

One day, I met members of a British mountaineering party who had summited some of the highest Argentinian peaks and were on their way home. They joked about all the free equipment they had been given — being sponsored, they hadn't paid for a thing. Even their passage by ship from the UK had been paid; and they'd been given cameras and dozens of other things, including 20 pairs of socks each, admitting they didn't need or use half the stuff. As they talked, I couldn't help thinking that travelling as Pete and I were doing under our own steam was much more challenging. And perhaps more rewarding.

A week or two after I appeared on television, a young Argentinian named Felipe tracked me down. Twenty-four years old and of average height and appearance, he spoke very little English, and wore a trademark green beret like the ones my pals Pete, Mike, Eddie and I had worn as young soldiers. Felipe was full of enthusiasm to travel. He was also a heavy smoker. No sooner had we met than he asked if he could accompany me on my onward journey — to the south — but I told him I wasn't going that way, as my intention on leaving Argentina was to go across country to Chile. He insisted that I should not miss a visit to the south and emphasized this over and over again, but I still said no, sure that I couldn't afford it.

For three weeks, Felipe telephoned or called on me personally every day and each time he spoke he implored me to take him along, to take his advice to travel south to the mountain region of Patagonia, a vast, spectacular region of glacial lakes, ancient forests, snow-capped mountains, and picturesque Andean villages. It sounded like a great destination, and the more I heard about it, the more I wanted to see it. Several other people insisted I would be missing something special and would regret not going. That did it. I would include it in my itinerary and agreed to take Felipe with me. It was a decision that I soon came to regret, for instead of bringing me luck, he did quite the opposite! This would be partly my fault, for I had not given enough thought to 'choose your travel companions wisely' as an older fellow traveller had once told me.

Without a doubt, I should have continued to travel solo, for I'd become well accustomed to travelling alone. Felipe saw the value of travelling with an experienced companion. He told me he could arrange lectures easily. Being Latin American and a fluent Spanish speaker, I naturally assumed he would be very helpful. How wrong I would be!

A week before departing from BA, Alberto invited me for a trip on board his yacht to Montevideo for a few days. It was another splendid opportunity which I immediately seized and, with two of his other friends, we were off. It was the first time since my boarding school days that I'd been on a sailboat. On our first day, we were becalmed in the centre of the Rio de la Plata and had to resort to using the outboard. It took three days to reach our destination after stopping at tiny fishing ports on the Uruguayan coast at night, awaiting the arrival of fresh, strong winds. On our second day, the wind blew up. While the others slept below, I happily took over the tiller. By then, I was convinced I'd finally acquired my sea legs, especially after my South Pacific voyage, but, alas, it was not to be. Quite suddenly, I began retching furiously and had to be relieved to go lie down. It was no good; it seemed my affliction was incurable.

We eventually sailed into Montevideo harbour after sailing over the wreck of the *Graf Spree*, a famous German battleship which had been blown up by the Royal Navy in the Second World War. There, we tied up alongside the yacht club. Having left only a few weeks previously, I never thought I'd be back there so soon.

The afternoon before leaving Montevideo, as Alberto and I sat in one of the many cafés sipping coffee, an attractive young woman seated at another table kept glancing our way. After a while, Alberto got up and went over to speak to her. No sooner had they exchanged a few words than he came back and casually said to me: 'She wants to go with you. I will meet you back here in two hours. OK?' Just like that! For a moment I was dumbfounded and didn't know what to say. The admiration between me and this young lady in her early twenties was obviously mutual. It turned out that Carla was a schoolteacher who had the day off and took me to her apartment nearby where both of us, hungry for sexual relief, became intimately acquainted for the best part of an hour and a half. This unexpected brief encounter couldn't have come at a better time, for my sex life in South America up until then had been almost completely non-existent. I was so grateful to Alberto for arranging it!

It was now time for me to be on my way, but this time I wasn't alone. As Felipe kept telling me it would be easy to get rides (he was full of assurances), we decided to hitch-hike. Our destination: Patagonia.

Luckily, we were soon picked up by a truck, which later stopped at the scene of an accident. Two fruit trucks had collided head-on, damaging both and littering the road with apples. Other motorists had stopped and, after checking to make sure that the drivers were uninjured, they started loading the fruit into sacks and the boots of their cars. It transpired that this was common practice, and that most motorists purposely carried sacks strictly for these occasions! Felipe and I joined in, eating and packing away as many apples as we could, as I'd forgotten by now my New Zealand apple-picking experience. As our original ride had left the scene, we thumbed a lift with another truck going our way, staying on board for most of the night.

I'd learned the hard way to always be alert while travelling, the point being driven home after several near misses including the accident in Northern Rhodesia. Because of our many adventures, friends had often declared that Pete and I had high survival skills (walking away unscathed from dangerous situations) which I put down to our soldiering days, but Felipe had not acquired the knack of being 'all eyes and ears.'

It had started to rain earlier. To make matters worse, although the road was in good condition, it was dead straight for hundreds of miles through the Pampas, providing an additional strain for the drivers who would sometimes fall asleep through sheer boredom. As we lay in the back of the truck, Felipe fell fast asleep, and I was dozing when I suddenly felt the truck sliding off the road. I leapt up, banging my head on the metal side in the process. I shouted to Felipe. Rather groggily, he sat up just as the truck left the road completely and ploughed into a field. After a few yards, we came to an abrupt halt incredibly still upright. Our driver had fallen asleep at the wheel. Luckily, no one was hurt.

The next day we passed yet another overturned fruit truck, this time carrying pears! We quickly joined the many motorists who had stopped to get their share of the booty. There was little of interest to be seen in that part of the country, except endless miles of tall grass, cattle, and the occasional gaucho on horseback with his dogs. 'No gauchos exist anymore,' some of my friends in BA had told me, but I saw dozens over the course of several days!

We eventually reached San Martin de los Andes, a lovely alpine village which reminded me of Switzerland, and where most of the locals had red cheeks, a result of the clear, cold air. Felipe and I checked into a small guest house and there met a young American couple named Ed and Fran. We would become good friends. They were making an extensive journey around South America and I would see them again later in La Paz.

As Felipe and I sat by the roadside awaiting a lift to San Carlos de Bariloche, Patagonia's principal town, several gauchos rode by and invited me to ride with them. Since we were in no hurry, I did so. It was exhilarating to join these flamboyant young men in one of my favourite pastimes. I marvelled at their horsemanship and was a mere novice by comparison. As Felipe looked on, I raced at breakneck speed with my new gaucho friends across the Pampas. 'Bueno, bueno,' they cheered loudly, impressed that I could keep up with them.

I was greatly taken with what I observed as we continued further south, for I had never dreamt that such beauty existed in that part of the world. Similar to Britain's Lake District, but on a much grander, wilder scale. It was Mother Nature at her most savage and pristine and a paradise for lovers of the great outdoors, be they hikers, mountain climbers or birdwatchers. Happy as a lark, I felt completely at home as I always did when surrounded by majestic scenery. I couldn't resist whistling and breaking into song and wishing we could stay longer.

One evening, after a truck dropped us near a roadwork site, we strolled through a forest to a beautiful lake. There, we chatted to a priest who was staying in one of the only two buildings in the vicinity, a small hotel. After Felipe told him we were looking for a cheap place to stay, the priest excused himself for a few minutes before reappearing with the keys of an empty guesthouse. The hotel manager had put it at our disposal for free after a word from the priest. We were absolutely delighted and moved in immediately. Situated right on the shore of beautiful Lake Espejo, we revelled in our comfortable overnight accommodation and the stunning scenery all around. From the veranda, we admired the view. It was heavenly! We had it all to ourselves and couldn't believe our luck.

As hitchhiking was usually easier alone, we decided to split up the next morning and then meet in Bariloche, where I hoped to do some travel talks. It wasn't long before a young couple gave me a ride and for the next four hours, we stopped several times to admire the high-country vistas, arriving in Bariloche around 5 p.m. I left my rucksack for safekeeping at the local police station and wandered around the town, awaiting Felipe's arrival. He eventually arrived two hours later having walked several miles and had his pullover swiped by a young boy on board a truck which he flagged down.

We stayed in the scenic alpine town of Bariloche for a few days, during which Felipe and I visited various clubs, hoping to arrange some lectures. But no luck. If we stayed a week or two, something could definitely be arranged, we were told, but hanging around meant spending more money. It was wonderful to linger in such a gorgeous setting, but we had to make a living.

On leaving Patagonia we took a launch to Puerto Blest across Lake

Nahuel Huapi, and then transferred to a bus, en route to Chile. Over the course of two days, we took boats and buses several times, it being the only way to travel where we were headed. The boat rides turned out to be quite expensive, something else Felipe had overlooked when considering how much money he would need to join me.

Once or twice we walked the distance between launch rides, stopping on the way to pick blackberries and to gaze at the scenery. It turned out that Felipe tired very easily. I, on the other hand, with my years of experience walking for lengthy periods at a time, could go for hours without ill effect. Also, Felipe was quite hopeless at getting up in the morning. I always had to shake him awake.

Finally, we entered Chile after crossing the border at El Hito and taking the launch across the crystal-clear green glacial water of Lake Frias. Throughout this period, despite living mostly on bread, salami, cheese and fresh fruit, Felipe's money began to dry up. So now I had to cover the costs for both of us, which meant my money would go much faster than anticipated.

After a few days in the picturesque town of Peulla with its stunning mountain views, we booked a train for Santiago de Chile, the lovely capital city in a valley surrounded by the snow-capped Andes. Like BA, it is a metropolis of museums, theatres, restaurants, and bars. After a 24-hour journey on hard wooden seats, we arrived in the city and were lucky enough to find comfortable accommodation in a students' hostel.

Many times in my travels, I was befriended by students. Like the police and some of the truck drivers who gave me a ride, they could seldom do enough for me. Our new Chilean friends were no exception, making Felipe and me feel immediately at home and inviting us to dine with them. Such acts of kindness by strangers never failed to impress me. New friends always seemed to understand how challenging life can be in foreign cities, knowing no one and having to find somewhere to stay and eat.

As I was now impatient to continue the trip, I decided against teaching as it was too time-consuming, hoping that I could give enough lectures to cover my expenses. Someone at the hostel suggested I contact the principal of Grange School, a well-known all-boys private school in Santiago, which I did. There, I met the quietly spoken Headmaster George Lowe — New Zealand born educator, explorer, and famous Everest mountaineer.

On the successful 1953 Everest expedition, he had been selected along with a sherpa to attempt to reach the summit. Alas, they were forced back due to adverse weather. The following day, Sir Edmund Hillary and Sherpa Tenzing Norgay were chosen, and they succeeded.

George admired my initiative and immediately suggested I give a lecture to the senior boys at the school, which was staffed mostly by British teachers. With its large green playing fields, architecture and classes, it was modelled after some of the finest schools in England. After lunching with George, his wife, some of the teachers, and a number of the senior boys, I made my presentation for which I was paid something like $25.

A few days previously, while visiting a newspaper office, I was advised to visit one of the local commercial radio stations, there being 22 in Santiago alone! I called at Radio Balmaceda, where I met a young Hungarian reporter named Gabor who seemed very pleased to meet me and wrote a story about my travels to be read over the air the following day.

Gabor insisted on showing Felipe and me much of the city, stopping once or twice for 'ace kreem'. He reminded me very much of Hugo Brown. Tall, dark-haired, he was a qualified pilot and former scuba diver, who had given up his underwater activities because of the recently much-publicised stories of aggressive sharks. He had interviewed many people from De Gaulle to other heads of state, so I felt quite privileged when he spoke with me. It seemed that he had friends in every part of the city and when I explained about needing a clearance certificate from the police prior to leaving the country, he said, 'Don't worry. Interpol. I have friends there too.' And he did! As an Anglophile, his greatest ambition was one day to visit the British Isles.

It was now autumn in Santiago; the trees were losing their leaves and, although the days were still quite hot, the nights were becoming very cool. So I bought a handsome Indian-made sweater.

Back at the hostel, my student friends had arranged a *conferencia* and invited some 30 or 40 others, all of whom willingly paid a small fee. Amusingly, I found myself with three willing interpreters who translated into Spanish what I said, but kept contradicting one another which kept everyone in stitches, including me. The evening ended with a party.

In all, I succeeded in giving about half a dozen lectures in different parts of Santiago, but the money I earned was only enough to cover our daily expenses. I wasn't saving anything. But we saw much of the city and its environs.

One afternoon, while travelling on a *colectivo* (inexpensive local bus), several small street urchins sneaked aboard through the rear entrance and proceeded to make nuisances of themselves. After riding for free for several blocks, they all jumped off but, just as the bus pulled away from the stop, one of the youngsters spat in my direction. Bingo. It hit me squarely on the cheek. My immediate anger quickly subsided as I had to admit it was a great shot.

After spending some weeks in Santiago, it was time to move on. Not surprisingly, by now I'd come to the conclusion that with Felipe by my side, I was getting nowhere fast. He wasn't pulling his weight and we were also getting on each other's nerves. In a nutshell, I'd had enough of my lackadaisical companion and his lack of stamina, patience, initiative, grit, and guts, all attributes required for the nomadic life. Felipe couldn't organise anything. Not even, in soldier's language, 'a piss-up in a brewery'. He'd become a liability to me rather than an asset. It was time to say 'Adios' and go our separate ways. Feeling greatly relieved, I headed north to Mendoza on my own.

I purchased a train ticket to Los Andes and, upon arriving there, I visited the police station to enquire whether or not any *colectivos* went to the border. They said 'no' but advised me to visit the nearby army garrison a couple of miles from the town where I might catch a ride.

No military vehicles were available so, for the next two and a half hours, I stood in the open back of a truck, en route to a road maintenance site high in the Andes. The night was freezing cold and, despite wearing my pullover and anorak, the icy-cold wind penetrated every part of my body. As I gritted my teeth and hoped for the best, I felt that I'd never been so cold in my life. Finally, we stopped. Greatly relieved, I thawed out with a cup of hot black coffee, hunched over the night watchman's open fire. I was then shown to a stuffy room full of cigarette smoke which, I was told, I could share with several of the labourers for the night. Shortly after, I fell asleep exhausted on one of the spare bunks, oblivious to the loud snoring, farting, and the combined odours of tobacco, alcohol and sweat.

Up early, I started to walk the four miles to the Argentinian frontier. It was still a bone-chilling cold, and I was unprepared for such exceptionally frigid weather. I don't remember the exact altitude, but it must have been at least 11,000 to 12,000 feet. After an hour and a half of walking, during which I nearly froze to death, and afraid of stepping inside the customs building to warm up only to miss a lift, I lucked out. A small bus with a Chilean football team on board came along and offered me a lift, which I gratefully accepted. Throughout the ride, the soccer players peppered me with questions in their halting English.

Arriving in Mendoza, a city with wide, leafy streets and small plazas in the heart of Argentina's principal wine region, I checked into a small hotel and then did some sightseeing.

In an effort to organise something quickly, I visited a local newspaper where, as was my practice by then, I produced some of my previous clippings to a young reporter. When he saw them, he remarked, 'But where do I start? What can I ask you? This is tremendous.' Although the newspaper article did get printed, it was of no help to me for it appeared after I'd left Mendoza.

It was around this time that I heard the tragic news of Martin Luther King's assassination in Memphis, Tennessee.

Less than five years earlier, in August 1963, while I was living in East Africa, Martin Luther King had made his stirring 'I have a dream' speech on the steps of the Lincoln Memorial in Washington, DC. That speech echoed around the world. I could remember much of it to this day, and I would recall some of it a few months later when I would be living, lecturing and doing odd jobs in the United States:

> *I have a dream that my four little children will one day live in a nation where they will not be judged by the colour of their skin but by the content of their character. I have a dream today ... I have a dream that this nation will rise up and live out the true meaning of its creed, 'We hold these truths to be self-evident; that all men are created equal'... I have a dream.*

As prospects for making some money in Mendoza seemed bleak, I decided to head north to Bolivia and Peru.

When I purchased my bus ticket to the town of Tucaman, I was compelled to pay for my rucksack. I had never paid extra for it before, even on planes, and it came as a shock, especially as my money was, once again, fast dwindling. I could, of course, have stayed and worked, except by then I was restless to continue the trip. Already, Pete and I were far behind our original plan. It hadn't bothered either of us until then; for it had been a really great trip so far. Indeed, the time of our lives. But both of us were well aware that our travels could easily stretch into 10 or 20 years like some of the old guys we'd met who couldn't settle down. We had both reached the age and stage in our lives where, we knew it would soon be necessary to become established in a career.

If I was to continue on my way, I'd have to watch my spending very carefully. After a two-day bus ride, I reached Ju Juy, but had missed the late-night bus connection so once more resorted to staying at a police station. The young inspector was delighted to see me and bade me sleep on a camp bed outside the prison cells after asking me lots of questions about *Inglaterra* over coffee.

The next morning, I walked the mile and a half to the railway station to catch a train to the Bolivian border. There, I joined a long queue of local Indians waiting to buy tickets. The train pulled in. Like in other third-world countries, it was immediately besieged by everyone jostling and shoving to get a seat. I was one of the last to clamber on board, where I was lucky enough to be able to stand outside on the bottom step at the end of the carriage. I strapped my pack onto a rail and stood next to it for the first hour, holding on tightly!

Finally, I managed to force myself into a crowded carriage where for a period of eight hours I remained standing, crushed between Indians and their children and luggage and vegetables. To them, it was extremely amusing to see a gringo travelling this way, and they couldn't contain themselves from giggling with their hands over their mouths to not make it too obvious. I, too, saw the funny side, but as the ride was so uncomfortable, I simply couldn't laugh.

While staring out the train window at the passing scenery, I was clapped on the back by a fellow who had recognised me from BA. José was a journalist heading north on an assignment. He and his photographer colleague, Antonio, invited me to dinner, which was especially welcome, for I hadn't eaten properly for two days. At La Quiaca on the Bolivian border, they kindly suggested that I take advantage of their sleeping compartment as they would be spending the night in a hotel. The timing of such an invitation couldn't have been better for I was dead tired at the time. Climbing into one of the two empty berths, I was pleased as punch, and slept like a log.

CHAPTER 18:

My Life in their Hands

> *'Twenty years from now*
> *you will be more disappointed*
> *by the things you didn't do*
> *than by the ones you did do.'*
> — Mark Twain

I had anticipated problems at the Bolivian border, having heard from fellow travellers that it was a difficult country to enter and that I certainly required an onward ticket. But I didn't have one. I would have to do some fast talking. So it was with a feeling of apprehension that I walked up to the border.

There, to my relief, my passport was flicked through, my pack received only a slap, and I was allowed to proceed after walking over the bridge into the small adobe border town of Villazon. Some ten minutes later, I sat on a station bench awaiting the train to take me to La Paz, the highest capital city in the world at slightly over 11,000 feet.

I had thought that mountainous Ecuador was in many ways similar to Ethiopia, and so was Bolivia. Incredibly rugged and beautiful and with an almost completely indigenous population, it had some of the same problems as its sister Latin American countries — bad roads, bandits, and strict customs officers who patrolled the trains and buses spot-checking and confiscating clothes, liquor, etc. As my train pulled out of the station, most people busied themselves hiding or sitting on their newly acquired goods, hoping that they would pass undetected. One short fat man sitting opposite me calmly proceeded to stack bottles of cognac and Cinzano all along the windowsills then sat back adopting the most innocent, angelic expression. What's more, it worked as none were confiscated.

Once again, the gringo was asked to help. This time, instead of being asked to wear two sombreros, I was requested to wear a brown bowler hat, the popular headdress of most Bolivian women. The journey from the Argentine border to La Paz took two days and a night on hard wooden seats, during which time we had our ticket punched at least six times, leaving barely anything to discern or show. As usual, I was travelling in budget class. To stretch my limbs, I placed some newspapers down and slept on the floor under the seats, much to the amusement of my fellow passengers.

When the train stopped and women came along the track selling food, I was introduced to two national Bolivian dishes — *apie* (sweetened maize, brown in colour and made with hot water) and warm *pastellas* (puffed pastry filled with sweetmeat or cheese). For the first time, I sampled coca leaf tea, which the locals use to alleviate altitude sickness.

Gradually, our train passed great plains, deserts, and mountains. At times the scenery was so breathtakingly beautiful it made me want to shout out with pleasure. Again I saw huge herds of grazing llamas and alpacas with their shepherds. Soon we were surrounded on all sides by immense snow-covered peaks as we continued at an altitude of around 15,000 feet and, although the sun was strong, the early mornings at that height were mind-numbingly cold. Even my hardened and brown as berries fellow passengers couldn't resist applauding when we passed another train loaded down with people in open goods wagons, oohing and aahing when they saw the almost frozen occupants. I'd faced many extreme conditions while travelling, but even I was amazed what they would do for a free ride.

Throughout those two days and a night on the train, *El Gringo* became very popular, for I played with the children and laughed and joked with everyone else. As we neared El Alto above an immense valley, my fellow passengers urged me to look at the view. There before me, in a gigantic crater and dominated by the shadow of Mount Illimani, I gazed down upon the city of La Paz, a collection of fine, imposing buildings and hundreds of tin and wooden shacks. That night found me in a small, stuffy hotel room in the city centre after spending an hour searching for a cheap place to stay.

At a snack bar in town, I asked two young fair-haired women if they knew of somewhere I could change my traveller's cheques. No sooner did I say, 'Excuse me,' when they both greeted me with, 'You're David Skillan! You gave us a slide show in the Te Anau Youth Hostel in New Zealand.' It transpired that Anne and Vivian, both from Australia, had been in the audience of the crowded room many months previously. No sooner had we said 'hello' than Anne excused herself with a severe headache and went to sit on a park bench while Vivian and I went together to purchase some medicine for Anne's pounding head.

A middle-aged German businessman entered the *farmacia* and I asked him if he wouldn't mind interpreting for us which he kindly did. We then all stepped outside where Mr Scharzenburger invited us to join him and his wife in their Mercedes-Benz for a drive to El Alto to view the city at night. The ride proved very worthwhile. Apart from admiring the magnificent view, we also gleaned some interesting facts from this charming couple.

Mr Scharzenburger had lived in Bolivia for over 30 years, during which time he established Bolivia's first export business. His wife was a native Indian and lawyer who no longer practised her profession, saying, 'I now only help my husband wiz hees beesness. Before, whenever I took on a case, nobody was ever prepared to take me on in thee court.' With her strong, fiery temperament, I could understand why.

According to their story, the previous president, Victor Estenssoro, had been a ruthless dictator who had introduced Gestapo-like methods into his government, and imprisoned people who didn't support him or whom he simply didn't like. Including Mrs Scharzenburger. Being who she was, she rose to the occasion, assembled her entire family — dozens of nieces, nephews, uncles, and aunts — and instigated the successful military coup of 1964. Her cousin (Barrientos) then became president and her brother, Minister for Foreign Affairs. She continued, 'Now the people are happy, poor but free, but if there is another revolution' — Bolivia being infamous for changing presidents and governments at a phenomenal rate — 'we of course must flee.'

She also told us — by now she had our full attention — that, on a visit to Miami two years previously, a fellow was taken off the plane by CIA agents. She kicked up a fuss at seeing the fellow manhandled, but was quickly informed that the fellow was a communist and would probably have hijacked the plane to Cuba. Had this occurred, she and her husband may well have been murdered for her brother was the colonel commanding the troops who captured and killed Che Guevara, the charismatic Argentinian guerrilla leader who supported Cuba's Fidel Castro in overthrowing President Batista's military junta. On the orders of President Barrientos, Che had been executed on October 9, 1967, just several months earlier in the small Bolivian village of La Higuera. He was 39. President Barrientos would die in a helicopter crash in 1969.

The next day, I heard my name called from across the street. Looking around, I saw the young American couple, Ed and Fran, whom I had met in San Martin de los Andes. They were amazed to find me still going strong, for when we had originally met, I was suffering from one of my frequent tummy upsets. Typically helpful and generous, they insisted on buying me a really tasty, substantial meal.

Just before that I'd dropped my camera while taking photographs in one of the town squares but, with Ed's help, I got it repaired. Despite the fact that my money was limited, I continued to buy film and take photographs, convinced even more after my recent lectures that it was well worthwhile. Sooner or later, I would benefit from it. Together we explored the black market, openly referred to by that name and boasting all manner of goods, including many European foodstuffs.

The night before leaving La Paz was a restless one, for I was afraid of missing the 5:30 a.m. bus to the Peruvian border. Struggling up a steep hill, my breath came in short gasps due to the altitude as I made my way to the station. There, I was relieved of my pack by the driver's assistant while I watched to see that it was safely loaded.

I then thought of Vivian, who had failed to take this necessary precaution in Peru, and when told, 'OK,' by the driver's sidekick, she left her rucksack on the pavement and got on the bus. Upon reaching her destination, she discovered her pack had gone astray. In fact, it hadn't been loaded and, although she complained to the police, they merely shrugged, as if to say, 'Too bad.' As far as they were concerned, that was the beginning and the end of it. Vivian subsequently purchased an enormous wicker basket which she somehow carried together with a fully packed kitbag. She herself was a petite five feet but lugged around quite the most baggage as anyone I'd ever seen.

It took the best part of five hours to reach the small town of Cococabana and, for most of that time, we travelled along a dangerous, narrow mountain road running alongside Lake Titicaca. At 12,506 feet above sea level, it is the highest lake in the world and at 120 miles long and about 50 miles wide, the largest freshwater lake in South America. Not far past Cococabana, we drove aboard a small, flimsy reed-made craft on which two men hoisted a large tattered sail then produced two long oars (one each) and proceeded to row us across the narrowest section of the lake, while a third stood in the stern using a giant paddle as a tiller. So much for the local ferry service! It was no surprise to me to hear that those small boats occasionally sank in rough weather.

As I walked beside the lake, the weather was perfect with blue skies and golden sunshine which reminded me of a beautiful English summer's day. After a short stop at a little lakeside restaurant for a large bowl of tasty sweet potato and cabbage soup, I set off on foot to walk the five miles to the Peruvian frontier. It wasn't long before I started humming as I walked along the dirt road by the side of the lake. The air was cool and fresh, and birds were singing as I waved to the peasants working side by side in the fields. Once more, I felt as though I didn't have a care in the world. I hadn't walked far when a truck stopped to pick me up, and I clambered into the back, which was already filled with local Indians and sheep.

At the border town of Yunguyo in Peru, I reported to the immigration authorities, paid a nominal amount for a visa, and had my passport stamped. There I spied Vivian, who had left La Paz the day before me. Anne, her travelling companion, had decided to travel south while Vivian, like Pete, was headed for Canada.

It was in Yunguyo that I once again joined up with Ed and Fran. By then, they had been away from home and on the road some eight months and had covered a lot of ground. When they recognised me, they were amused at what Ed described as 'this stooped, laden-down figure, wearing a cravat, shouldering a pack and carrying a "swagger" stick.'

Ed was a tree surgeon and an inveterate traveller when time and money permitted. He was tall and rangy and normally very fit and healthy. Fran was very attractive with long straight blonde hair and an ex-Peace Corps worker. Both were in their late twenties and hailed from Boston, Massachusetts. It came as a great surprise to them that this skinny, wiry guy had managed to travel for so long, they told me. I too wondered how I was still alive and well. Their respect for my endurance increased even more, when earlier they both had become so ill with altitude sickness they could hardly stand, let alone travel.

The trip to Puno was extremely uncomfortable as the bus had broken springs and the road was full of potholes. However, the stunning panoramas compensated for the discomfort of the journey. Once, we stopped at an army post to have our passports checked, and being rather thirsty, I asked the CO (a colonel) if I could have a glass of water. 'Venga' (come), he said, escorting me to his office where he dispatched an orderly to fetch me a bottle of Fanta Orange which he just gave me. At the town of Puno, Ed, Fran, and I all checked into a small hotel, followed by a meal and a walk around the town.

All three of us wanted to visit Machu Picchu, the lost city of the Incas. This extraordinary feat of engineering built five hundred years ago was one of my must-sees while in South America and, if necessary, I would have walked there.

We all boarded a train to Cusco. Inside our compartment, we met two young Frenchmen, Jean and Phillipe, also en route to the world-famous ruins in the Peruvian jungle. It seemed natural for the five of us to join forces for company and to share food. That first train journey took ten hours and was pleasant enough with spectacular scenery.

For our one night in Cusco, the five of us stayed in a small cheap hotel, and then boarded the train to Machu Picchu early the next morning. That second rail trip lasted five and a half hours, stopping at every tiny station and hamlet as the railway wound through the great long valley, dominated by lush jungle and very steep, high mountains. We were the only Europeans on board and, while we bartered with fellow passengers for fresh cheese and oranges, most of the Indians looked at us with amusement.

All five of us agreed that we would rather climb than travel up to the ruins by bus. When we reached the station at Aguas Calientes, the base for visiting Machu Picchu, we set off almost immediately, crossing the iron bridge over the river and starting to hike straight up. It was blazing hot and by the time we had gone halfway, we were too tired to continue via the direct route, so we took the main dirt road, which was narrow and perilous and could only handle minivans. After hiking for another two hours, the effort was certainly worthwhile, for the view of the ruins set high on a mountain was breathtaking. After wandering around, fascinated and intrigued by what we saw, we decided to spend the night in one of the newly restored stone huts. The hut had a straw roof and would house llamas that sometime in the future would be introduced as an added tourist attraction.

Not far from the main entrance to the ruins stood the one and only Machu Picchu hotel, which was quite expensive, catering primarily to wealthy tourists. And which none of us could afford. As darkness fell and before turning in, we went to the hotel where we took advantage of the washrooms and joined hotel guests sitting in the lounge. There, we celebrated our visit to this amazing and unique place with ice-cold Peruvian beers.

Inside the lounge hung a photo of Hiram Bingham, the American explorer who had discovered and explored the ruins in 1911. While camped in the valley below, he had heard of a 'lost city' from some Indians and, after hacking and climbing his way through dense jungle, had found the ruins — completely overgrown. Ultimately, the Peruvian government took over and cleared the area, making one of the wonders of the world more accessible and into a hugely popular commercial enterprise.

Sleeping rough in the ruins was memorable, and we awoke the next morning to bright sunlight streaming through the window openings. No sooner were we out of our sleeping bags than we were spotted by the keen-eyed, one-legged watchman brandishing his crutches and shouting at us for trespassing! We hurriedly departed.

After a welcome cup of coffee back at the hotel, more exploring the ruins and picture taking, we took one of the twice-daily minibuses back down to Aguas Calientes. At the station — only a bare platform with no buildings — we discovered that the train was five hours late, so we whiled away the time chatting, taking a siesta, and watching well-to-do tourists stepping out of the first-class express train.

Back in Cusco, we spent another night in our previous little hotel; then the three of us boarded a bus early the next day for Lima. The journey took three days and two nights and was terribly hot during the day and bitterly cold at night. By then, I'd been aggravated and bitten many times by bugs and fleas, but they were even worse on the dilapidated bus. To keep expenses to a minimum, we munched bread, bananas, and large juicy, prickly pears which we purchased at regular stops. Once again, the scenery in this remote area was simply awe-inspiring as we continued our journey high in the Andes.

Arriving in Lima early in the morning, we walked and walked, looking for a cheap place to eat. Eventually, we found somewhere we could order some fried eggs and where we were befriended by a lone *Peruano* drunk who insisted on buying us all breakfast. Only when we finished did he change his mind. And left us to pay — including for his meal!

By this time, I had only a few dollars to my name, so I decided to look up Rolf and Marianne, whom Pete and I had first met on board the *Lauro Sodre* Amazon steamer. At the time, they had insisted that I wasn't to pass through Lima without contacting them. So I telephoned Rolf, hoping one of them might be able to help me to arrange some lectures. Fortunately, they were delighted to hear from me and asked me to meet them at a favourite restaurant.

I left Ed and Fran in town, where they were busy writing a cable to friends in the US asking for a loan to buy a ticket to get home. Unbeknownst to us, we would meet one more time towards the end of my odyssey.

Meanwhile, I caught a *colectivo* to 'O Que Bueno' (OK Good), a restaurant near the Colegio Pestalozzi where I was reunited with Rolf and Marianne. It was a very happy occasion, and they invited me to stay at their flat. As they had a maid and a spare bedroom, they assured me that it would be no inconvenience. After I'd taken my first shower in days, I tumbled into the comfortable bed and, needless to say, slept very well.

Both Rolf and Marianne assured me constantly over the next ten days to relax and make myself at home, comforting words indeed. While Rolf enjoyed practising his English with me, Marianne fussed over my meals and general comfort. I couldn't help imagining where I would be if it hadn't been for them and so many others who had befriended me.

As I relaxed, I thought about what I'd been through the previous couple of weeks: arduous travelling conditions; no decent washing facilities; almost non-existent funds; a broken zip in one of my two pairs of pants and a hole in the bum of my other pair; and, finally, the loss of my only comb. In perfect contrast, I was now ensconced in a comfortable home equipped with all the 'mod cons' as well as excellent food, a proper bathroom, classical music and, by no means last, fine friends. I awoke the next morning to the gentle strains of Beethoven's *Fifth Symphony* coming from the living room, showered and shaved, then sat down to a breakfast of bananas, grapefruit, eggs, toast and coffee.

Calling at the offices of the British Council trying to arrange a lecture, instead I was offered a teaching position for six months by the friendly director. I thought about the offer but then turned it down. Later, I thought that I should have accepted, for I was to pay dearly for my impatience.

At the office of *Caretas*, a well-known Peruvian magazine, I met the managing director and editor, Señor Don Enrique, who was only too happy to purchase some of my photos. As he was fairly influential in his own right, he also promised to arrange a free flight for me to the Ecuadorian border. Fortunately, Señor Don Enrique kept his word and presented me with an air ticket the day before I left.

I was amused by the *colectivos* in Lima, which had to be seen to be believed. Often vintage models, rusty and falling to pieces, somehow they managed to keep moving, and it was a ridiculous but amusing sight to often see elegant women and smartly dressed men climb in and sit down, frequently on exposed seat springs. As they drove, the drivers threw up one, two or three fingers for all to see, indicating how many seats were available.

One afternoon, I recognised an American fellow whom I'd seen on the Machu Picchu train. He had recently returned from construction work in Vietnam and was also travelling around the world. We went into a restaurant where he ordered a large meal, and I ordered only a cool drink. As we sat there, he boastfully spoke of the $21,500 he had earned in the last 18 months and then complained that he had only about $19,500 left. I made no comment about my worldly wealth but noticed that he didn't offer to pay for my drink, which made me even more determined to help fellow travellers whenever I could.

On my last morning in Lima, Rolf woke me at 6:00 a.m., and all three of us sat down to another fine breakfast. Marianne thrust some fruit, bread and cheese into a brown paper bag, 'for your journey.' At the airport, Rolf excused himself for a few minutes, then reappeared and presented me with three bars of chocolate. 'Also for your journey,' he said. Feeling as usual incredibly humble, I stepped aboard the turbo-prop, knowing that I carried their blessings and good luck wherever I went.

The plane was scheduled to leave at 8 a.m. but, by the time all the passengers were on board, it was 8:30. Only then was a technical fault discovered, which took 40 minutes to repair.

We landed twice at airports whose names I don't remember, and the second time was very scary indeed. As we came in to land, the undercarriage failed to engage, and the plane suddenly had to climb. For a while I had my heart in my mouth. To make matters worse, there were mountains and low clouds in the vicinity. Twice we circled, and each time the undercarriage failed. It was then that some of the passengers became visibly agitated and I was becoming convinced that we would have to make a crash landing. Luckily, the third time, the wheels dropped, everyone breathed a huge sigh of relief, including the crew, and we landed safely. Needless to say, I was thankful to stand on *terra firma*. I soon forgot the frightening episode, but at the time it reminded me how often my life was in other people's hands.

At Tumbes on the border of Peru and Ecuador, I presented myself at a police station dead tired. After asking permission, I fell asleep on the concrete floor despite the fierce humidity and mosquitoes.

The next morning I met Gunther, a German backpacker whom I had originally seen on the train to Machu Picchu. Like me, he was travelling on a shoestring. We chatted for a while, then parted, fully expecting to meet again as we were headed in the same direction.

My passport checked, I walked across the metal bridge back into Ecuador then boarded another *colectivo* going to Machala. There I transferred to another vehicle to reach Puerto Bolivar in order to catch the night ferry across the Gulf of Guayaquil. I spoke to a police sergeant in the street who, wanting to practise his English, immediately befriended me and took me to his quarters where I washed and left my rucksack prior to exploring the town.

The ferry left that evening. Quite ancient, it had seen better days and took all night. For the equivalent of 75 cents, you could travel third class in a dirty hammock or, as I did, in a second-class clean one for a dollar. Arriving early on the sergeant's advice, I quickly chose my hammock and tied my handkerchief there to claim it — a well-known practice at the time. The crossing was very calm, but I didn't sleep too well due to the noise and the heat. In contrast to the interior of Ecuador, where it can be bitterly cold in the mountains, the coast is tropical and very humid.

In the dirty, smelly port of Guayaquil, I ran into Gunther again. As he was heading the same way as me, we teamed up to travel together for a while. Like me, he was fascinated with Latin America; unlike me, he hadn't found it too frustrating. A month later, in Mexico City, his attitude would be quite different.

While Gunther took a siesta beneath the statue of San Martin, I was accosted for money by a young man. During our brief conversation, he showed me some pornographic photos and told me he could find me 'a good girl'. I couldn't resist laughing at the proposal for I barely had enough money for myself, let alone to spend on prostitutes.

Gunther and I successfully hitch-hiked to Santo Domingo, passing through great banana plantations on the way and filling ourselves up with bananas as we travelled. At one time we were picked up by a very talkative Ecuadorian businessman. While we sat almost petrified travelling at around 90 miles per hour, our garrulous driver sang the praises and virtues of the Ecuadorian women, who according to him were 'hot, passionate and loving'. He also proudly told us that his country had the greatest banana industry in the world which we were in no position to doubt.

En route back to Quito, our final destination together, the bus broke down high in the mountains in drizzling rain and mist. After waiting an hour or so, we transferred to another, which also broke down on the outskirts of the city. Finally, a local bus took us into the centre where we returned and checked into one of the many small hotels in La Onza, a colourful cobbled street, a place Pete and I had discovered on our previous visit.

Always trying to plan ahead, I investigated my options for the next leg of my journey. Should I fly directly from Quito to Panama? This would be a faster but more expensive option on my limited budget. Alternatively, I could travel to Colombia by bus, and then fly. The bus would take four days and cost very little.

A helpful young man at an airline office contemplated my situation and proposed my best option was to travel by bus to Medellín, Colombia, and from there fly to Panama. He himself had taken that route and assured me that it would be far less expensive, but I had to hurry to catch the once-weekly plane. He worked out the bus timetables and connections, which I asked him to double check. Yes, he was quite sure it could be done.

Did I have to pay airport tax? No, he assured me. Having by this time become familiar with many aspects of the Latin way of life, I requested his assurance on this no less than three times to which he seemed to take exception. But I knew what I was up against, and although not rude (I greatly appreciated his cooperation), I made it quite plain that I couldn't afford to take chances.

Now in quite a hurry, I boarded the bus for Tulcán at the Colombian border. After a harrowing nine-hour drive in extremely cramped conditions, we reached our destination. This was due entirely to luck and the good Lord, for the driver was atrocious, driving very fast on the mountain roads with absolutely no consideration for his passengers. Not content with speeding and racing other buses, the driver chose to overtake on bends at 50 miles per hour. We eventually reached our destination at 4 o'clock the next morning. While I waited for the immigration office to open, a tall Peace Corps volunteer, Anthony, approached me and invited me to join him and his wife, Audrey, for breakfast. It was such incidents that always seemed to crop up most conveniently that made my life easier.

Another nightmarish ten-hour bus ride got me to Medellín. Not knowing anyone in the city, I went to the police station where the duty officer, a detective inspector, was delighted to see me. Like so many people keen to practise English, he invited me to spend the night at his home before flying to Panama City the next day. Alonso Angel was his name and, after a very pleasant overnight stay with him and his family, they drove me to the airport where I discovered it was, in fact, necessary to pay airport tax. I was fuming, not because I had to pay the tax but because I'd been misinformed even after being assured three times to the contrary. I realised I had saved absolutely nothing in my headlong dash to reach Panama quickly.

No sooner was I seated on the plane than I spotted a familiar figure walking down the aisle. It was Vivian, still clutching her enormous basket. After getting over our mutual surprise, she sat down next to me and told me all about her extended stay in South America. Shortly after we had last seen each other in Bolivia, she had fallen ill and had to alter her itinerary and rest up for some time in Bogotá.

Outside Panama City airport, I discovered just how heavy Vivian's worldly possessions were when I offered to carry her kitbag to the bus stop! There, a young fellow offered us a ride to the city centre. A Scot, he lived and worked in Venezuela and spoke with an American accent! He very kindly dropped us off at the door of the Hotel Centrale where I discovered to my chagrin when checking in that prices had gone up considerably since I was last there.

I had little to do in Panama which, as I wasn't too fond of the place, suited me fine. I made enquiries about buses to Mexico. Fortunately, there was one leaving in two days via various Central American cities. I was now focussed on making my way to the United States to make and save some much-needed money.

Walking around town, I noticed large groups of menacing-looking characters roaming the streets. When I saw them assembling at a corner, I was curious and decided to take a closer look. The next thing I knew, bullets were whistling past my ears and over my head, so I ducked behind a concrete pillar. Peering around discreetly, I saw a mob stopping cars in the streets.

I quickly learned that the commotion was related to the imminent elections. If the occupants were not supporters of the group who stopped their vehicle, they were dragged out and their vehicle was torched while they watched, horrified and powerless to do anything about it. It wasn't long before the street was littered with ransacked, burning cars. As I watched from a safe distance, a police Land Rover screeched to a halt and riot police, complete with shields and rifles, leapt out and let loose a volley of shots into the air. Everyone scattered.

This terrifying period lasted three days during which time many people were injured, and some killed. After things quieted down, and in my quest for interesting and unique photographs, I couldn't resist taking a few shots of a burning car in front of a Shell petrol station.

One evening in Panama City as I walked along the street, I instinctively felt someone's eyes on me. Someone was giving me the once-over. Standing on the corner, obviously up to no good, stood a dodgy-looking character with a malevolent glare. I had no doubt that he was after the contents of my shan bag, which I casually carried over my right shoulder. He definitely didn't want to practise his English on me. I moved away, and the fellow began to follow me. I stopped, he stopped. I walked further along the street and so did he. By now, sure of being robbed, I looked around for somewhere to dart, for I realised only too well that this fellow could easily knife me and steal my valuables on such a crowded thoroughfare. So I did the only thing I could think of, which was to walk quickly, weaving in and out of the crowd in the hope that the would-be assailant wouldn't follow me. Much to my relief, when I looked back, he was gone — no doubt in search of easier prey.

My stay in Latin America continued to be unpredictable and interesting. I was very glad to have come this far, still very much alive and in one piece!

Fording river at border of Peru and Ecuador

One of many truck drivers who gave author a ride

Amazon River activity with SS Lauro Sodre (top right)

Teresa and Breta travelling down the Amazon River

With children on board Amazon steamer

Indian children near Iquitos, Peru

Children window shopping in Ecuador

Bolivian children with author's swordstick

Children in small town, Bolivia

Mother and child near Cusco, Peru

Travel companion, Felipe, and trucker, Argentina

With gaucho in Lake District, Argentina

Fisherman in reed boat, Lake Titicaca, Bolivia

Andean musician above Cusco, Peru

At Machu Picchu, Peru

On Copacabana Beach, Rio de Janeiro

Balloon sellers, Mexico City

Guitarist and food vendors at train station, Mexico

Wyoming, USA

Bob overlooking Grand Tetons

American cowboy

In Grand Teton National Park

Mount Rushmore National Memorial, South Dakota

Chief Ben Black Elk of the Sioux Tribe at Mount Rushmore

Teenagers on park bench in Washington, DC

CHAPTER 19:

Mariachis and Robbers

*'Travel makes you modest.
You see what a tiny place you occupy in the world.'*
— Gustav Flaubert

I was disappointed to be going through Central America without being able to explore the region. I wasn't used to travelling so fast. But time was of the essence.

Throughout my six-day bus journey north to Mexico, Central America was a blur of scenic countryside and small towns stopping only for meals, connections, immigration, and overnight stays.

Passing the vast Lake Nicaragua, I heard from a Canadian missionary on board that one of his colleagues had died there a few years before, a victim of a shark attack. It is probably the only freshwater lake in the world with sharks.

In Managua, the Nicaraguan capital, I met a 20-year-old American serviceman from the army who had gone AWOL, a punishable offence if he was ever caught. He boasted he'd left the US with 97 cents and had arrived in Nicaragua with 95. 'You just have to play it cool, man,' he said when I asked him how he had done it. I later learned that he had knocked on people's doors on the way asking for food and shelter and, with no passport in his possession, had entered the different countries by walking, often a few miles, around the various border posts.

There were several other Europeans on board the bus for Guatemala City, none of whom knew any Spanish and, rather amusingly, I found my limited services required quite often when we stopped at restaurants. Since I'd always considered myself a hopeless linguist, nobody was more surprised than me!

A fast and most uncomfortable journey by long-distance bus finally brought me to the Guatemala-Mexico border, the driver having raced along yet another dangerous road through the Guatemalan jungle, throwing his passengers around in an effort to reach our destination quickly. When we arrived at 1 a.m. in the morning, the immigration and customs office was closed, so we were all compelled to sleep in the bus until 5:30 in the morning when it opened. Why, I wondered, couldn't the driver have taken all night at a safe and comfortable speed? Perhaps he had a girlfriend to meet and was in a hurry to see her. Like so many things in Latin America it left me mystified.

Fortunately, there were no special formalities required for holders of British passports entering Mexico and, at the border town of Tapachula, I caught another bus to Mexico City. It rained for 20 of the 22 hours of the journey, creating flooding in the streets and slowing us down considerably. To make matters worse, there were two leaks in the bus, and I happened to be seated under one of them!

We stopped several times en route to eat and stretch our legs. The more I saw of Mexico, the more it was as I'd imagined. Men, women, and even children worked side by side in the fields with their patient donkeys. Nearly everyone wore straw sombreros, and maize, Mexico's staple food, was grown everywhere.

Arriving in Mexico City, my home for what would be four weeks, I checked into the Hotel Zamora, a small, inexpensive three-storey building in Avenida Cinco de Mayo, half a block from the cathedral in the heart of the city. I quietly celebrated yet another birthday, my 29th, away from home. Like most of the others during my travels, this one also passed without fanfare.

On my first day in the vast, teeming, bustling metropolis that is Mexico City, I was walking around as usual with my camera and getting my bearings, when I spotted a colourfully clad Indian woman seated on the pavement selling her wares. I attempted to take her photograph. No sooner did I produce my camera than the woman was on her feet shouting and threatening to break my camera. As she was brandishing a bottle, I put my camera away. Prior to this incident, I'd never had any difficulty taking portraits, including those of the many indigenous people I had visited. It's true that it sometimes took time but, using tact and discretion, I had always succeeded in getting the desired photo. Except in this case. However, this was not an isolated instance for it would happen twice more in Mexico City, probably due to too many tourists attempting to do the same!

As was my routine now in large cities, I visited some of the newspaper offices. By this time, I was almost penniless, and it was essential that I make some money as quickly as possible. Unfortunately, a spate of travellers had passed through recently, most of who had related their stories to the press. From what I gathered, most of them had travelled comparatively short distances; the majority of them had experienced a few adventures and travelled solely in the Americas. It seemed too that their intention was to send the cuttings home to impress their family and friends while mine was to seek new opportunities. As a result, the field had become somewhat overcrowded as I discovered when I presented myself to some very cynical and blasé reporters. Fortunately, they seemed genuinely pleased to meet me each time after I produced some references and photographs. All admitted that I had something to say and indeed something to sell, although I was never ever paid by newspapers for providing my story.

While being interviewed by a reporter at the offices of *The News*, then Mexico's only English newspaper, I was told about a German traveller who was travelling the world by bicycle. I immediately guessed it must be Heinz Stücke, whom I'd first run into in Tanzania almost four years ago.

Walking back to my hotel, I noticed that someone had attempted to relieve me of the contents of my back pocket. Fortunately, whoever it was had only succeeded in tearing off the button. I shrugged it off as part of the travelling game.

The following day, my story appeared in the *Excelsior* and *El Universal* newspapers, and I received several calls at the hotel, including one from a doctor who wanted me to stay with him and his family for a few days. That was the only call that I didn't receive personally, for at the time I was eating breakfast in the little restaurant next door. The good doctor left his telephone number, which regrettably the hotel receptionist mislaid. This annoyed me no end as I couldn't return his call. He didn't call again, presumably thinking that I wasn't interested in his very kind offer. Several students also called; they had read that I wanted to arrange some *conferencias* and all wanted to be of service. I duly joined several of them for coffee, and they discussed how they would try to assist me. The best way, they decided, was to organise a show themselves and get all their friends and relatives to attend, which is exactly what they did.

Later, back at *The News* where I had gone to seek information as to the whereabouts of various magazines, I was told that a German cyclist had been looking for me. As I sat in the restaurant next door to the hotel, I spotted the familiar figure of Heinz crossing the street to his bicycle, which was chained to a lamp-post. I jumped up and raced after him, saying when I reached him, 'I met you in Africa,' whereupon he countered with, 'You must be David Skillan.' We immediately took coffee together and reminisced about our travels, the first of several such meetings during my stay in Mexico City. He was still selling brochures about his travels, and appeared to be quite well off as a result.

At Plaza de Garibaldi, there were many mariachi bands. Dressed in their full regalia of matching high-heeled leather boots, black, brown or white outfits and massive wide-brimmed sombreros, the seven- or eight-man groups of musicians congregated there night and day to serenade couples, lovers and tourists. It was, I thought, a very romantic way of telling your partner, wife or fiancée that you loved them. One day, Heinz and I sat in a small restaurant at the plaza listening to the music and enjoying oyster cocktails and beer with a dash of salt and lemon Mexican-style. After strolling around, we ate tortillas and hot mashed maize and butter followed by delicious pancakes filled with dulce de leche.

An engineering student called on me looking for a travelling companion. Having been through my experience with Felipe, I told him I wasn't interested. Under no circumstances would I travel again with someone I barely knew.

Another call I received was from a young woman named Marta who wanted to meet the *trotamundos* (globetrotter) and practise her English. She kindly volunteered to act as my guide, driving me around the city as she did some errands. At 9 a.m., she picked me up in her fiancé's red Volkswagen. After that, we met on a few other occasions to visit parks and markets, but I only met her fiancé, Enrique, a few days before I left when we all drove to the Cathedral at Guadalupe. There, we watched throngs of pilgrims slowly moving across the vast square on their knees to enter the cathedral.

One young woman called from the swanky Hotel de Prado, and I arranged to meet her. Juanita turned out to be a voluptuous, 20-year-old Cuban who was the public relations officer for the hotel. She, too, wanted to practise her English, and we met a few times for coffee.

One Saturday afternoon, Juanita invited me to join her sister and boyfriend Carlos on a drive to the pyramids of Teotihuacan. No sooner had we driven a mile and a half in the city when we collided at speed with another vehicle. Fortunately, no one was hurt, only shocked. After returning home and changing his damaged vehicle for another car, Carlos continued to take us to the pyramids. I'd sensed immediately after stepping into the original vehicle that we were travelling with a young and reckless driver. He seemed to delight in showing off to me and the two girls, but he paid handsomely for it later when the bill for the damage arrived. No doubt he also received a reprimand from his father for he was more subdued and careful at the wheel of the second car. Hopefully, he'd learned a lesson.

It was a glorious sunny day when we arrived, and we spent several hours climbing up and down the pyramids, arriving back in the city at sunset. We then drove to see the house where Castro had held some of his clandestine meetings, plotting and planning his successful revolution. At the Anglo-Mexican institute, I was informed that, as much as they would like to have me speak, all dates were fully booked several weeks in advance. Once again, my timing was off. It was now imperative that I make enough money to facilitate my crossing the US border.

The first money I earned in Mexico City was by selling some photographs to the editor of *Siempre*, Señor José Llego. I called at his office as arranged, showed him some of my photographs and he agreed to purchase several saying, 'You must sell your photos in the States. Whatever I give you will be robbery for there you can make so much money.' But I was in no position to choose my markets, and when I looked somewhat crestfallen, he suddenly said, 'Look, how long do you intend to stay here?' I told him probably two or three more weeks. He then asked how much I would need to cover my expenses. I told him some ridiculously low figure whereupon he produced twice as much from his wallet and handed it to me saying, 'Here, take this for now and come and see me tomorrow. I will give you the money. I will not need the photographs.' Thanking him profusely, I assured him I would get the photographs duplicated and give them to him which is exactly what I did. They were included in a later issue of the magazine.

In the small Chinese restaurant where I often ate, I became very friendly with the young proprietor named Richard Wang, whose family had migrated from Hong Kong. As I ate in his restaurant regularly, it wasn't long before Richard started giving me a discount on my meals. He didn't stop there but also invited me a few times to his parents' restaurant. When his older brother got married, I was invited to the wedding, which was a rowdy and most entertaining affair.

Borrowing a projector from a student friend, I gave one of my few lectures in Mexico, which was well attended and for which I received a small fee. All of my lectures received excellent responses, and were arranged by students, except for one, which was arranged by a young secretary named Irene who was president of her young people's social club. I later met her family. They'd spent several years in the US, and all spoke excellent English. They virtually adopted me, and I was invited several times for meals at their luxurious home. Not content with that, Señor Remirez kindly suggested that I use his office and telephone. He liked me so much he even took me to dinner one night and then to a game of Jai Alai, one of the world's fastest ball games, something I thoroughly enjoyed having never seen this unusual game before.

With friends, I visited Xochomilko where, for a small fee, we hired a punt with a boatman who poled us leisurely along some of the water-lily filled canals and waterways. Over a period of two hours, we were approached no fewer than half a dozen times by others in boats intent on selling us souvenirs or singing us a song. As was usual with new-found friends, I always tried to pay my own way, but my offer was usually refused with the words, 'No, let me. You need dee mon-ee.'

Several times I came across festivals, including the famous annual Corpus Christi in the Plaza del Zócalo where an entire market was set up, and indigenous dancers went to pay homage and dance. They were colourfully clad with feather hats and satin cloaks. In addition, hundreds of small children were dressed in the national costume. Strapped to the back of each little boy was a *huacal*, a kind of square basket made from bamboo reeds, filled with fruit or vegetables — a tradition from colonial days when the natives presented baskets of fruit, etc. to the Spaniards instead of money. It was a wonderful opportunity to take some fine photographs of the children in their native costumes. These I would later sell.

Late one night, I befriended an English traveller as he walked down the street looking for an inexpensive hotel and suggested that he stay at my digs. Derek was very grateful, being a complete stranger to the area, on a limited budget and knowing almost no Spanish. He was an exchange student in the US currently on holiday in Mexico. I knew what it was like to be hungry and more or less broke.

By now, as much as I liked Mexican food, I was getting tired of rice, beans and burritos, the inexpensive country-wide dishes in their many different forms. At least sticking to the same diet had allowed me to continue living on a strict budget. Some things I refused no matter how well cooked as, for example, in Peru where I was offered roasted guinea pig and iguana soup.

When I wasn't trying to make money or gallivanting around, I would meet Heinz over tacos and beans or a cup of coffee. One evening, after he went back to his hotel, I walked to the Plaza de Garibaldi to buy a pancake. It was about 9 p.m., drizzling with only a few people about. As I strolled down Avenida Juarez, a drunk veered towards me and demanded 'un peso.' Within seconds, two others quickly approached me, and all three surrounded me, blocking my path. They were shady-looking characters and quite inebriated. Preferring to lose one peso than perhaps my life, I dug into my pocket and pulled out some bills. Incredibly quickly, one snatched a ten-peso note from my hand, and together all three disappeared into the crowded street.

I went straight to the nearest policeman who accompanied me to the spot of the crime but the robbers had vanished into the night. Some 15 minutes later and annoyed as hell, I met Heinz and related what had happened. I asked him to accompany me back to the same place thinking perhaps the drunks might return to rob some other unsuspecting victims. Heinz was quite sure they wouldn't be there but agreed to join me.

It was still drizzling when we arrived back at the scene where I'd been accosted. Lo and behold, there were the same three men laughing and joking and knocking back more alcohol. It was my intention to simply go up and demand my money back, but Heinz had other ideas. As soon as I pointed them out, Heinz went up to one of them, snatched a bill from his top pocket and punched him with full force in the face. As he fell, Heinz punched him again then kicked him as he fell on the ground. I was flabbergasted as Heinz looked around for the other two who, having seen everything, walked quickly and conveniently into a restaurant. But within minutes they came out, no doubt thinking we had gone. Heinz then walked up to another one of them and began hitting him. By this time, a small crowd had gathered as Heinz and the other fellow rolled on the pavement. A couple more punches, and he declared he was done, having given the culprit a good hiding he'd never forget.

Heinz had obviously learned a thing or two during his travels about dealing with such offenders. He was utterly fearless and liked nothing more than a good punch-up. Although not a big guy, Heinz was powerfully built and as tough as they come thanks to the years of travelling and pedalling his bike. He had his own philosophy about life, no doubt learned through harsh experience. He told me, 'Every punch that doesn't kill me toughens me.' Heinz would continue to travel the world on his bike for the rest of his life.

Almost everywhere I went, be it Chihuahua Park or any large square or public place, there was no escape from balloon sellers, organ grinders or ice cream carts. No sooner did one organ grinder move on, than another took his place. I enjoyed the music but took exception when the piece being played was never completed. A few turns of the handle, a passing of the hat and then they were gone!

As I waited for friends who had invited me to a show at the Palace of Fine Arts, I spotted Gunther again. He had been quite sick and was now almost broke, frustrated and greatly fed up with Latin America. I bought him a meal and then introduced him to Heinz, who kindly took him under his wing.

It was June 6, 1968, while sitting in a restaurant that I heard the shocking news of Robert Kennedy's assassination, just three months after Martin Luther King had been killed. Customers suddenly stopped eating and huddled around a radio. I could tell the news was not good from the expressions on their faces and when some of them started sobbing. The newspapers were full of the story the next day. It left me reeling. This heartbreaking event caused upset and dismay among the Mexican people as with many others around the world. It made me think back to November 1963 in Napier, South Africa, when we'd first heard of President John Kennedy's assassination, while Pete, Mike, Gavin and I were living there.

Although I'd made a little money in Mexico City, I was frustrated with my circumstances. So I decided to head to the US and take a chance at getting in, no matter how strict the entry formalities.

Prior to moving on, my new friends showered me with gifts — books, handkerchiefs and other things — which, as usual, touched me and made me slightly reluctant to leave. All wished me 'Hasta la vista' (Until your next visit) and 'Buen viaje' (Have a good trip).

Regardless of all the ups and downs I'd faced travelling in Latin America, I'd seen and done a lot and met many delightful people of all ages and occupations who worked, played, and made their home there. And I would never lose my new and everlasting affection for them.

The train ride to Mexicali on the US border took two days and nights, and during that time we passengers sweated by day and nearly froze at night. For the first part of the journey, we were accompanied by a six-man military guard because of the risk of bandits who might hold us up. We stopped umpteen times at towns, villages, and hamlets, sometimes for a couple of hours during which time we were besieged by womenfolk selling beans and tortillas wrapped in banana leaves while blind men and beggars with hands outstretched wandered up and down both the inside and the outside of the train.

In my carriage was a young American hippie named Murray and we spent part of the journey in animated conversation. Apparently, he'd had a well-paying job as a shirt salesman in Los Angeles but, like many others in the late '60s, he decided to drop out of the rat race. After handing in his resignation, he grew a beard, donned 'way out' clothes and a beaded necklace, and went to Mexico with three friends. There they surfed, sun bathed, and drank tequila to their heart's content. When I met him, his funds had run out so he was heading back to the States to work again to sustain his carefree Mexican lifestyle as a hippie and beach bum.

Murray's girlfriend was meeting him at the border with a car, and he invited me to travel with them to Los Angeles. After disembarking from the train, I queued up at the immigration office with the many other aliens, mostly Mexicans, bracing myself for questions while Murray casually walked through, flashing his American passport. When I was called to step forward, the immigration officer checked my passport and newly acquired visa, asked me one or two questions, which I obviously answered to his satisfaction, and with that, I was allowed to enter the United States. For the umpteenth time, I breathed a sigh of relief. Entering the so-called 'Promised Land' had been easier than I thought. *It would all be different now in the US*, I told myself and, to an extent, it was. But not how I'd imagined! I was in for a rude awakening about certain things.

A few weeks after I left Mexico, full-scale riots erupted, and many people died. I had been warned of the imminent revolution from some of my student friends during my stay. Preceded by months of political unrest, the worst of the rioting occurred on October 2, 1968, just ten days before the opening of the Summer Olympics, and resulted in the killing of many students and civilians by police and army troops in Plaza del Zócalo, the same place where I'd admired and taken photographs of small children.

More or less simultaneously, student protests were breaking out all over the world: France, Germany, Italy and Argentina, even in Japan and, as I would see for myself, the United States. It seemed that 1968 was shaping up to be a year of demonstrations and dissent.

CHAPTER 20:

Welcome to America!

> *'Because in the end,*
> *you won't remember the time you spent*
> *working in the office or mowing the lawn.*
> *Climb that goddamn mountain.'*
> — Jack Kerouac

It was late June, and I couldn't have arrived in the United States at a worse time. I'd been terribly impatient to leave all the frustrations of Latin America behind me and to get settled as quickly as possible. But if I considered that I'd been plagued by bad luck before, even more was to come. I had long wanted to see the 'bastion of democracy' for myself, but I would find out very soon that it was not like Hollywood movies. For the moment, though, I was very excited about entering the 'land of opportunity' assuming that I'd saved the best for the last.

To say the US was going through a tumultuous time that year was a huge understatement, what with the Vietnam War and massive social unrest taking place. The American people were angry and wanted change. Over the next nine months, I would experience for myself much of their anguish and disillusionment as well as share many of their concerns.

Both Murray and Adele, his bosomy, tanned girlfriend, were organic food fanciers and twice on the way to Los Angeles, we pulled off the highway at drive-in road stops for a quick bite. I was introduced to cold date shakes, which were delicious, especially in the hot California sun.

They dropped me in downtown Los Angeles where I booked into a small hotel on 'skid row' not far from the Greyhound station. When they left me, Murray said, 'Play it cool, man. Don't flash money around. Get me?' then they drove off.

I'd been walking around the streets of the city for half an hour when two different individuals approached me — one a small black boy and the other a middle-aged white man — asking for a few cents to make up the difference in their bus fares. These were the first of many such encounters in the big American cities that I visited. I sat down in a restaurant and ordered a 'special' (hamburger, doughnut and Coke). Upon stepping outside, I was again accosted, this time by a shaggy, middle-aged white fellow for 'spare change'.

Feeling there was no time to lose, I called at the offices of the *L.A. Times*, where I met a friendly young reporter. He was very enthusiastic about my travels, but explained that the story would have to be approved by the editor's city desk. For whatever reason(s) the editor rejected it and only later did the reason dawn on me. With Hollywood so close by, no doubt the paper was besieged by people hoping to be discovered. I certainly had no aspirations in that direction. Besides, during my brief stint as an extra in Japan when even the stars seemed terribly bored hanging around for the next take, I had decided that acting wasn't the life for me.

During my stay in the Far East, I had noticed just about every man carried business cards, so I got some printed. Under my name appeared the words: Traveller-Adventurer. But, I was reluctant to hand them out. Silly me. Foolishly, I allowed my normally well-concealed bashfulness to get the better of me.

For two days I walked the streets of Los Angeles. At the Chamber of Commerce, I was given a detailed list of 300 clubs and associations in the metropolitan area but was told they were all closed for the summer months. Too bad, but I'd faced tough times before. *I'd find a way to keep going*, I told myself. However, I couldn't afford to sit at cafés waiting for windfalls. I needed to earn a living somehow and hoped to do illustrated travel talks.

My first impressions of the US, however, were not very favourable. I'd always thought of Americans as gregarious and generous, but many weren't. Even so, I found them to be extremely talkative. And opinionated — qualified or not! I'd never met so many 'experts'. Time and again, I saw people throw their hands up in despair and say, 'What's the good? No one listens anymore.'

Certain expressions had become popular at that time including 'Sock it to me, baby' (from Rowan and Martin's *Laugh-In*), 'Groovy, man', 'Where the action is', 'Doing your own thing', and 'Tell it like it is', to mention only a few. Regarding the latter, I promised myself I would.

Some months before, Pete and I had met a young Peace Corps couple from Santa Monica on the Amazon steamer, who had invited us to look them up at their beachside home. I took a 30-minute bus ride to see them. They were delighted and remembered both Pete and me well, especially having seen Pete save the Peruvian boy's life. Pat, a teacher, and John, a senior research consultant with Douglas Aircraft, had by then settled back into their normal North American routine, although they admitted to greatly missing their volunteer days.

After a delicious meal, they proudly showed me the local beach and supermarket in their new Thunderbird. We then drove to Hollywood to see the houses of various famous movie stars. We also drove down Sunset Strip past the Arthur discotheque before stopping for an enormous ice cream as 'a special treat.' We ended up outside the famous Grauman's Chinese Theatre on Hollywood Boulevard, where I read some of the messages on the Hollywood Walk of Fame. Gary Cooper, one of my favourites, left his mark with 'Syd, I got here at last 14/8/43.' Others read 'Syd, may you never die till I kill you. H. Bogart 21/9/42'; 'My time is your time, Rudy Vallee 21/7/41'; and 'Love to you all, Shirley Temple 14/3/35.'

My small, impeccably clean hotel was owned and managed by Mr and Mrs Takashima, a Japanese couple who always greeted me with a deep bow even though they'd lived in the States for over 20 years. I spoke with some of the residents there. Most were down-and-out older men and all very talkative. When I told one man that I hoped to give travel lectures in the US, he drawled, 'Son, what you need here is a "front".' He meant I should be well dressed, have an impressive address and also have a 'gimmick'. All of which, apparently, would give me credibility. Regrettably, most of my decent clothes were still back in Colombia; I certainly couldn't afford an expensive hotel; and I was much too naïve to be in possession of any gimmicks. What he didn't tell me, but I soon learned, was that it would help if I was brash and outspoken so people noticed me.

With few encouraging prospects in L.A., I decided to take a bus to San

Francisco. As I sat eating a doughnut in the Greyhound terminal café, I chatted to a middle-aged black waitress. I was unsure what to call people of coloured origin in the US and had no intention of committing a faux pas at their expense, so I asked her. 'Well,' she said, 'I dun care what I'se called. All people is just people to me, no matter where they come from.' My sentiments entirely.

On the bus, I sat next to a young married woman who had just waved a tearful farewell to her equally young husband. She complained bitterly about the 10-hour bus ride, which I found to be perfectly comfortable with delightful scenery. 'You should try buses in other countries,' I told her, but she didn't seem to get my point.

In San Francisco, often referred to as 'a rich man's city,' I was accosted again, this time by students seeking handouts as I walked to nearby Chinatown. Eventually I found the San Joachim Hotel, which had been recommended to me by someone in Mexico City. I checked in and being very tired, I went straight to bed. Thinking that everything would be considerably better in San Francisco, I was up early the next morning. After a breakfast of bacon and eggs, I went straight to the tourist bureau to enquire about the whereabouts of various newspaper offices, chambers of commerce and the like.

While in Mexico City, Irene Remirez had very kindly typed a summary of my adventures and experiences. At the *San Francisco Chronicle* reception desk, I asked for a staff reporter. A bearded man in his late thirties appeared, and I handed him my resumé which included a few highlights of my travels.

I understood that cynicism was part of a newspaperman's job, but I was flabbergasted when, at first, this hard-nose, veteran reporter, Michael Grieg, simply didn't believe my story. I'd never faced this reaction before; and even when I produced some of my references, he implied they were fake! It was then that I realised what enormous barriers I was up against. After some 15 minutes of questioning, my honesty finally won him over, and he agreed to write my story, which hopefully would give me some much-needed exposure and produce some positive results.

Two days later an article about my lengthy meanderings was featured in the *Chronicle* entitled, 'A World for Pennies,' with a photograph and caption. Fortunately, the article produced some enquiries. KGO, a local radio station, called me at the hotel and asked me to appear on one of their programmes. I was delighted. Once again, things were looking up.

After a meal of spring rolls and chop suey at Sam Weh's, a popular restaurant in Chinatown, I set off to canvas various clubs and organisations in the city. But everywhere I went I was told the same thing as in L.A., 'Sorry. Closed for the vacation.'

In the three weeks during my stay, I covered almost every part of the beautiful city of 'San Fran', on foot, by bus and cable car. I decided that, like other port cities I'd been fortunate enough to see, it was one of the most beautiful in the world. Much of the time the weather was fine except early mornings when it was often misty and quite cold. Several times, usually at sunset, I walked to the statue of Christopher Columbus at the top of Telegraph Hill from where I had a great panoramic view of the harbour and surroundings.

The time soon came for me to appear on Owen Span's daily KGO Talk Radio show. On arriving at the studio, I nearly turned tail and ran when I saw my name on the noticeboard welcoming me to the station. As usual, my apprehension was unwarranted. I was questioned by Owen on the air for almost 30 minutes, then for the next hour the lines were opened and people called in. It was another novel experience as I sat wearing earphones, hunched over the mic and, although I didn't get paid, I was able to advertise my services. The programme could have been a little longer, but it was interrupted by commercials and breaking news about the church minister who had gone into hiding since giving a ride to Sirhan Sirhan, the man who shot Robert Kennedy. Sirhan had not then been captured, but the Reverend had received threats to his life.

Many of those who called in asked questions about my travels; others suggested who best to approach regarding lectures and what to see in the US. One young woman asked about travelling in Europe, and a middle-aged man greeted me in Amharic, then reminisced about his days in Ethiopia until he was cut off! Yet another woman asked how I liked L.A. I didn't and told her so — she seemed quite upset. Some called just to hear the sound of their own voice. A Californian named Bob, whom I would meet in due course, called in with some general travel questions.

Finally, the last caller, an Australian woman invited me to dinner to meet her family. Two nights later, I dined as arranged with Alison Bennett, her daughter, Nancy, and her husband, Mike, a marine sergeant who had recently completed a tour of Vietnam. They took me for a drive around the beautiful campus of the world-famous Berkeley University. It boasted every amenity and facility imaginable, yet later it would be the scene of the most violent demonstrations, as were many other campuses.

Another day, I met Charles Felix, an advertising executive, and one of those who'd telephoned after reading the newspaper article. After picking me up at my hotel in his Buick, he introduced me to his delightful wife, Pauline, and family of three children at his Tudor-style home in Mill Valley across the Golden Gate Bridge. Then we all drove to Buena Vista and picnicked at a winery where we sampled some of California's famous wines.

Curious to know more about my travels and adventures, Charles suggested that he organise a lecture at his home. A week later and after I had spent a weekend with the family, I put on a show to some 40 of his friends, mostly couples, and all professional people. They appeared to enjoy it immensely, and I received a nice payment after Charles produced a small wicker basket as a collection box, tossed in a dollar bill, and passed it around.

Back at the hotel, I received two letters from people requesting some travelling information. Both enclosed self-addressed envelopes and one contained two crisp single dollar notes 'in appreciation'. I replied to both immediately.

As a result of the radio show during which I mentioned I was looking for work, Barry Eagle, a tradesman from the suburb of Oakland, called me up and offered me a job painting an old house. By then I'd had enough of painting, which, like fruit picking, I found incredibly monotonous and boring. Nevertheless, I was in no position to be fussy and was very anxious to make some money. So, for five days, I commuted by bus for 30 minutes to the nearby port town, leaving my hotel early in the morning and arriving back around six in the evening. The dreary job turned out to be a blessing in disguise, for one day I met an attractive young woman at the bus stop.

As I worked alongside Barry, I listened to the popular radio talk shows which had mushroomed over the last few years and were similar to the ones on which I had appeared. Many were hosted by self-declared 'experts' on anything and everything, and dealt with controversial subjects such as drugs, sex, racism, and, of course, politics. People phoned in simply to vent, to air their point of view, or to receive advice from these experts, very few of whom were qualified but had the 'gift of the gab'; they were more entertainers than experts in anything. Freedom of speech covered everything, it seemed, but I was not impressed!

On the last day of my painting job, I met Shirley, a 22-year-old English Lit student at Berkeley University. As I stood at the bus stop at the end of the working day, I rather shyly struck up a conversation with her. Much to my delight, she responded, and we chatted happily together on the bus into the city. She was softly spoken, of average height, with short, brown curly hair, very large hazel eyes, and a face that was made to adorn the covers of women's magazines. That chance meeting at a bus stop was the beginning of a pleasant, albeit entirely innocent and very brief, interlude. Over the course of just a few days, we explored much of the city together, visiting Fisherman's Wharf and riding from Union Square to Nob Hill by cable car.

One Sunday afternoon as we sat sipping tea in the Japanese garden in Golden Gate Park, Shirley leaned over and kissed me lightly on the cheek. 'Welcome to America,' she said. Suddenly I felt like a new man. With that personal and intimate gesture, she had unwittingly given me renewed confidence in myself. Regrettably, my friendship with Shirley was to fizzle out as quickly as it started when I got the impression there was somebody in the background, and no room for me.

Sometimes in the evenings, I strolled through North Beach, which reminded me vaguely of Sydney's King's Cross and London's SoHo and which was only a stone's throw from my hotel. There, in an effort to sample the flavour of the place, I mingled with the seamen, tourists and hippies some of whom disappeared into the many strip joints, San Francisco being the home of the original topless shows. I always walked warily and often passed some scruffy-looking youth or girl who whispered from the corner of their mouths, 'Pot?' or 'Grass?'

A couple of times I strolled up and down the streets of Haight-Ashbury, home of the original hippies whose mantra was, 'Turn on, tune in, drop out.' There was no doubt in my mind that the flower children were extremely idealistic declaring universal peace for all. However, instead of the colourful, interesting scene that I expected, I found the area to be dreary and depressing, full of scruffy pseudo-hippies and weirdoes, some stoned out of their minds. Once I picked my way over an attractive young woman semi-prostrate on the sidewalk, leaning up against a building to prop herself up, quite dopey from drugs. I left what had been a famous district but was now a sad, pathetic scene projecting a loss of faith and hope.

One evening I went to help an old man to cross a street. However, no sooner did I try to assist him, than he pulled away and shouted, 'Don't touch me! Don't touch me!' I was incredulous. He was afraid that I was going to rob or beat him! Apparently a regular occurrence on the streets of American cities.

Yet another person whom I met as a result of the radio show was Earl Brown. A middle-aged, genial Californian with a very loud voice, he'd been a fighter pilot during the Second World War flying Spitfires with the Royal Canadian Air Force. As one who still enjoyed a somewhat dangerous life, he operated a giant crane on the top of a new 40-storey skyscraper under construction. At Earl's invitation, I visited him at work after travelling by lift the first 32 floors, then climbing the stairs the rest of the way up. What a view it was. From his lofty perch, we could see almost 360 degrees of the city and suburbs and well into the distant Coast Mountains.

Earl also invited me to join him, his wife and some friends one weekend at a well-known beauty spot not far from the city where they frequently went to get away from it all. He'd met Hank, one of his best friends, at the same place years earlier when they both discovered they were sharing the same mistress! Most of the time was spent in conversation, dominated by Earl, knocking back beer and firing at logs and bottles with the revolvers that everyone seemed to have brought. This made me think long and hard about the people of the United States. They liked their firearms.

It was now mid-July, and, although I could have had bookings for lectures in September, time was money and I couldn't afford to hang around. So, when I heard that Bob, the guy who'd called in to the radio station with questions, was driving to New York City, I seized the opportunity and asked if I could accompany him, considering it to be a wonderful chance to see much of the country. 'Sure,' he said without thinking twice. He was delighted to have me, especially as I offered to pay half the cost of petrol, food, and miscellaneous expenses.

At 32, Bob had done a multitude of different things. He was, by profession, a qualified engineer but didn't like working in that capacity. 'Too competitive,' he said. For experience and variety, he had been a cab driver and a nightclub dancer. He wasn't gay or weird. He simply wanted to learn professional dancing and did so. At one time he had worked with Carol Doda, the first and most famous of the city's topless dancers, until one evening before a packed house, he tossed her into the air and she landed in the wrong position, breaking two ribs — putting her out of action for a while, and himself out of a job! Articulate and knowledgeable, he punctuated most of his conversation with 'Groovy, man, groovy.'

Around about that time, I received a letter from Pete suggesting I join him in Canada. But I was too intent on doing my own thing. I wanted to organise myself my way, choosing to do so in the US.

Bob would prove to be the ideal travelling companion for a thrilling road trip crossing the entire country from west to east. Likeable, easy-going, and well-informed, he was one of those people who did, in fact, know his own country well. Even though he'd not actually crossed it yet, he had mapped out an excellent itinerary for us to follow which included a number of national parks as well as places off the beaten track. We would be travelling in his spacious and weathered Volkswagen 1500 station wagon.

It was in very high spirits that we set off to drive across the nation on, of all roads, Route 66. Bob and I were familiar with the popular TV programme of the same name, and we couldn't help feeling as we drove that we resembled the characters in some way. Driving on the freeways maintaining a constant speed of 70 miles per hour was disconcerting to me at first. But, like so many things I'd done for the first time, I concentrated on the job at hand, and before long, became reasonably comfortable with highway driving. I couldn't fail to be impressed with the excellent roads which invariably made motoring 'out West' a pleasure.

We first drove to L.A. where we called on Bob's sister, then drove down Hollywood Boulevard and past various places of interest before getting onto the highway to head east. With a 'gung-ho,' another of Bob's favourite expressions, we were off. Two young guys eager to make the most of what for many would be 'the ultimate road trip'. *New York, here we come.*

Throughout the two-week journey to New York City, we camped out.

We either slept in our sleeping bags on a groundsheet or inside the car. Fortunately, the weather was glorious most of the time.

After passing Palm Springs and motoring through the Coachella Valley desert and San Bernardino Mountains, we left California behind and entered the state of Arizona. Once, we stopped to look at some fossilised footprints of dinosaurs that had roamed the earth millions of year ago. As we drove, I marvelled at the great wide-open spaces and the sweeping vistas in all directions. A visual feast that took one's breath away.

Seeing the Grand Canyon for the first time was a jaw-dropping experience. We stayed and camped in the area for two days to take it in from all angles. One evening, we watched mesmerised as the sun set over the canyon rim — a never-to-be-forgotten experience and, like so many others, one to be cherished. No doubt about it, when it came to natural wonders, the United States was richly blessed.

Travelling through the Painted Desert, we stopped to give a lift to a young Navaho Indian named Joe. He was the first of several hitchhikers we picked up. Like many Native Indians, Joe was a sheepherder and resented the government's attitude towards his people, for all too often they were regarded as and felt like aliens in their own land. At more or less 10-mile intervals along the original Navajo Trail, now a fast highway, we passed trading posts patronised mostly by Indians and the occasional motorists stopping for cool drinks or ice cream.

Picking up our fifth of seven or eight hitchhikers and making small talk with him made me reflect on my own experiences when I'd relied on my thumb and clean-cut appearance to get to countless places. No doubt about it, my inner voice told me, hitchhiking, and for that matter rough travelling was fraught with danger. And definitely not for the faint of heart.

Despite the fact that we were travelling at the height of the summer season, often our vehicle was the only one in sight. Several times, we drove along dirt side roads. Then we only saw the occasional farmer usually heading in the opposite direction.

We arrived in Arches National Park, Utah, just in time to see the extraordinary rock formations at sunset; then we swam in a safe, secluded spot in the Colorado River and camped on the bank. There, I was buzzed and attacked by mosquitoes. No matter where I was, they still loved me!

Welcome to America!

To minimise expenses, we purchased food from roadside stores and supermarkets and occasionally cooked a meal over open fires. But for much of the time we existed on enormous hamburgers and root beer from roadside diners, or peanut butter and strawberry jam sandwiches which we made as we drove. As always when travelling, I felt like a local, eating and living as they did and relishing the lifestyle. Frequently, we saw wildlife — deer, squirrels, rabbits, and hares. And once, ambling along a quiet backroad, a black bear.

After visiting Arches National Park, we picked up a young hitchhiker and took him with us to Jackson Hole, Wyoming, an old miners' and prospectors' town. We then drove into Grand Teton National Park where we tossed stones into one of the crystal-clear lakes and gazed up at the high rocky peaks. The fresh mountain air was invigorating and, as always in such a natural and magnificent environment, I felt completely at home. Sadly, that blissful feeling of euphoria would be short-lived.

In Yellowstone National Park, we were fortunate to spot no less than three grizzlies at very close quarters. The famous Old Faithful geyser erupted just as we arrived as if expecting us. Wandering around the hot springs, we saw moose and elk grazing in the nearby marshland.

I had thought that the days of the Wild West had been over for a long time, but in Wyoming, Montana, and the Dakotas, it seemed that everyone catered to the cowboy. Most of the men and women in towns wore Stetsons, checked shirts, and jeans while country music came from jukeboxes in the many bars. There seemed to be western outfitters and saddlers everywhere, and the accents were as familiar as in western films. Twice, we stopped to chat to and take photographs of cowboys who seemed to love posing for the camera. Passing through Indian Territory, we met people of the Cheyenne, Shoshone, and Blackfoot tribes. This was the US I had often read about and seen in films, I realised, as we drove leisurely and happily through the Badlands and Black Hills.

Our next destination was Custer State Park and Mount Rushmore National Memorial where we gazed in wonder at the massive faces of the four presidents — George Washington, Thomas Jefferson, Abraham Lincoln, and Theodore Roosevelt — carved out of granite in an extraordinary feat of engineering. There, we fell into conversation with Ben Black Elk of the

Sioux tribe. Often called the fifth face on the mountain, he sat at the foot waiting to be photographed for which he earned considerable amounts of money almost every day which he distributed, so he told us, to his people. A pamphlet about him that I picked up declared that he was a philosopher and a man of great wisdom who, 'bridges the gap between the world of the white man and the world that the white man has forced on these long-suffering people.' Once, when testifying at a Senate hearing, he talked about the confusion in all Indians' minds; that the young people were living too fast and were becoming neither white nor Indian, forgetting and gradually losing their identity and history, a worldwide problem for indigenous people no matter where they lived.

While Bob took a nap inside the car in the hot afternoon sunshine, I stood and watched great herds of buffalo grazing on the plains. Later that day, we pulled up and camped on the banks of the Missouri River. There, surrounded by the incomparable grandeur and beauty of it all, I couldn't resist bursting into song. It was just so good to be alive!

On we went through the agricultural state of Minnesota, dotted with picturesque farms and fields and exceptionally flat for miles and miles. Passing through the city of Indianapolis, I never dreamt that I would be there again a few months later addressing a luncheon audience of 120.

We entered West Virginia via a bridge over the Potomac River. The state reminded me of parts of the British Isles with its beautiful green hills, forest and farmland, boasting large mansions with white painted fences and cattle and horses grazing in the well-kept fields. Later the same day, we crossed over the wide Shenandoah River, the tall green trees along the banks with plenty of 'No trespassing' and 'No fishing' signs nailed to them.

A few times as we drove, I suggested to Bob that we call on farmers in the hope of being allowed to sleep in their barns. But he was strongly against that idea. 'Too dangerous,' he said. 'We'll be shot.' He was convinced, and so was I after talking it over with him, that because of the current 'explosive' situation in the US, we wouldn't be too warmly received.

Throughout the sixties, the United States had seen tumultuous times with massive social upheaval, assassinations, riots, demonstrations, and protests of all kinds. It was also the time of the Black Movement, the hippie era, and the Vietnam War, making 1968 particularly memorable. Due to escalating violence throughout the country, many American people were afraid. And no longer were strangers to be trusted, especially in the more remote, rural areas. 'What a bloody shame,' was my reaction at the time. But, upon reflection, completely understandable. Like it or not, it was the way things were.

A few miles past Middleburg, Virginia, along the Lee Jackson Highway, we stopped at a small antique shop to ask about a place to stay. A young man tinkering with his car overheard us and invited us to his home for the night. 'I'd welcome the company,' he said. This stop ended up being one of the highlights of what was proving to be a very interesting road trip.

John Tiller was a quiet, reserved university student who lived on the family farm. The beautifully maintained grounds were more a retreat than a farm for the only animals there were a few horses and several dogs owned by the keeper whose job was to patrol the property and look after the gardens. John's mother lived there most of the time but was away when we visited. The most interesting fact about the property was that it had formerly been leased by President and Jackie Kennedy as a family getaway.

Bob and I spent the night in one of the very comfortable cottages which had housed the secret service agents during the Kennedys' stays. That evening, the three of us swam in the outdoor pool. As we relaxed over a few beers, we listened to our young host's stories about the Kennedys. Like many of his countrymen, he idolised them both. Lying in bed that night, I reflected on all the misfortune the Kennedy family had endured, then fell asleep to the noisy chirping of crickets.

The next morning was slightly misty as John showed us around the sprawling grounds. On the lawn outside the main house were a slide, swings and some toys which the Kennedy children, Caroline and John-John, had probably used. The scene brought a lump to my throat as I imagined the whole family laughing and playing together.

Back on the road, we encountered very heavy traffic, mostly cars driven by government people on their way to work. As we entered Washington, DC, the nation's capital, the dome of Capitol Hill came into view followed by several other famous landmarks: the Washington Monument, Lincoln Memorial, Pentagon, Supreme Court, Smithsonian Museum, and White House. All these impressive buildings with which I had been only vaguely familiar from movies and occasional glimpses on television suddenly came to life. It was thrilling to see them for the first time.

Since it was my intention to try to sell some photographs to National Geographic, I called at their offices. After waiting quite a while for someone to see me during which time I browsed around the famous building, I was introduced to one of the photo editors. He was polite but quite snooty when he explained to me, 'You know, it's very seldom that we accept photographs from people that we don't know. Unless they are well known or we commission them to do a job on our behalf. Sorry.' I felt crushed for, on the advice of several people, I'd felt sure that some of my photographs would be accepted. No doubt my sights were set much too high as National Geographic represented the finest photography in the world.

The weather was insufferably hot and humid as Bob and I wandered around, taking in the sights. This was apparently normal for summer, and our shirts soon clung to our backs. But New York City would be just as bad or worse!

As we sat in the car munching sandwiches, we asked a passer-by if he knew of a nearby campsite. Scratching his head, then stooping down to catch a better look at us, he suddenly said, 'Come to our place. You'll be very welcome.' His name was Michael Strothers. He and his wife, Patsy, occupied a large house in the suburbs where we happily stayed for two nights. Michael was a young budding politician, an intern who had decided that the best place to do his training was the capital. It seemed ironic that, despite most being very leery of strangers, there were still some people who took us at face value, and realised that we meant no harm.

Bob had to conduct some business during our brief stay, and I accompanied him on some of his appointments. In between, we strolled past the White House, guarded by steely eyed police officers. We also looked around the Treasury Building, an imposing structure and home of the Secret Service. We then joined a party of tourists for an escorted tour of the FBI headquarters. We ended the afternoon by looking around the National Art Gallery then spent half an hour in a small park talking to two black teenage boys. They seemed pleased to chat and to have their picture taken. Another great shot!

The next morning after some last-minute sightseeing and a lunch of hamburger and fries grabbed from a drive-in, we drove through a very heavy deluge which, luckily, had stopped and dried up by the time we pulled off the road to camp. Up early the next morning, we took the I-95 and, after five hours of speeding along the freeway and skirting Baltimore and Philadelphia, the famous Manhattan skyline came into view. After negotiating our way through dense traffic jams, we arrived smack in the centre of New York City, which I didn't know at the time would end up being my home for the next several months.

After a dozen enjoyable and interesting days crossing the country from coast-to-coast and covering slightly over 3000 miles, Bob and I had arrived in 'the Big Apple'. By now, I'd had plenty of highs and lows not to mention remarkable experiences on my marathon journey, but New York City was destined to complete my global education.

CHAPTER 21:
Down and Out in New York

*'Plunge boldly into the thick of life,
and seize it where you will,
it is always interesting.'*
— Johann Wolfgang Von Goethe

I'd known for some time — both from the favourable reactions from the newspapers who wrote about my peripatetic life and those who attended my travel talks — that my story was of interest to many people. Now was the time to try to exploit it to my advantage. Such was my attitude on arriving in New York City. But as I'd already heard and would witness for myself, at the time the United States was a turbulent and volatile country. And nowhere more so than in New York City which in the late sixties was a violent, noisy, crime-ridden place. And I had a front row seat!

With our wonderful road trip over, Bob and I booked into the International Student Association hostel in a tenement building on the west side near Central Park. It was filled with foreign students on holiday, and the only available space was on the floor where we threw down our sleeping bags. The bathroom facilities were totally inadequate, and there were cockroaches running around. But we didn't mind; we had a roof over our heads. Bob stayed for two days, then drove to Buffalo to visit friends. I was left on my own, not knowing a soul in the city.

My first Monday in New York City dawned, and I was full of apprehension for I'd heard the most terrible reports about the city, most of which turned out to be true. Once, I asked a bus driver where 75th Street was. 'One block from 74th,' was his sarcastic and glum reply. For around two weeks, I wandered in a slight daze, calling at offices, as my mind whirled with ideas on how to make money.

Like most big cities, New York was quite bewildering at first but, armed with a city map, I soon got to know it as well as anyone else. Over the space of a few months, I visited all five boroughs of the city, travelling by bus, train, underground and on foot. I saw the city from the top of the Empire State and Chrysler buildings; viewed the Manhattan skyline from the Staten Island ferry; and wandered along Fifth, Park and Madison Avenues, through Rockefeller Center, and down Broadway.

In Times Square, I saw huge billboards advertising *Finian's Rainbow*, a new musical starring Fred Astaire, Petula Clark and Tommy Steele. At the same time, the movies *2001: A Space Odyssey* and *Barbarella* were showing. Judy Garland was playing at the Palace Theater, *Hair* at the Biltmore, and *Hello Dolly* at the St James. But I had no time or money for shows. I had to find a way to make a living.

Once or twice, I stopped to watch demonstrations, but with so many occurring, hardly anyone paid attention. People picketed and protested for the slightest reason. It was also a time for 'love-ins' and 'sit-ins'. The previous night, two policemen had been ambushed and gunned down when they went to investigate a false alarm. Watching the police officers keeping the crowds moving, they seemed very tense. They resembled walking arsenals with their weapons, night sticks, hand cuffs, and walkie-talkie radios.

I'd changed my casual travelling clothes for something more formal — grey flannels, a lightweight jacket, a white shirt and tie — and, although I wasn't the smartest man in town, I was at least respectable. The temperature was 97°F and, like all New York summers, it was very humid and sticky. Travelling on the subway was quite dreadful. Even with air conditioning, it stank of urine and body odour. It was also a dangerous place, especially at night when armed transport police patrolled up and down the train and on the platforms — an essential service introduced three years previously to reduce the growing pattern of murders and beatings. Time and again I would be warned by friends not to take the subway at night; when I did I was always extra vigilant and always looked for the car with an officer. I had no intention of my world tour ending abruptly by being murdered on the subway or on the mean streets of New York. I was never beaten up, thank God, but when travelling around the city, the thought often occurred to me.

Intent on getting some publicity, I called at the offices of the *New York Times* where, after explaining briefly to the receptionist why I was there, I waited for a staff reporter. A tall bearded fellow appeared and asked me what I wanted. I began to explain my story to him, but he obviously didn't believe one word, giving me precisely two minutes of his time. 'Sorry. We aren't interested in that sort of thing,' he said and walked away while I just stood there speechless. It was the first of a number of similar dismissive encounters.

It had taken me several years of intermittent rough and tough travelling to get to this point, and I wasn't going to be put off by some office jerk who maybe had never left the city. I'd faced all kinds of challenging situations in my travels, yet time and time again in New York, I would be dismissed after only two or three minutes. I received a slightly warmer reception at the *New York Post* but was told that the time was not right. The features sub-editor at *The News* did in fact express great interest but said, 'not for a while.' He suggested that I call on the 'British boys' who were in an office upstairs.

With nothing to lose and intent on meeting as many people as possible, I did just that. In the offices of the *Daily Express*, two secretaries, Barbara and Lois, were thrilled to meet me and hear a little about my world travels. They must have considered me very undernourished, for within minutes both invited me to dinner at their respective homes on different evenings. Lori was single and, like me, in her late twenties. Barbara was in her early thirties and living with her boyfriend. As events unfolded, she kindly took it upon herself to telephone some radio and television stations on my behalf, and a few radio stations were only too pleased to have me as a guest.

The *Daily Express* reporters were impressed by my life on the road and seemed very pleased to meet me. 'We may be able to help you when you get home,' said one when he took me for a drink in a nearby bar. 'But I'm afraid there's nothing we can do at this end.' A fairly understandable conclusion. *However, I'd made my first friends and suddenly everything was looking up.*

That evening, I dined with Barbara and her partner, Bill, a well-built, slightly balding fellow in his mid-thirties with a nervous habit of twirling his moustache while talking. During the meal Bill whispered to me, 'We're living in sin.' 'Who cares,' I replied. 'It's none of my business.' After all, it was the Swinging Sixties.

More than anyone else, this generous and delightful couple would play a very important role during my five-month stay in New York. With funds rapidly dwindling, I could hardly afford to stay at the student digs for long despite its reasonable price. At just the right time and with typical American hospitality, Bill and Barbara invited me to stay with them despite the fact they were living under the constant threat of eviction and could be turfed out at any time. The main reason they lived where they did, they told me, was not due to a lack of money, but because their home was so spacious it allowed Bill to pursue his passion for photography.

They occupied a loft on the first floor of an old apartment building in Canal Street, a seedy major east-west thoroughfare in lower Manhattan not far from The Village and just a few blocks from Chinatown, SoHo, and Little Italy. Years later, it would be gentrified and become a fashionable, trendy place with high-end boutiques and elegant coffee shops. But in those days, most of the old buildings in the area housed shops and offices on the ground floor while upstairs were huge lofts occupied by artists, writers and musicians. For a long time, Greenwich Village had been their favourite haunt but, when that area became too expensive, many people moved to the Canal Street area or further afield where the rent was still reasonable.

When Bill and Barbara made it clear that I was more than welcome to stay as long as I liked, I moved into one of their spare rooms and made myself at home. Luckily for me, their studio was huge by most apartment standards, probably 1,200 square feet. My room had little furniture and my sleeping arrangements consisted of a mattress placed on an old wooden door, which was placed on two high wooden chairs, leaving me about three feet above the floor. Basic and spartan though my new digs were, they provided a welcome shelter over my head for the time being. And I was both relieved and extremely grateful. In fact, I couldn't thank my new hosts enough.

As soon as I moved in with Bill and Barbara and to earn my keep, I helped to clean and paint the place. I also made friends with Dum-Dum the family dog, a big brown mongrel which I soon started taking out for daily walks.

When my confidence reached its lowest ebb in New York City, I found comfort in being with the dog, realising why some people resort to animals for understanding. Dumb they may be, but better any animal than cynical, suspicious, and indifferent people were my thoughts at the time.

Disposing of the trash was easier said than done. Like many people living in unregistered tenement buildings, they had no access to garbage bins, so it was necessary to drop rubbish into the street litter bins. Many times, under cover of darkness, I found myself with large smelly bags cradled under each arm, walking several blocks to drop them off away from the eyes of police prowl cars.

Upstairs on the next floor lived an eccentric Korean electronics engineer. Above him lived a young couple (the husband a budding cinematographer) and above them was a Japanese-American couple whom we seldom saw. At the entrance on the ground floor was a pizza parlour on one side and Rosie's Bar on the other. A frequent visitor to the place was Chris, a strapping, bearded artist-cum-sculptor who lived a few doors down the block and occasionally came to play chess with Bill.

Bill was a full-time firefighter with the New York Fire Department, but he regarded the job as a means to an end. His principal interest was photography, and there was no doubt that he excelled in his chosen field. However, like literally thousands of other hopeful photographers, artists, authors and poets, he was always striving and hoping for 'the big break' which, for most people living in New York, never comes. He told me he would have died to shoot the cover for either *Life* or *Look* magazine, his ultimate goal.

To add to his considerable portfolio, he recruited Lori as a nude model. She was tall and slim with long flowing fair hair. Barbara made sure these photo sessions were 'all perfectly decent and respectable'. Sometimes on a shooting day, usually a Saturday, I got a sneak peek. Like most men, I greatly admired the female form, especially when exposed in all its naked glory.

Bill also moonlighted as a barman downstairs at Rosie's Bar. Once or twice I worked with him or stood in for him, but very quickly I discovered that I was useless at mixing drinks.

After I'd been in New York for some weeks, a long weekend interrupted my plans. Americans seemed to celebrate frequent public holidays, which more than once caught me unawares. Lori invited me to spend a weekend at Fire Island. I'd never heard of the place until then, but I was soon to find out for myself the reputation it deserves. The island, a long narrow piece of land known for its pristine beaches, state park and lighthouse, is an hour-and-a-half drive from the city centre followed by a half-hour ferry ride. We drove in Lori's mother's Pontiac which had power steering. Lori steered with one solitary finger most of the time, making me very nervous as we cruised at 70 miles per hour along the four-lane highway, listening to Diana Ross and the Supremes on the radio.

Fire Island was then a summer resort frequented by young singles who descended on the place in droves. Lori and I joined several other couples to share expenses and occupy one of the many beachside cottages. Of the dozen or so who stayed in that cottage, all of us were under 30. All, except for Lori and I, had been married and were now divorced and all, despite it being illegal, smoked marijuana. Or used much stronger drugs, such as heroin.

Our short stay was a welcome break from the city, and Lori and I spent most of the time swimming and beachcombing along the golden sands, during which time my back and shoulders were badly sunburned. It was obvious that New Yorkers seized every available opportunity (most weekends) to get away from the crowded and bustling metropolis. Contending with all the noise, traffic, and pollution was almost too much to bear. People just needed to escape. Now living there, I realised how much I preferred the natural world and wide-open spaces to city life.

Back in the city, Barbara had arranged for me to appear on Jim Gearhart's WNBC radio show. Jim immediately made me feel at home and, at 2:15 p.m. on August 13, 1968, I made my New York radio debut. Jim obviously took to me and introduced me as, 'smart and good-looking' (which made me smile) and just the person he'd choose to take him on safari. As in San Francisco, the interview was followed by opening the phone lines for questions. Some of the calls were amusing with one wise guy suggesting that I sign up for welfare. A woman who didn't give her name suggested that I join the FBI or CIA. A Mrs Roundhouse called in and suggested that I do some dishwashing, and another woman suggested that I register at an employment agency. Yet another lady with a snooty voice offered me a job as a salesman in her art gallery saying, 'You just smell of success,' and someone else suggested that I join a private detective agency. Others asked questions about my travels. Several asked when they were going to see me on TV. I had no answer to that.

That programme lasted two enjoyable hours, but no one proposed any substantial ideas or contacts for lectures, probably because it was still summer. A few days later, I was invited to appear on Radio WHO's John Wingate Show. John interviewed me briefly and included such questions as, 'What was the world's attitude towards the Pope's decision on birth control? What was your funniest form of transportation and where? When sleeping in Malaysia, Thailand and Rio, did you sleep by yourself?' I don't recall my answers but everyone seemed satisfied with my replies. Again, no offers came my way.

Another day, I got a call from WNEW radio asking me to appear on Alan Burke's show. Alan was a well-known host who had his own syndicated radio and television talk shows. A controversial character, he was known for putting down his guests. I duly turned up at the TV studio, which had a live audience. Waiting in the wings alongside me was none other than world heavyweight boxing champion Muhammed Ali, formerly Cassius Clay before his conversion to Islam. There he was, towering above me, muscles rippling through his Saville Row suit as he acknowledged me with a cursory nod. At 26 and six feet, three inches tall and roughly 200 pounds compared to my puny 135 pounds, I was glad not to be boxing him.

To this day, I can't remember what I said during Alan Burke's interview, for I had a bad case of stage fright. But I do remember that I was just a 'filler' which meant I was put in the show for just a few minutes to give it a little variety. And thank God he wasn't unkind to me. Mohammed Ali, then known as the greatest boxer of all time, got slightly different treatment. He certainly wasn't shy; in fact, he was an extreme extrovert. I'd seen him frequently on television and admired him for his boxing skills and Muslim cause, but his boastfulness irritated me. Once I was done, I didn't hang around for long.

Well-known guests who appeared on Burke's show around that time included Norman Mailer, prolific author, novelist, filmmaker and controversial journalist; and F. Lee Bailey, a prominent and aggressive lawyer who 25 years later would become famous as the lead defence in the O.J. Simpson trial.

Other famous figures of the day would make a profound impression on their times, and appeared regularly on one of the Big Three television channels — NBC, ABC and CBS, then still in black and white. Dr Spock, renowned paediatrician whose advice, much of which he would later admit was erroneous, was followed faithfully by millions of American mothers.

Andy Warhol was a well-known illustrator and modern, contemporary, and abstract artist. Timothy Leary, a Harvard professor, urged everyone to try LSD, the 1960s being the heyday of illegal drug use. John Lindsay was the charismatic and handsome mayor of New York who faced more than his share of city and community problems during his tenure.

Then there was Allen Ginsberg, poet and philosopher who, along with William S. Burroughs, writer and novelist, and Jack Kerouac, famous for his novel, *On the Road*, formed the Beat Generation. Much as I would have liked to have met French-Canadian Jack Kerouac, I never got the chance. He was something of a restless soul who, like me, enjoyed the freedom of the open road.

Influential women of the time included Jane ('Hanoi Jane') Fonda, actor, political activist and outspoken critic of the Vietnam War. Gloria Steinem, journalist and political activist, along with Bella Abzug, a lawyer, and Betty Frieden, feminist and author of *The Feminine Mystique*, were founders of the newly created Women's Liberation Movement. Australian Germaine Greer, feminist firebrand and author of the soon-to-be published *The Female Eunuch*, was particularly strident in her views.

The hour-long news show *60 Minutes*, with the tough and outspoken Mike Wallace, was just beginning its long and illustrious life span. Most of the nightly and weekly shows were watched by millions and were hosted by such luminaries as Johnny Carson, Phil Donahue, Dick Cavett, Merv Griffin and Mike Douglas. I watched and listened to them all, wondering how some had acquired such lucrative, high profile jobs. Being a good talker, I decided, seemed to be their main qualification.

Having little social life, there were plenty of TV shows to provide me with a good laugh and a welcome distraction from the everyday trials and tribulations of big city life. Once a week I watched Lucille Ball and husband Desi Arnaz in *I Love Lucy*, and such stalwarts as Jack Benny, Dick Van Dyke, Red Skelton and, of course, Bob Hope.

Borrowing Bill's trusty Remington typewriter, I set to work teaching myself to type so I could send letters to TV stations offering to share some of my experiences. What surprised me was that I received hardly any replies. This perplexed and annoyed me, for I knew that there was no excuse for that anywhere in the world. It was just bad manners and very unprofessional. But, as I learned, typical of the New York scene.

Throughout my travels, I'd always been an early bird, but in New York City I became a night owl — brainstorming with Bill and Barbara about potential sources of income, typing until the early hours of the morning, or catching glimpses of late-night TV. Now down to my last few dollars and jobless, I was feeling the pinch but lived in constant hope that it would all change.

New York, known as 'Fun City' in the summer, is a delightful place for a few days and with pockets full of money but a very different story when you live there and are broke! However, one moonlit evening, I was thrilled to join hundreds of New Yorkers in Central Park listening to a free concert by the New York Philharmonic Orchestra. It was the last entertainment I enjoyed until I had made a little money.

At most magazine offices that I visited, I couldn't get past the receptionists without an appointment. I did manage to see the editor of a true adventure magazine. Unfortunately, I was pipped to the post by a young New Zealander who had been there before me. His story, 'Around the World on $1 a day,' was on the front page and a little misleading. Upon reading the article, I discovered that he had travelled only through the Americas and spent considerably more than the title suggested. No similar travel story would be printed for some time, I was told.

I tried photography agencies, hoping to sell some of my pictures, but I was told the same thing, 'Sorry, our photographs are taken by people we know.' To me, this seemed ludicrous. To achieve almost anything in New York, you had to know somebody, or the doors were closed.

I also called at some women's magazines, confident that I could at least sell some of my photographs of women and children taken in different parts of the world. The response was the same — I was unknown. It was a vicious circle with the same writers, photographers and TV personalities getting all the business and exposure.

Time and time again, my six years travelling the world were dismissed as though they hadn't happened. I felt I had something to share but was up against a wall of indifference. It hurt that no one, except good friends, seemed to believe me. As Bill would say, 'Nobody gives a shit!' He also reminded me more than once, New York City was a selfish 'dog eat dog' and 'everyone for themselves' place.

Those days I hardly ate anything, afraid of spending money, except for a snack during the day (burgers, hot dogs, or pizza). Barbara cooked an evening meal which she insisted I eat and which I greedily gobbled up.

One day, I noticed a middle-aged invalid lady attempting to cross Fifth Avenue. Unlike my experience with the elderly man in San Francisco, she seemed to be only too pleased to get a helping hand from me, ever the Boy Scout. Over the months, I lost count of the number of old ladies I helped to cross busy streets. Most people ignored or neglected them despite posters on the buses extolling, 'Give a damn.' There was no room for courtesy in New York as I was fast finding out.

All too often, as before, I was accosted in the streets for 'spare change' or 'a cigarette'. Indeed, hardly a day passed when I wasn't approached at least once or twice by panhandlers, sometimes demanding or even threatening. I felt they were parasites, just as capable of doing a good day's work as anyone else. As I was constantly out and about, I needed a solution to the problem. The last thing I wanted was to be mugged, or worse. When I mentioned this to Bill, he advised me to always walk with one hand in my pocket, which suggested I might be carrying a knife or gun. Luckily, the pose worked. From then on, I was seldom pestered for anything. It was also a welcome addition to my growing (and never-ending) 'street savvy' education.

In the midst of my New York stay, I suddenly remembered Mike, the Peace Corps worker we'd met during our fun days in Ethiopia. Having been so preoccupied, I'd forgotten all about him until now. He was thrilled when I called, convinced that both Pete and I had forgotten him. His immediate reaction was, 'Come and stay.' I didn't but was reminded that I had no shortage of friends. At least now, for a while, I could hang out and drown my sorrows with a fun-loving pal.

Mike and I met often over the weeks for coffee or pizza or to stroll around the Village or Washington Square. His company was refreshing at a time when my social life was almost non-existent. He'd been very impressed hearing our travel stories. In much of his spare time, he was involved in the Biafran relief campaign, but gradually lost heart as support for the cause waned. Little did he or I know then that some years hence a similar, tragic occurrence — another terrible drought — would befall Ethiopia. Since returning to the United States, he had become a social worker. One day he invited me to have a look around his workplace, Unit 89, not far from Harlem on the West Side. It was disheartening to see the large waiting room full of black Americans and Puerto Ricans waiting to be interviewed or to receive their welfare cheques.

One day I particularly recall feeling so thoroughly disillusioned after walking around much of the city from office to office and street to street. I was incredibly weary, my feet were sore and my eyes were gritty from the dust and grime. I returned to the apartment almost in tears, the closest I'd come to crying since I was a kid. It was then I realised that, having come to New York City to achieve my immediate goal of making money, I was in the wrong place at the wrong time.

No matter how hard I tried or which avenue I pursued, my powers of persuasion fell on deaf ears. It seemed that circumstances were against me. More than once I yearned to be back in my beloved Africa. Indeed, the more I thought about it, the more I realised that I was at my happiest when surrounded by magnificent scenery or on the road as free as a bird.

After doing everything that I and others could possibly think of such as contacting TV stations, magazines, newspapers, and photo agencies to no avail, I followed the only option left to me. I swallowed my pride and resorted to doing temporary, menial jobs to cover my daily expenses.

Finding odd jobs wasn't easy until someone referred me to Jack Skully, the manager of Everything for Everybody, a small, somewhat unique employment agency on Eighth Avenue in the West Village. Boards outside and inside the office displayed ads, including: 'W103081 needs two attractive young ladies between 24-28 as escorts to black-tie affair'; 'E70-658 needs ride to California as soon as possible'; 'W82-249 needs small bathroom painted'; 'F24-193 wants to converse with someone in French'; 'F72-894 needs hip chick to take care of apartment'; 'E71-722 boy needs girl'; and dozens more. Jack took 15% of whatever was paid for the job, and I got some painting and cleaning work. Fortunately nobody bothered to ask if I had a work visa.

For two weeks, I commuted daily to Brooklyn to help renovate an old house. The job of scrubbing and cleaning woodwork was hard, and the hours were long. At another place, on the East Side, I scraped paint off bathroom tiles. I lost count of the number of apartments and windows I cleaned from Fifth Avenue to the Bronx and Queens. And from the Lower West Side to the Upper East Side. Some were old and shabby places and others were incredibly luxurious with armed security guards posted at the main doors.

Over a period of time, I did jobs for a variety of people who assumed that I was a struggling student trying to work my way through college. When I attempted to tell them exactly what I was doing, none of them seemed to believe me. It seemed to be totally beyond their comprehension. I was 'putting them on', I read in their faces. Once for two days, I distributed posters and brochures in shops and supermarkets for The Living Theatre, an appropriate name for New York! It certainly was a human comedy — of the worst possible kind.

In my travels around the city, I met many people who had been nowhere and done nothing in particular yet acted like they knew everything. 'Bullshitters of the first order', as my old drill sergeant would say. And as James Reston, a highly respected journalist, wrote in the *New York Times*:

> *We're in a mood of indifference. Something is happening to the American mind. It is being drugged by facts and diverted from reality. Every hour of every day, it hears and reads the most astonishing things, but it is not astonished. It is given the facts of human condition but it doesn't feel them or doesn't think it can do anything about them.*

I couldn't have agreed more.

At that time I found an escape from all the negativity by listening to music of all kinds. Bob Dylan, Joan Baez, Pete Seeger and Canadians Neil Young, Gordon Lightfoot and Joni Mitchell were up and coming folk singers in North America. And I always enjoyed a long-time favourite of mine, the incomparable Frank Sinatra.

Before the long, hot summer came to an end, I began phoning around to try and arrange lectures. Unlike the vast majority of Americans who conduct just about everything on the telephone, I disliked using it, preferring to speak face to face. Nevertheless, I obtained the names of clubs and associations from the phone directories and called 40 or 50 organisations over the next few days.

Most of the people I called hung up as soon as I spoke or simply didn't believe me. I was staggered, but I quickly figured out why. So many people had been pestered by obnoxious callers that many were afraid of answering the telephone. Those who didn't hang up immediately were so suspicious they asked how I'd obtained their numbers. Of the few people I did speak to, some were downright rude.

Despite these strange reactions, I persevered and eventually spoke to a minister who invited me to speak to the women's society at his church. But not for another three weeks. At least, it was a start.

Bill and Barbara, although seasoned New Yorkers themselves, had to admit that what was happening to me was simply 'unbelievable' and they expressed this time and time again. Barbara continued to make spasmodic calls on my behalf but, as the reaction was always negative, her initial enthusiasm waned. 'What's the good,' she said, 'when nobody believes anyone anymore?'

However, slowly and surely and through sheer persistence, I gradually made some money, and a tremendous feeling of relief surged through me. It was also time to repay Bill and Barbara's hospitality, but they wouldn't hear of it and assured me I had nothing to pay back.

As usual, Pete and I kept in touch by mail, and his latest missive informed me that he had a job on the *SS Prince George*, a passenger ship owned by the Canadian National Steamship Company. It was one of only two ships that plied up and down the British Columbia coast during the summer season, sailing via Prince Rupert through the Inside Passage to Ketchikan, Juneau and Skagway in Alaska. His job was jack of all trades, but mostly stewarding — cleaning cabins, making beds, and serving drinks and meals to the roughly 300 passengers. Most importantly, the pay was good.

Fortunately, despite facing a few hurdles that were to be expected in every country we visited, Pete was in no way disappointed by the vast and diverse country to the north. In fact, to the contrary. He told me gleefully in his letters that Canada was, in fact, everything he hoped it would be.

Summer finally came to an end, and I began to give lectures. Most of them were arranged by church ministers and women's societies. All were well-received, and I was usually inundated with questions afterwards. Thankfully, as more people learned about me, I received more bookings and spoke in a variety of places.

After giving a lecture to the West End Women's Society, the following writeup appeared about me in one of their newsletters:

He has seen, recorded and remembered much, interested less in scenery than in the people of the world. His pictures are beautiful, his delivery excellent and he impressed us with his sincerity, his ability to make the best of a situation and to profit from past experience. He is certainly an inspiration to our youth who could not but thrill to the challenge of such a life.

Naturally, such recognition bolstered my spirits and my wavering confidence.

The most amusing aspect of all this was that I was leading a regular Walter Mitty existence, racing back to the studio from painting and cleaning jobs in dirty, well-worn jeans to shower and change into more suitable speaking attire.

Nineteen sixty-eight in New York was known as 'the year of the strikes'. That year, the teachers and garbage collectors went on strike, as did the parcel delivery companies, the utility (electricity) services, and subway personnel. Even the firefighters and policemen staged a one-week slow down.

Intermittently over a long period, I gave a hand to a carpenter friend named Larry whom I'd met through Bill and Barbara. Aged about 45 and self-employed, he operated out of a small workshop along the Bowery, the infamous street that New Yorkers prefer to forget. A long straight road, it had hardly changed in 80 years. Dreary, dirty and depressing, the Bowery was the home of street people who were referred to, unkindly I thought, by the majority of the populace as 'bums', 'drunkards', and 'derelicts'. They were down and out, yet nobody seemed to care. It was a pathetic scene as I picked my way over the many prostrate bodies in the space of only one or two blocks. My problems were nothing compared to what I saw there.

I met and spoke to many. Although some were men who had allowed their weaknesses to get the better of them, others had suffered from sheer bad luck through no fault of their own including one who had lost his entire family, house, and belongings in a fire. With almost nothing left to live or hope for, he had resorted to drinking. Seeing him in such a bad way made me shudder. My heart went out to him and all of those impoverished, destitute individuals. Frequently, I dug into my pocket but, before handing over some small change, I insisted they spend it on a sandwich or hot drink. Most had the same, dirty, and unshaven appearance and that hopeless, distant look. As if being down and out wasn't enough, buses would sometimes stop while the passengers took photographs of them. One day, one old man with whom I frequently chatted stood in his usual spot and quipped, 'I've been photographed more often than the president. I'll be a movie star one of these days!'

By now, we'd been away from the UK for slightly more than six years, and Pete was ready to return home. I wasn't yet, making it one of the few times that I wasn't prepared to move on with him. He flew in from Vancouver, and I met him at the airport. Bill and Barbara kindly invited him to stay and he slept on the floor in my room. He was surprised and impressed at how well I knew my way around New York, for he found it quite intimidating. I didn't bother telling him there had been over 500 murders in the city that year!

Pete told me on more than one occasion that the trip had changed his life. I felt the same way. He thanked me for persuading him to come along and I was so glad that he'd decided to join me. Our mutual love of travel and shared adventures had cemented our friendship. For various reasons, we had often followed our own paths, joining up again every so often to share some new adventure or experience. Now he was heading home without me.

Over a period of a few days, Pete tried very hard to persuade me to go back to England with him. As he pointed out, we'd begun the trip together and been through so much together that it made sense to go home together. I agreed, but couldn't leave, at least not yet. I still needed to make some decent money. He understood but disliked the idea of returning home alone. He'd missed many of my extraordinary experiences, but he'd had quite a few remarkable ones of his own.

After Pete left, I put on my stiffest upper lip and did my best to remain optimistic, convinced that things would turn around. I may have been down, but I wasn't completely out!

As luck would have it, I had a middle-aged, widowed aunt named Doris Robinson who was a health care worker then living in Boston. She had left England for the United States 20 years previously in search of a change of scenery and a better life after her husband died of a lengthy illness. Her one claim to fame (apart from Simon and Garfunkel's song, 'Mrs Robinson', which was a #1 hit in 1968) was being chatted up by none other than Clark Gable on a Miami park bench years before, she once laughingly told me.

She had written to me a couple of times, inviting me to visit her. Although I had every intention to go and see her at some point, I felt reluctant to do so until my financial position had improved. In each letter, she tried to persuade me to come to the slower-paced capital of Massachusetts because it was 'far nicer' than New York. So, one weekend I bought a round-trip ticket to Boston and headed there on a Greyhound bus. I, too, was badly in need of a change of scenery. *To hell with New York!*

CHAPTER 22:

Where There's a Will . . .

'Anything you can imagine you can make real.'
— Jules Verne

It was so nice to see Aunt Doris again. She had been visiting her sister, my Aunt Mabel, in England when our original group of four had left and they were the last ones to see us off. Even though she was only vaguely aware of some of my exploits, she fussed over me and was very proud of my achievements. As a keen traveller herself, she was quite certain that my story was well worth telling and had been in touch with one or two newspapers on my behalf, including the *Boston Globe*. There I was interviewed by Christina Tree, a young and attractive Globe Travel Writer, who, after our meeting, took me to lunch and showed me around for the rest of the day.

Again, after a few days of sightseeing, my aunt urged me to stay in Boston, so I decided to do so, but not until after my story had appeared. That way, my name might ring a bell in people's minds. I stayed five days in Boston at the YMCA. As I'd discovered in Australia and elsewhere, such hospitable establishments were usually located in the heart of downtown and always offered clean and comfortable accommodation at a reasonable price.

When I returned to New York, I continued to work at dreadfully boring and mundane jobs, still hoping that a break would come my way. After all, I reasoned, I'd just about finished my epic journey, my material (photographs and references) was readily available, and I had an endless supply of interesting stories to tell. I merely wanted the opportunity to present them publicly.

Why I bothered to stay in New York, I'll never know. Yet I am convinced now, just as I was then, that nobody could have tried harder to succeed than me. I left no stone unturned to organise myself. I was out and about all day every day, meeting and talking to people, trying to make money. Somehow I survived, helped by good friends and my own determination to make a living, no matter how meagre and no matter the job. I could have thrown in the towel any time I liked but, when I considered doing that, the encouraging words of 'Stick with it, lad,' from John, my old building site boss in Gladstone, rang in my ears.

Fortunately, life in one of the most well-known and spectacular cities in the world, while greatly disappointing, was not all drudgery for me. I got to see all the major sights and, because I was constantly on the go, I met many people. One was Ellen, a friend of Bill and Barbara, a plump young woman in her late twenties and, as I was soon to find out, as horny as hell. It had been far too long since I had enjoyed intimate female company, and I craved the smell and feel of a woman's body.

For some weeks, Bill and Barbara had been trying to fix us up, and one weekend they finally arranged a date that was mutually convenient. I was to meet Ellen at her apartment on a Saturday evening where she would be waiting to serve me dinner. I turned up at her place near Central Park, clutching a bottle of wine. No sooner had we drunk a little, chatted about this 'n' that, and finished the first course of mushroom soup, than Ellen made it apparent, by turning the lights down and the music up, that the second course, supposedly spaghetti bolognese, was going to be me. She plied me with wine and, as I became more and more inebriated, hurriedly and methodically undressed me and did what she wanted with me. I don't remember much of the evening nor precisely what took place as I was too drunk. But I do recall that I was a willing partner and did what was expected of me. Later, Bill and Barbara told me Ellen was quite wealthy from a tidy inheritance, and wanted to pursue a relationship with me. But I just wasn't interested. I had no special feelings for her.

One of my lectures was held in the palatial Biltmore Hotel opposite Grand Central Station. I made a point of arriving early as I always did whenever I made arrangements of any kind. For some inexplicable reason, I lost my way in that massive hotel and went back four times to the front desk to enquire exactly where the meeting was going to be held. The receptionist must have thought me quite mad for, despite having acquired an uncanny sense of direction, I'd never been lost in my life — until then!

By now, I thought I'd experienced my share of misfortune, but the final humiliation came one night when I fell out of bed and knocked my nose out of joint. It happened when I accidentally rolled out of my makeshift bed and smashed my face on the arm of an adjacent wooden chair. Staggering up, and in excruciating pain, I made my way to the bathroom where I carefully and quietly bathed my bloody face in warm water, then tried to go back to sleep. In the morning, Bill and Barbara were horrified to see my condition, admonished me for not waking them, and insisted that I go to a hospital.

At St. Vincent's, one of New York's largest hospitals, I joined the line in Emergency. A mix of people stood around, scared and bewildered, suffering from various serious injuries such as broken arms, stab wounds, and acid burns. I felt such a fool standing there for two hours waiting to be seen with my minor issue. The young doctor who examined me took just four minutes before declaring, 'You're through,' followed by, 'Just let Mother Nature take her course.' The cashier then presented me with an $11.00 bill. I had witnessed some of the workings of a busy New York hospital! And forfeited some of my hard-earned cash.

To make matters worse, I had a lecture that day and gave it with a handkerchief covering my swollen face. My explanation brought the house down. Even I couldn't resist laughing despite the pain.

One evening, Bill invited me to his fire station in the district of Brownsville, one of the toughest neighbourhoods in the city. Donning fire-fighting gear, including an 'auxiliary' yellow helmet, I went to three alarm calls during the night — yet another unique and interesting experience — but I decided it was an unenviable job. A third of all the emergency calls were false alarms, and the calls come in so fast and furiously that they, the firefighters as well as the city's police and ambulance services, are the busiest in the world. Three times was enough to give me a vague idea of the life of a firefighter. Later, I just slept while the others were called out several more times that night. Once, Bill told me when he stood in his usual position on the back of the engine, someone fired a shot which whizzed past his left ear. He returned home after that shift somewhat shaken. No wonder Barbara worried constantly about his dangerous job.

One family that took me under their wing after I delivered a lecture to members of their Methodist church was the Trell family. Reverend Trell was president of the church society and, after enjoying my presentation, he arranged for me to give a second one at another church. Besides that, he kindly invited me several times to spend weekends with his family. He and his wife were concerned whether I was eating properly, which I wasn't. A few times, I took them up on their offer and relaxed in pleasant surroundings at their suburban home where they always treated me as a member of the family.

I first heard of Richard Nixon's closely contested victory as I stepped out of an apartment building. I was tired at the end of a tedious and fairly hard day scraping off wallpaper. In the street, three small boys were playing baseball chanting, 'Nixon, Nixon, he's the one. If he can't do it, no one can.' He would become president early in the new year, taking over from Lyndon Johnson. Within a few years, he ended the draft and opened the door to China, Watergate happened, and he would resign in disgrace.

I'd stayed in touch by mail with Breta and Teresa, the two young European women Pete and I'd met on our trip down the Amazon. As winter approached and seeking respite from my life in New York City, I decided to visit them in Toronto, Canada's largest city. I made the trip by train which took all day, stopping at every hamlet and town along the way. It was snowing when my taxi arrived at their apartment from the station, and it continued throughout my three-day stay. After I borrowed a thick sweater and some gloves, we ventured outside a few times to see some of the sights, hang out in coffee shops, and skate at the public ice rink by City Hall.

As I'd found out on our thrilling Amazon River journey, Breta and Teresa were stimulating company. We continued where we'd left off, laughing and talking about our adventures and mutual love of travel. They too were free spirits who, after meeting and becoming friends in the Azores, had taken a year off work to travel. Little did I dream then as we chatted that several years later, I too would be living in Canada.

Back in New York, when Thanksgiving arrived, I made the most of it like everyone else and stuffed myself with turkey. It seemed that American families ate turkey the whole week long. It reminded me of the many repetitive meals of steak in Australia and unlimited apples in New Zealand. It was almost too much of a good thing. It went without saying I appreciated both the celebration and the invitation to participate.

The winter of 1968–69 in New York City turned out to be icy cold with piles of snow bringing the city to a halt. That winter would record the biggest snowstorm since 1926. Fortunately, during the worst of it, I had already moved to Massachusetts.

Christmas 1968 arrived, and I received no fewer than six invitations to join various families for the holiday period. I chose to stay with the Trell family at their holiday home in the snow of the New England countryside. While staying with them, I joined a family group of 15- to 30-year-olds shovelling snow and singing carols to shut-ins. And as the tumultuous year drew to an end, the spaceship Apollo 8 brought something positive for the American people to talk about, when it successfully completed its lunar orbit. Several months later in July 1969, they would have even more to celebrate when Neil Armstrong and Buzz Aldrin became the first men to walk on the moon.

Two days before New Year's Eve, I met two young Welsh nurses named Alice and Jean, who were visiting the city for the weekend. They'd come from Kingston, Ontario, for a few days, didn't know a soul in the city, and had become very alarmed when they heard shots ring out in the streets their first night. For two days, I took it upon myself to be their guardian and guide as we travelled together by bus and subway all over the city. On New Year's Eve, we joined the swarms of revellers in Times Square. Crushed and pulled by the crowd, we celebrated the occasion by watching the spectacular fireworks display, looping arms and singing 'Auld Lang Syne'.

The *Boston Globe* eventually published Christina Tree's article about my travels, 'One way to see the world'. Quite unexpectedly, Ed and Fran — with whom I'd travelled on and off through South America and explored Machu Picchu — spotted the article, tracked me down, and telephoned me from their home in Boston. They were 'thrilled and amazed' that I was still around. They'd had enough travelling for a while. *I've had a bellyful of New York!* I told them. That same day, after saying farewell to Bill and Barbara, who had been so kind and helpful to me, I left by Trailways bus for Boston.

Throughout the four-hour journey, I reflected on my life in New York. I thought about other young and eager hopefuls who, like me, flocked there to try their luck, only to face rejection time and time again, and my heart went out to them. I knew I wasn't the first to be disillusioned and by no means would I be the last. Earlier, I'd watched the Miss America Contest on television, and I was reminded of the song performed by the lovely runner-up. A teacher of deaf children, she sang 'What the world needs now is love, sweet love,' an appropriate song for everywhere, especially New York.

Ed and Fran met me at the bus terminal and immediately rushed me off to a party where they introduced me to another delightful couple, Maurice and Betsy DiMarco. 'You're welcome to be our guest for as long as you like,' offered Maurice in his clipped New England accent after learning I was looking for a place to say. This confirmed what I had always believed: American people are some of the most hospitable, generous and warm-hearted in the world.

Like most American couples, they both worked. Maurice was a real estate broker and Betsy, a secretary. They were an odd-looking couple, for Maurice was short and tubby while Betsy stood six feet. But there was no denying their fondness for one another. They typified the 'All-American' family and occupied a large house on Chestnut Hill Avenue in the suburb of Brighton with an in-law suite and separate entrance, which suited me perfectly. A happy and devoted couple, they had two adorable young daughters, aged two and four, who attended daycare five days a week, a large dog called Buffy, and a cat called Kate which gave birth to six kittens during my stay. A big event when you live in the suburbs.

Boston, home to Harvard University and the birthplace of Ralph Waldo Emerson, the famous philosopher and writer, was far nicer to me than New York had been. I made more friends in two weeks than during five months in New York. There, my social life had been virtually non-existent; in Boston, I was invited to numerous parties, some put on by my hosts.

While settling in and finding my way around, I did lots of odd jobs around the house, and just like Bill and Barbara, Maurice and Betsy couldn't do enough for me. While exploring the city, I travelled on the ancient (now obsolete) streetcars; walked the Freedom Trail; gazed down on the metropolis from the top of the John Hancock building, then the tallest in the city; and peered into every nook and cranny of *Old Ironsides*, a wooden-hulled three-masted frigate of the US Navy built in 1794 and named by President George Washington.

From New York, I had written to various clubs in the Boston area and received a few replies. Nevertheless, it took a few more letters and umpteen telephone calls before things started to happen.

Maurice helped by introducing me to the editor of a local newspaper, and an interview was arranged. One of the editor's protégés, a rookie reporter called David Ramsay, asked me questions and took notes. A couple of days later, an article with a photo of yours truly appeared with the captivating caption, 'Name: Skillan; Address: The World'. I read on, intrigued at what the youthful reporter had to say about my life on the road:

> *Ever dream of leaving it all? Of going around the world giving lectures and taking photographs of breathtaking picturesque landscapes? Of shouldering a pack [and] . . . climbing Mount Kilimanjaro?*

The article continued:

> *For most, it's a dream never fulfilled. But for a very special kind of man, one who calls himself 'a dreaming realist,' that dream of world travel and adventure came true.*
>
> *David W. Skillan, a 29-year-old British traveller and adventurer, who lists his home address as 'the world,' completed his military service in England, talked three friends into accompanying him, and set off to see every far away paradise he had ever dreamt of, and working every job he wanted to. It was a boyhood dream, one which this 'dreaming realist' made come true.*
>
> *Skillan, a trim 140 pounds and a slight 5' 9½", has seen more places, worked more jobs, lived with more diverse types of people, experienced more adventures and lived more of life than men twice his age and physical stature.*

Unbelievable though it may have seemed years before, there I was in the United States reading a rather flattering story about myself in a Cambridge, Massachusetts newspaper. And yes, it was all true.

Almost immediately after that article appeared, I was contacted by all kinds of people and began giving illustrated lectures at colleges and various cultural and professional associations, for which I was nicely rewarded, both by the response and financially. Delighted as I was, I couldn't help thinking how different everything might have been had my story appeared in one of the New York dailies.

After a talk to members of the International Institute in Boston, the Activities Director wrote a nice review which added greatly to my confidence:

> *This modern Odysseus personally recounted experiences about his epic wanderings to members of our international Professional circle. It was enormously exciting to meet this young man now ending his world trek before returning to his native England. The earth which our gallant astronauts saw as a beautiful blue sphere in space has felt his intimate, boyish, eager affection in its remotest places.*

Having been frustrated and impatient in New York, I began to unwind in Boston. The more I relaxed, the more the telephone seemed to ring with requests for my speaking services. And thanks to my adopted family, I ate well and regularly.

One evening, the local Baptist minister called on my gracious hosts, drumming up support for the production of the annual children's play. When he heard about some of my adventures, he became very enthusiastic and insisted that I speak to his congregation — which I subsequently did and for which he arranged a special collection.

Another of my talks was to members of a nationwide organisation with branches in every state. The English-Speaking Union was an international charity fostering global understanding, communication and goodwill between people of different cultures and languages. *I fitted in nicely*, I thought.

Briefly, I was asked to give lectures in three different parts of the country. My expenses would be covered, and I would be adequately remunerated.

This organisation was used to 'famous' speakers and, although rather nervous at the prospect, I was delighted to be following in their footsteps. Suddenly, between lecturing at various clubs in Boston, I found myself travelling first class by jet to not one but three American cities. On arrival at the airports, I was whisked by limousine to a prestigious hotel to address a luncheon or dinner meeting! Often, during my travels, I had declared half-jokingly and half-seriously, that at the time I was roughing it and 'One day I'd be doing it in style.' It was finally happening!

I always saw the amusing side when presenting my story to such successful and distinguished audiences for, although I remained far from well off, I frequently sat next to and spoke to well-to-do and prominent people. The whole affair had a dream-like quality and provided an ironic twist to everything that had happened.

One particular highlight on my brief and memorable speaking itinerary included an overnight stopover in Charlotte, North Carolina, in the 'Bible Belt'. There, it was my pleasure to be hosted by an elderly couple named Mr and Mrs Duncan. Exuding Southern charm, Mr Duncan was a God-fearing man and successful lawyer who idolised his wife; after 40 years of marriage, he still stood up whenever she entered the room. Mrs Duncan's main interest was in birdlife. She had written several books on the birds of the region and was a recognised authority in the avian field. Over the years, their vast garden had been transformed into a bird sanctuary called 'Wing Haven'. Living so close to nature, this contented couple were at complete peace with the world.

All the lectures I gave in the United States (including those in New York) were well-received, assuring me that I did have something to offer, which made me feel happy and warm inside. A few people told me how much they loved my infectious enthusiasm while others said I was like a breath of fresh air. As far as I was concerned, I was providing a service: educational, entertaining and inspirational talks about people and places, life and adventure in different parts of the world.

The most enjoyable part of the meetings for me was always the question-and-answer sessions which followed. Often, especially with students, I answered questions for up to two hours — and even then some people seemed reluctant to let me leave! Although I appreciated the recognition and praise, it was embarrassing and a little overwhelming. Once, one man leapt up from his seat and turning to face the audience loudly proclaimed, 'This man is made of steel!' Which made me blush. Besides, I knew only too well that I was just an ordinary man — who had led an extraordinary life.

To would-be travellers, I offered practical tips, advice and suggestions based on what Pete and I had learned from our enjoyable and sometimes unpleasant experiences. Once or twice, the odd sceptic suggested we must have been sponsored, declaring, 'No way could you do it alone.' To comments like that, I could only laugh. While I was careful to emphasize it wasn't necessary for everybody to travel for long periods, there was no doubt in my mind that travel was a sure-fire way of bringing people together. Just about everyone told me I must write a book to which I replied, 'I hope to one day.' Only a book could tell the whole story. Many said they'd like to read it. And some even predicted it would be a success. For my part, I definitely wanted to record all the triumphs and tribulations of my epic travels, if only for the sake of posterity.

Summing up, I pointed out that not having quite completed the trip — I had yet to cross the Atlantic — I hadn't yet organised my thoughts. The passage of time would put them into perspective. But of some things, I was already sure. People were much the same everywhere; most merely wanted a decent living for themselves and their families. I was much more appreciative of the little things that so many of us take for granted — good food, a daily bath, and a decent bed. I knew that very little could have been accomplished without the assistance of so many kind people; that one can get used to anything if necessary; and that anyone can have whatever they want if they persevere long and hard enough. I had proven this time and time again. It was with great satisfaction that I noticed members of the audience leave with beaming smiles, hopefully determined to do things they, too, never thought they could.

Then it happened. Boston experienced a record snowfall of 50 inches, the worst in 97 years! For several days, I helped dig out and clear the path to the house while the city was virtually at a standstill. Making the most of it, I joined the family ice-skating and tobogganing on the ponds and nearby slopes.

Time slipped by. Soon, my visa was due to expire, and I would go back to England. Although I had written home faithfully at least once a month, I owed it to my family to see them soon. Besides, I'd been living by my wits long enough. It was time to get a proper job and settle down — at least for a while.

The evening before departing from the United States, I lectured at the Harvard University Library Club. The person responsible for arranging it, Mr John Depta, president of the club and librarian by profession, had attended one of my previous presentations. Prior to my talk at the university, he had distributed flyers advertising it. Beautifully designed and printed, the flyers were bordered with various cartoons depicting some of my adventures. Under my name, the following words from Tennyson's *Ulysses* appeared:

> *I cannot rest from travel; I will drink life to the lees: all times I have enjoyed greatly, have suffered greatly, both with those that loved me, and alone, on shore and when thro' scudding drifts the rainy Hyades, Vext the dim sea ... For always roaming with a hungry heart, Much have I seen and known: cities of men, and manners, climates, councils, governments ... I am part of all I have met.*

Somebody was very perceptive, I thought, *and got it. Yes, it was definitely very nice to have my efforts and achievements, such as they were, recognised.*

After my talk, I was presented with a couple of books, one of which was entitled *Travellers in Disguise*. It was about two of Italy's most famous explorers, Poggio Bracciolini and Ludovico De Varthema, who had travelled extensively throughout the Middle East and Asia in the early fifteenth century. And faced every imaginable hardship. I regarded this kind gesture as the supreme accolade. For the umpteenth time, I felt very humble — but proud.

The next day, after packing my few possessions, including my precious slides, Ed, Fran, Maurice and Betsy accompanied me to Boston's Logan Airport to see me off. It was the evening of March 22, 1969. My plane was due to take off at 9 p.m. My US visa expired at midnight. I had spent exactly nine months in the United States. Now I was going home in style in a BOAC VC-10.

After giving my swordstick to Ed, it was with considerable regret and nostalgia that I threw away my original rucksack — now torn and tattered —and replaced it with a new, inexpensive suitcase. That pack had served me well and, like a faithful companion, had accompanied me everywhere I'd roamed.

'God speed,' 'Don't forget to write,' and 'Come back soon,' my friends chorused as we hugged one another and said our goodbyes.

The big jet took off on schedule, and while we were airborne, I didn't sleep a wink — still apprehensive about flying — but sat lost in thought. I felt almost indifferent, as if it was all a long dream. It had taken me more than six-and-a-half years, nearly 80 months, to accomplish what I'd set out to do, and now I was returning to where I had started in only six-and-a-half hours! With this flight, my long, long journey was coming to an end.

As the plane droned through the night, I reflected on the last several years. Incredibly, what had begun merely as a boyhood dream had become a reality, and its success had surpassed all the hopes I'd entertained. Certainly, I'd known great moments of euphoria and exhilaration as well as my fair share of problems, suffering and despair. I'd witnessed and experienced things that most people only dream about.

With Pete and often solo, I'd seen many of the wonders of the world, eaten every kind of food imaginable, rubbed shoulders with people from all walks of life, and endured fluctuating changes of fortune. I had tasted true freedom, gone where I wanted and done what I pleased and, to an extent, been master of my own destiny. Yet I hadn't become blasé and, despite everything, didn't feel any the worse for wear. I had no idea how many miles I'd travelled but with all my different jobs and diversions it must have been at least 250,000. Had I changed? Undoubtedly. The trip had left its mark, both emotionally and mentally.

I hoped I'd become wiser, more understanding and tolerant than before — in other words, a better man. No, I hadn't been to university. But I'd had the best education possible — at the school of hard knocks.

Probably the greatest satisfaction stemmed from knowing I'd done it my way. Thanks to so many generous and kind souls and through my own enthusiasm, determination and perseverance, I'd achieved what I had set out to do. Along the way, numerous challenges had been met and faced, and because

I had dared, I had succeeded. There was no doubt about it — I had truly laughed, loved and lived.

Looking back, the hardships and frustrations I had experienced were insignificant compared to the sense of satisfaction and accomplishment I now felt. And thanks to God's good grace, I had survived and lived to tell the tale.

No amount of money could have bought so many hard-earned experiences, both good and bad. Nor could they ever be taken away from me. As my worldwide odyssey approached its end, I couldn't wait to share my travels and adventures with others and, at the same time, inspire them to follow their own passions and interests whatever they may be. Maybe in my own limited way, I could help make the world a better place.

As I continued to reminisce, a parade of familiar faces appeared before me, and I thought of them all with gratitude and affection. *What had become of Shadarak and Elijah?* I wondered. *Were they still involved in the monkey business? And what of Naas? Was he still building giant masts? And what about little Mohammed in Algeria, Slippery of the outback, Alberto and his beloved city of Buenos Aires, and some of the lovely women I had known? Would I ever see any of them again?* I asked myself. I sincerely hoped so. My life had been enormously enriched by knowing them. They had all meant so much to me and would never be forgotten. That much was the least I owed them.

I hoped all the effort I'd put into something I loved doing would ultimately pay off. It remained to be seen just how much my passion for travel would influence my life in the future, but I was confident that my never-to-be-forgotten wanderings and extraordinary experiences would stand me in good stead. Time would tell. I also reflected on an oft-used quote by philosopher and writer, Henry David Thoreau:

> *If one advances confidently in the direction of his dreams, and endeavours to live the life which he has imagined, he will meet with a success unexpected in common hours.*

Those prophetic words had often danced in my head, for they came to epitomise my vagabond life. Certainly, I regarded that period of my life as my Everest, the hardest thing I had ever done. Nothing, I felt sure, would equal it or even come close.

My sombre mood brightened as I told myself I could return to any of those distant lands at any time and immediately feel at home.

The plane descended from the early morning sun into occasional cloud, and I felt strangely numb as I looked down on England's green fields and caught a glimpse of Windsor Castle. We landed at Heathrow and taxied up to the terminal. Soon after, I stepped onto British soil. Mission accomplished. My epic journey had come to an end. It was March 23, 1969. I was just a few months shy of my 30th birthday.

It was typical English winter weather, overcast, blustery, and cold. I turned up the collar of my new raincoat and walked across the tarmac with my book, *Travellers in Disguise*, under my arm. With only a cursory glance at my third passport, an immigration officer greeted me with the words, 'Good trip?' and waved me on. Seconds later, a customs inspector asked me, 'Anything to declare?' *Just memories*, I thought, shaking my head.

Just beyond the airport exit doors stood a smiling Pete and Hoshiko waiting to welcome me home. They'd recently returned from Paris, where they'd tied the knot. 'Hi, pal. Good to see you,' exclaimed Pete, with a thumbs up as hugs and warm handclasps were exchanged all around. Soon after, we were all on a bus headed towards London, lost in conversation as we caught up on each other's news.

Earlier, after getting home and working at a few dead-end jobs, Pete had made plans to return to Canada. The very next day after my arrival, he left with Hoshiko by his side. As a self-made entrepreneur, he would carve out a successful career for himself and his family in his adopted country.

From London's always busy Waterloo Station, a two-hour train ride took me to my parents' home in the suburbs of Southampton where they had moved during my long absence abroad. There, for the next six months in my small, cramped bedroom, I would relive my memorable marathon journey while writing the original manuscript for this book.

Still firmly bitten by the travel bug and with my passion for discovering new places as intense as ever, I would be employed for the next few years by a travel company where my hard-won, international experience would be put to good use. I'd return to many of the places originally visited as an itinerant wanderer — but this time staying in fine hotels with all expenses paid. But that's another story to be told another time...

ABOUT THE AUTHOR

As a young traveller and trailblazer who wandered the world on a shoestring throughout the 1960s, David Skillan vowed the next time he travelled it would be in greater comfort. This ambition would be realized when he was employed by a London-based Swiss travel organization seeking a foothold in the British travel market.

When he joined the company as a dynamic 30-year-old year old and led his first tour — to Japan in March 1970 — he openly declared that he would put them 'on the map'. During the five years he spent with the company, they went from obscurity to the UK's most successful tour operator winning numerous awards and accolades. For David, it was the beginning of a lengthy and varied 45-year career in the travel industry during which he escorted well over 200 long-haul tours to many of the places he'd first visited as a youthful globetrotter.

In 1976, he moved with his young family to Canada where he ran his own successful travel company. He also published a travel newsletter for more than thirty-five years. An incurable romantic and self-described free spirit, David counts himself lucky to have seen much of the world from both the 'roughest and smoothest' of angles.

A long-time resident of Vancouver, BC, David is a retired tour operator, tour manager, writer and photographer. He maintains strong ties with the British Isles, the land of his birth, and likes nothing more than visiting his home country every so often.

CPSIA information can be obtained
at www.ICGtesting.com
Printed in the USA
BVHW051341230921
617332BV00004B/8